T0269365

Adaptation, Learning, and Optimization

Volume 15

Editors-in-Chief

Meng-Hiot Lim
Division of Circuits and Systems, School of Electrical and Electronic Engineering,
Nanyang Technological University, Nanyang 639798, Singapore

Yew-Soon Ong
School of Computer Engineering, Nanyang Technological University, Block N4,
2b-39 Nanyang Avenue, Nanyang, 639798, Singapore

For further volumes:
http://www.springer.com/series/8335

Serkan Kiranyaz · Turker Ince
Moncef Gabbouj

Multidimensional Particle Swarm Optimization for Machine Learning and Pattern Recognition

 Springer

Serkan Kiranyaz
Moncef Gabbouj
Department of Signal Processing
Tampere University of Technology
Tampere
Finland

Turker Ince
Department of Electrical and Electronics
  Engineering
Izmir University of Economics
Balcova, Izmir
Turkey

ISSN 1867-4534   ISSN 1867-4542 (electronic)
ISBN 978-3-642-43762-5   ISBN 978-3-642-37846-1 (eBook)
DOI 10.1007/978-3-642-37846-1
Springer Heidelberg New York Dordrecht London

ACM Computing Classification (1998): I.2, I.4, J.2

Printed on acid-free paper

Springer is part of Springer Science+Business Media (www.springer.com)

Preface

The definition of success—To laugh much; to win respect of intelligent persons and the affections of children; to earn the approbation of honest critics and endure the betrayal of false friends; to appreciate beauty; to find the best in others; to give one's self; to leave the world a little better, whether by a healthy child, a garden patch, or a redeemed social condition; to have played and laughed with enthusiasm, and sung with exultation; to know even one life has breathed easier because you have lived—this is to have succeeded.

Ralph Waldo Emerson

The research work presented in this book has been carried out at the Department of Signal Processing of Tampere University of Technology, Finland as a part of the MUVIS project. This book contains a rich software compilation of C/C++ projects with open source codes, which can be requested from the authors via the email address: MDPSO@email.com.

Over the years the authors have had the privilege to work with a wonderful group of researchers, students, and colleagues, many of whom are our friends. The amount of our achievements altogether is much more than any individual achievement and we strongly believe that together we have really built something significant. We thank all of them so deeply. Our special thanks and acknowledgment go to Jenni Raitoharju and Stefan Uhlmann for their essential contributions.

Last but not least, the authors wish to express their love and gratitude to their beloved families; for their understanding and endless love, and the vital role they played in our lives and through the completion of this book. We would like to dedicate this book to our children: the new-born baby girl, Alya Nickole Kiranyaz, the baby boy, Doruk Ince, Selma, and Sami Gabbouj.

Tampere, December 2012

Prof. Dr. Serkan Kiranyaz
Doc. Dr. Turker Ince
Prof. Dr. Moncef Gabbouj

Abstract

The field of optimization consists of an elegant blend of theory and applications. This particular field constitutes the essence of engineering and it was founded, developed, and extensively used by a certain group of creative people, known as *Engineers*. They investigate and solve a given real world or theoretical problem as *best* they can and that is why optimization is everywhere in human life, from tools and machinery we use daily in engineering design, computer science, IT technology, and even economics. It is also true that many optimization problems are multi-modal, which presents further challenges due to deceiving local optima. Earlier attempts such as gradient descent methods show drastic limitations and often get trapped into a local optimum, thus yielding a *sub-optimum* solution. During the last few decades, such deficiencies turned attention toward stochastic optimization methods and particularly to Evolutionary Algorithms. Genetic Algorithms and Particle Swarm Optimization have been studied extensively and the latter particularly promises much. However, the peculiar nature of many engineering problems also requires dynamic adaptation, seeking the dimension of the search space where the optimum solution resides, and especially robust techniques to avoid getting trapped in local optima.

This book explores a recent optimization technique developed by the authors of the book, called *Multi-dimensional Particle Swarm Optimization (MD PSO)*, which strives to address the above requirements following an algorithmic approach to solve important engineering problems. Some of the more complex problems are formulated in a multi-dimensional search space where the optimum dimension is also unknown. In this case, MD PSO can seek for both positional and dimensional optima. Furthermore, two supplementary enhancement methods, the *Fractional Global-Best Formation* and *Stochastic Approximation with Simultaneous Perturbation*, are introduced as an efficient cure to avoid getting trapped in local optima especially in multi-modal search spaces defined over high dimensions. The book covers a wide range of fundamental application areas, which can particularly benefit from such a unified framework. Consider for instance a data clustering application where MD PSO can be used to determine the true number of clusters and accurate cluster centroids, in a single framework. Another application in the field of machine intelligence is to determine the optimal neural network configuration for a particular problem. This might be a crucial step, e.g., for robust and accurate detection of electrocardiogram (ECG) heartbeat patterns for a specific patient. The reader will see that this system can adapt to significant inter-patient variations in ECG patterns by *evolving* the *optimal* classifier and thus achieves a high accuracy over large datasets.

The proposed unified framework is then explored in a set of challenging application domains, namely data mining and content-based multimedia classification. Although there are numerous efforts for the latter, we are still in the early stages of the development to guarantee a satisfactory level of efficiency and accuracy. To accomplish this for content-based image retrieval (CBIR) and classification, the book presents a global framework design that embodies a collective network of evolutionary classifiers. This is a dynamic and adaptive topology, which allows the creation and design of a dedicated classifier for discriminating a certain image class from the others based on a single visual descriptor. During an evolution session, new images, classes, or features can be introduced whilst signaling the classifier network to create new corresponding networks and classifiers within, to dynamically adapt to the change. In this way the collective classifier network will be able to scale itself to the indexing requirements of the image content data reserve whilst striving for maximizing the classification and retrieval accuracies for better user experience. However one obstacle still remains: low-level features play the most crucial role in CBIR but they usually lack the discrimination power needed for accurate visual description and representation especially in the case of large and dynamic image data reserves. Finally, the book tackles this major research objective and presents an evolutionary feature synthesis framework, which aims to significantly improve the discrimination power by synthesizing highly discriminative features. This is obviously not limited to only CBIR, but can be utilized to synthesize enhanced features for any application domain where features or feature extraction is involved.

The set of diverse applications presented in the book points the way to explore a wide range of potential applications in engineering as well as other disciplines. The book is supplemented with C/C++ source codes for all applications and many sample datasets to illustrate the major concepts presented in the book. This will allow practitioners and professionals to comprehend and use the presented techniques and adapt them to their own applications immediately.

Contents

1 Introduction 1
   1.1 Optimization Era 2
   1.2 Key Issues 4
   1.3 Synopsis of the Book 7
   References 10

2 Optimization Techniques: An Overview 13
   2.1 History of Optimization 13
   2.2 Deterministic and Analytic Methods 29
      2.2.1 Gradient Descent Method 29
      2.2.2 Newton–Raphson Method 30
      2.2.3 Nelder–Mead Search Method 32
   2.3 Stochastic Methods 33
      2.3.1 Simulated Annealing 33
      2.3.2 Stochastic Approximation 35
   2.4 Evolutionary Algorithms 37
      2.4.1 Genetic Algorithms 37
      2.4.2 Differential Evolution 41
   References 43

3 Particle Swarm Optimization 45
   3.1 Introduction 45
   3.2 Basic PSO Algorithm 46
   3.3 Some PSO Variants 49
      3.3.1 Tribes 51
      3.3.2 Multiswarms 53
   3.4 Applications 55
      3.4.1 Nonlinear Function Minimization 55
      3.4.2 Data Clustering 57
      3.4.3 Artificial Neural Networks 61
   3.5 Programming Remarks and Software Packages 74
   References 80

4   Multi-dimensional Particle Swarm Optimization   83
   4.1   The Need for Multi-dimensionality.   83
   4.2   The Basic Idea.   85
   4.3   The MD PSO Algorithm   87
   4.4   Programming Remarks and Software Packages   92
       4.4.1   MD PSO Operation in *PSO_MDlib* Application   92
       4.4.2   MD PSO Operation in *PSOTestApp* Application.   94
   References   99

5   Improving Global Convergence.   101
   5.1   Fractional Global Best Formation.   102
       5.1.1   The Motivation   102
       5.1.2   PSO with FGBF.   102
       5.1.3   MD PSO with FGBF   104
       5.1.4   Nonlinear Function Minimization   104
   5.2   Optimization in Dynamic Environments   116
       5.2.1   Dynamic Environments: The Test Bed   116
       5.2.2   Multiswarm PSO   117
       5.2.3   FGBF for the Moving Peak Benchmark for MPB. . . .   118
       5.2.4   Optimization over Multidimensional MPB.   119
       5.2.5   Performance Evaluation on Conventional MPB   120
       5.2.6   Performance Evaluation on Multidimensional
            MPB.   124
   5.3   Who Will Guide the Guide?   128
       5.3.1   SPSA Overview   130
       5.3.2   SA-Driven PSO and MD PSO Applications.   131
       5.3.3   Applications to Non-linear Function Minimization . . .   134
   5.4   Summary and Conclusions.   141
   5.5   Programming Remarks and Software Packages   142
       5.5.1   FGBF Operation in PSO_MDlib Application   143
       5.5.2   MD PSO with FGBF Application Over MPB.   144
   References   147

6   Dynamic Data Clustering   151
   6.1   Dynamic Data Clustering via MD PSO with FGBF   152
       6.1.1   The Theory   152
       6.1.2   Results on 2D Synthetic Datasets   155
       6.1.3   Summary and Conclusions.   160
   6.2   Dominant Color Extraction   160
       6.2.1   Motivation.   160
       6.2.2   Fuzzy Model over HSV-HSL Color Domains   163
       6.2.3   DC Extraction Results.   164
       6.2.4   Summary and Conclusions.   170
   6.3   Dynamic Data Clustering via SA-Driven MD PSO   171

6.3.1 SA-Driven MD PSO-Based Dynamic Clustering
in 2D Datasets 171
6.3.2 Summary and Conclusions..................... 174
6.4 Programming Remarks and Software Packages 176
6.4.1 FGBF Operation in 2D Clustering 176
6.4.2 DC Extraction in *PSOTestApp* Application 179
6.4.3 SA-DRIVEN Operation in *PSOTestApp*
Application 183
References 185

7 Evolutionary Artificial Neural Networks 187
7.1 Search for the Optimal Artificial Neural Networks:
An Overview 188
7.2 Evolutionary Neural Networks by MD PSO.............. 190
7.2.1 PSO for Artificial Neural Networks:
The Early Attempts.......................... 190
7.2.2 MD PSO-Based Evolutionary Neural Networks 191
7.2.3 Classification Results on Synthetic Problems....... 193
7.2.4 Classification Results on Medical Diagnosis
Problems.................................. 200
7.2.5 Parameter Sensitivity and Computational
Complexity Analysis......................... 203
7.3 Evolutionary RBF Classifiers for Polarimetric SAR Images... 205
7.3.1 Polarimetric SAR Data Processing 207
7.3.2 SAR Classification Framework................. 209
7.3.3 Polarimetric SAR Classification Results 211
7.4 Summary and Conclusions.......................... 217
7.5 Programming Remarks and Software Packages 218
References 227

8 Personalized ECG Classification 231
8.1 ECG Classification by Evolutionary Artificial
Neural Networks 233
8.1.1 Introduction and Motivation................... 233
8.1.2 ECG Data Processing 235
8.1.3 Experimental Results 239
8.2 Classification of Holter Registers 244
8.2.1 The Related Work 245
8.2.2 Personalized Long-Term ECG Classification:
A Systematic Approach....................... 246
8.2.3 Experimental Results 250
8.3 Summary and Conclusions.......................... 253
8.4 Programming Remarks and Software Packages 255
References 257

**9   Image Classification and Retrieval by Collective Network
     of Binary Classifiers** 259
     9.1   The Era of CBIR 260
     9.2   Content-Based Image Classification and Retrieval
            Framework 262
            9.2.1   Overview of the Framework 263
            9.2.2   Evolutionary Update in the Architecture Space 264
            9.2.3   The Classifier Framework: Collective Network
                     of Binary Classifiers 265
     9.3   Results and Discussions. 270
            9.3.1   Database Creation and Feature Extraction 271
            9.3.2   Classification Results 272
            9.3.3   CBIR Results. 277
     9.4   Summary and Conclusions. 280
     9.5   Programming Remarks and Software Packages 281
     References 293

10   Evolutionary Feature Synthesis 295
     10.1   Introduction 295
     10.2   Feature Synthesis and Selection: An Overview. 297
     10.3   The Evolutionary Feature Synthesis Framework 299
            10.3.1   Motivation. 299
            10.3.2   Evolutionary Feature Synthesis Framework 301
     10.4   Simulation Results and Discussions 306
            10.4.1   Performance Evaluations with Respect
                     to Discrimination and Classification 307
            10.4.2   Comparative Performance Evaluations
                     on Content-Based Image Retrieval 309
     10.5   Programming Remarks and Software Packages 314
     References 321

Acronyms

2D	Two Dimensional
AAMI	Association for the Advancement of Medical Instrumentation
AC	Agglomerative Clustering
ACO	Ant Colony Optimization
aGB	Artificial Global Best
AI	Artificial Intelligence
ANN	Artificial Neural Network
API	Application Programming Interface
AS	Architecture Space
AV	Audio-Visual
BbNN	Block-based Neural Networks
BP	Back Propagation
bPSO	Basic PSO
CBIR	Content-Based Image Retrieval
CGP	Co-evolutionary Genetic Programming
CLD	Color Layout Descriptor
CM	Color Moments
CNBC	Collective Network of Binary Classifiers
CPU	Central Processing Unit
CV	Class Vector
CVI	Clustering Validity Index
DC	Dominant Color
DE	Differential Evolution
DFT	Discrete Fourier Transform
DLL	Dynamic Link Library
DP	Dynamic Programming
EA	Evolutionary Algorithm
ECG	Electrocardiogram
ECOC	Error Correcting Output Code
EFS	Evolutionary Feature Syntheses
EHD	Edge Histogram Descriptor
EM	Expectation-Maximization
ENN	Evolutionary Neural Networks

EP	Evolutionary Programming
ES	Evolution Strategies
FCM	Fuzzy C-means
FDSA	Finite Difference Stochastic Approximation
FeX	Feature Extraction
FF	Fundamental Frequency
FFT	Fast Fourier Transform
FGBF	Fractional Global Best Formation
FT	Fourier Transform
FV	Feature Vector
GA	Genetic Algorithm
GB	Global Best
GLCM	Gray Level Co-occurrence Matrix
GMM	Gaussian Mixture Model
GP	Genetic Programming
GTD	Ground Truth Data
GUI	Graphical User Interface
HMM	Hidden Markov Model
HSV	Hue, Saturation and (Luminance) Value
HVS	Human Visual System
KF	Key-Frame
KHM	K-Harmonic Means
KKT	Karush–Kuhn–Tucker
KLT	Karhunen–Loéve Transform
kNN	k Nearest Neighbours
LBP	Local Binary Pattern
LP	Linear Programming
MAP	Maximum a Posteriori
MDA	Multiple Discriminant Analysis
MD PSO	Multi-dimensional Particle Swarm Optimization
ML	Maximum-Likelihood
MLP	Multilayer Perceptron
MPB	Moving Peaks Benchmark
MRF	Markov Random Field
MSE	Mean-Square Error
MST	Minimum Spanning Tree
MUVIS	Multimedia Video Indexing and Retrieval System
NLP	Nonlinear Programming
P	Precision
PCA	Principal Component Analysis
PNR	Positive to Negative Ratio
PSO	Particle Swarm Optimization
R	Recall
RBF	Radial Basis Function
RF	Random Forest

RGB	Red, Green and Blue
SA	Stochastic Approximation
SAR	Synthetic Aperture Radar
SCD	Scalable Color Descriptor
SIFT	Scale-Invariant Feature Transform
SLP	Single-Layer Perceptron
SO	Stochastic Optimization
SOC	Self-Organized Criticality
SOM	Self-Organizing Maps
SPSA	Simultaneously Perturbed Stochastic Approximation
SVEB	Supra-ventricular Ectopic Beats
SVM	Support Vector Machines
TI-DWT	Translation-Invariant Dyadic Wavelet Transform
QP	Query Path
UI	User Interface
VEB	Ventricular Ectopic Beats

Tables

Table 2.1	Pseudo-code for generic line search method.	30
Table 2.2	Pseudo-code of the simulated annealing algorithm	34
Table 3.1	Pseudo-code for the bPSO algorithm.	48
Table 3.2	A sample architecture space for MLP configuration sets $R_{\min} = \{9, 1, 1, 2\}$ and $R_{\max} = \{9, 8, 4, 2\}$	66
Table 3.3	The Overall Test CE Statistics	73
Table 3.4	The data structure, **PSOparam**.	75
Table 3.5	Initialization of the PSO swarm	77
Table 3.6	The main loop for (MD) PSO in **Perform**() function.	77
Table 3.7	Implementation of Step 3.1 of the PSO pseudo-code given in Table 3.1	78
Table 3.8	The termination of a PSO run.	78
Table 3.9	Implementation of Step 3.4 of the PSO pseudo-code given in Table 3.1	79
Table 4.1	Pseudo-code of MD PSO algorithm	91
Table 4.2	Implementation of Step 3.4 of the MD PSO pseudo-code given in Table 4.1.	93
Table 4.3	The callback function **OnDopso**() activated when pressed "Run" button on **PSOtestApp** GUI.	96
Table 4.4	Member functions of **CPSOcluster** class.	97
Table 4.5	The **ApplyPSO**() API function of the **CPSOcluster** class..	98
Table 4.6	The function **CPSOcluster::PSOThread**().	99
Table 5.1	Pseudo-code of FGBF in *bPSO*	103
Table 5.2	Pseudo-code for FGBF in MD PSO	105
Table 5.3	Benchmark functions with dimensional bias.	106
Table 5.4	Statistical results from 100 runs over 7 benchmark functions	112
Table 5.5	Best known results on the MPB.	123
Table 5.6	Offline error using *Scenario* 2	123
Table 5.7	Offline error on extended MPB	128
Table 5.8	Pseudo-code for *SPSA* technique	130

Table 5.9   Pseudo-code for the first *SA-driven PSO* approach   132
Table 5.10   PSO Plug-in for the second approach   133
Table 5.11   MD PSO Plug-in for the second approach   134
Table 5.12   Benchmark functions without dimensional bias   135
Table 5.13   Statistical results from 100 runs over seven
             benchmark functions   136
Table 5.14   Statistical results between full-cost and low-cost modes
             from 100 runs over seven benchmark functions   139
Table 5.15   *t* test results for statistical significance analysis
             for both SPSA approaches, A1 and A2   140
Table 5.16   t table presenting degrees of freedom vs. probability   140
Table 5.17   Implementation of FGBF pseudo-code given in Table 5.2 . .   144
Table 5.18   The environmental change signaling from the main
             MD PSO function.   145
Table 5.19   MD PSO with FGBF implementation for MPB.   146
Table 5.20   The fitness function **MPB()**.   147
Table 6.1   Processing time (in msec) per iteration for MD PSO
             with FGBF clustering using 4 different swarm sizes.
             Number of data items is presented in parenthesis
             with the sample data space.   159
Table 6.2   Statistical results from 20 runs over 8 2D
             data spaces   173
Table 6.3   MD PSO initialization in the function
             CPSOcluster::PSOThread().   177
Table 6.4   MST formation in the function **CPSO_MD
             <T,X>::FGBF_CLFn()**.   178
Table 6.5   MST formation in the function **CPSO_MD
             <T,X>::FGBF_CLFn()**.   179
Table 6.6   Formation of the centroid groups by breaking
             the MST iteratively.   180
Table 6.7   Formation of the *aGB* particle.   181
Table 6.8   The CVI function, **CPSOcluster::ValidityIndex2**.   182
Table 6.9   Initialization of DC extraction
             in **CPSOcolorQ::PSOThread()** function   183
Table 6.10   The plug-in function **SADFn()** for the second
             SA-driven approach, A2.   184
Table 6.11   The plug-in for the first SA-driven approach, A1.   185
Table 7.1   Mean (μ) and standard deviation (σ) of classification
             error rates (%) over test datasets.   203
Table 7.2   A sample architecture space with range arrays,
             $R_{min}^2 = \{13, 6, 6, 3, 2\}$ and $R_{max}^2 = \{13, 12, 10, 5, 2\}$ 204
Table 7.3   Classification error rate (%) statistics of MD PSO
             when applied to two architecture spaces.   204

Table 7.4   Summary table of pixel-by-pixel classification results
             of the RBF-MDPSO classifier over the training
             and testing area of San Francisco Bay dataset   212
Table 7.5   Overall performance comparison (in percent)
             for San Francisco Bay dataset   213
Table 7.6   Overall performance (in percent) using smaller training
             set (< %1 of total pixels) for San Francisco Bay dataset . . .   214
Table 7.7   Overall performance comparison (in percent)
             for Flevoland dataset.   215
Table 7.8   Summary table of pixel-by-pixel classification results
             (in percent) of the RBF-MDPSO classifier over
             the training and testing data of Flevoland   216
Table 7.9   Some members and functions of **CMLPnet** class.   219
Table 7.10   Some members and functions of **CMLPnet** class.   220
Table 7.11   **CDimHash** class members and functions.   221
Table 7.12   **CGenClassifier** class members and functions.   222
Table 7.13   The main entry function for evolutionary ANNs
             and classification.   223
Table 7.14   The function evolving MLPs by MD PSO.   225
Table 7.15   The fitness function of MD PSO process
             in the **CMLP_PSO** class.   226
Table 7.16   The constructors of the **CSolSpace** class.   227
Table 8.1   Summary table of beat-by-beat classification results
             for all 44 records in the MIT/BIH arrhythmia database.
             Classification results for the testing dataset only
             (24 records from the range 200 to 234)
             are shown in parenthesis.   242
Table 8.2   VEB and SVEB classification performance of the presented
             method and comparison with the three major algorithms
             from the literature.   243
Table 8.3   VEB and SVEB classification accuracy of the
             classification system for different PSO parameters
             and architecture spaces.   244
Table 8.4   Overall results for each patient in the MIT-BIH
             long-term database using the systematic approach
             presented. For each class, the number of correctly
             detected beats is shown relative to the total
             beats originally present.   252
Table 8.5   The overall confusion matrix   253
Table 8.6   The function **CPSOclusterND::PSOThread**().   256
Table 9.1   14 Features extracted per MUVIS database.   272
Table 9.2   The architecture space used for MLPs.   273

Table 9.3   Average classification performance of each evolution
method per feature set by 10-fold random train/test set
partitions in *Corel_10* database. The best classification
performances in the test set are highlighted. 274

Table 9.4   Confusion matrix of the evolution method, which produced
the best (lowest) test classification error in Table 9.3. 275

Table 9.5   Test dataset confusion matrices for evolution stages
1 (*top*), 2 (*middle*) and 3 (*bottom*) 276

Table 9.6   Final classification performance of the 3-stage
incremental evolution for each evolution method
and feature set for *Corel_10* database. 276

Table 9.7   Classification performance of each evolution method
per feature set for *Corel_Caltech_30* database. 277

Table 9.8   Retrieval performances (%) of the four batch queries
in each MUVIS databases. 278

Table 9.9   Retrieval performances per incremental evolution stage
and traditional (without CNBC) method. 279

Table 9.10   The CBNC specific data structures. 284

Table 9.11   The function: **RunFeXfiles(int run)**. 286

Table 9.12   The function: **TrainCNBC(int run)**. 287

Table 9.13   The class: **CNBCglobal**. 288

Table 9.14   The member function **TrainDistributed()** of the class
CNBCglobal. 289

Table 9.15   The class: **COneNBC**. 291

Table 9.16   The member function **Train()** of the class **COneNBC**. 292

Table 10.1   A sample set of $F = 18$ operators used in evolutionary
synthesis. 303

Table 10.2   A sample target vector encoding for 4 classes, $c_1, \ldots, c_4$. . . . 305

Table 10.3   Discrimination measure (DM) and the number
of false positives (FP) for the original features. 307

Table 10.4   Statistics of the discrimination measure (DM), the number
of false positives (FP) and the corresponding output
dimensionalities for features synthesized by the first
run of EFS over the entire database. 308

Table 10.5   Statistics of the DM, FP and the corresponding output
dimensionalities for features synthesized by the first run
of the EFS evolved with the ground truth data over
45 % of the database. 308

Table 10.6   Test CEs for the original features 308

Table 10.7   Test CE statistics and the corresponding output
dimensionalities for features synthesized by a single run
of the EFS with 45 % EFS dataset. 308

Table 10.8   ANMRR and AP measures using original low-level
              features. "*All*" refers to all features are considered.   309
Table 10.9   ANMRR and AP statistics obtained by batch queries
              over the class vectors of the single SLP.   309
Table 10.10   ANMRR and AP statistics for features synthesized
              by a single run of the EFS when the output dimensionality
              is fixed to $C = 10$ and operator selection is limited
              to addition and subtraction.   310
Table 10.11   ANMRR and AP statistics for features synthesized
              by a single run of the EFS when the output dimensionality
              is fixed to $C = 10$ and all operators are used..   310
Table 10.12   ANMRR and AP statistics and the corresponding output
              dimensionalities for the synthesized features by a single
              EFS run using the fitness function in Eq. (10.4).   311
Table 10.13   Retrieval performance statistics and the corresponding
              output dimensionalities for the final synthesized features
              by several EFS runs using the fitness function
              in Eq. (10.4).   311
Table 10.14   ANMRR and AP statistics for the features synthesized
              by the best MLP configuration evolved by MD PSO.   314
Table 10.15   ANMRR and AP statistics for the features synthesized
              by the concatenated SLPs.   314
Table 10.16   The class **CPSOFeatureSynthesis**.   315
Table 10.17   The function: **CPSOFeatureSynthesis::PSOThread()**
              (part-1)   317
Table 10.18   The function: **CPSOFeatureSynthesis::PSOThread()**
              (part-2).   318
Table 10.19   The function: **CPSOFeatureSynthesis::FitnessFnANN()** . . .   319
Table 10.20   The function: **CPSOFeatureSynthesis::**
              ApplyFSadaptive().   320

Figures

Fig. 1.1   Some sample uni- and multi-modal functions.   4
Fig. 1.2   Sample clustering operations in 2D data space using K-means
            method where the true K value must be set before..   5
Fig. 2.1   Iterations of the fixed (*left*) and optimum $\alpha(k)$ (*right*) line
            search versions of gradient descent algorithm plotted
            over the Rosenbrock objective function, $x(0) = [-0.5, 1.0]$ . . .   31
Fig. 2.2   (*Left*) Iterations of the Quasi-Newton method plotted over
            Rosenbrock function, $x_0 = [-0.5, 1.0]$. (*Right*) Iterations
            of the gradient descent (*red*) vs. Quasi-Newton (*black*)
            methods plotted over a quadratic objective function,
            $x(0) = [10, 5]$ 32
Fig. 2.3   *Left:* Iterations of the Nelder–Mead method plotted over
            Rosenbrock function with $x(0) = [-0.5, 1.0]$. The vertices w
            ith the minimum function values are only plotted. *Right:*
            consecutive simplex operations during the iterations 3–30. . . .   33
Fig. 2.4   The plot of 1,532 iterations of the simulated annealing
            method over Rosenbrock function with $x(0) = [-0.5, 1.0]$,
            $\varepsilon_C = 10^{-3}$, $T_0 = 1$, $\psi(T) = 0.95T$, $N(x_0) = x_0 + 0.01x_{range}r$
            where $r \in N(0, 1)$, and $x_{range}$is the dimensional
            range, i.e., $x_{range} = 2 - (-2) = 4$ 35
Fig. 2.5   The plot of 25,000 iterations of the FDSA method over
            Rosenbrock function with $x(0) = [-0.5, 1.0]$, $\varepsilon_C = 10^{-3}$,
            $a = 20$, A $= 250$, $c = 1$, $\upsilon = 1$, and $\tau = 0.75$ 37
Fig. 2.6   A sample chromosome representation for the two problem
            variables, a and b   38
Fig. 2.7   A sample crossover operation over two chromosomes g_i
            and g_j after $L = 10$ bits. The resultant child chromosomes
            are c_i and c_j.   39
Fig. 2.8   The distributions of real-valued GA population
            for generations, $g = 1, 160, 518$ and 913 over Rosenbrock
            function with $S = 10$, $Px = 0.8$ and σ linearly decreases
            from x_{range} to 0.   40

Fig. 2.9   A sample 2-D fitness function and the DE process
            forming the trial vector   42
Fig. 2.10   The distributions of DE population for generations,
            g = 1, 20, 60 and 86 over Rosenbrock function with
            S = 10, F = 0.8 and R = 0.1.   43
Fig. 3.1   Illustration of the basic velocity update mechanism in PSO . . .   47
Fig. 3.2   Particle velocity tendency to explode without
            velocity clamping   49
Fig. 3.3   Local PSO topologies   54
Fig. 3.4   Some benchmark functions in 2D   56
Fig. 3.5   The plots of the *gbest* particle's personal best scores
            for the four sample non-linear functions given in Fig. 3.4.
            a Sphere. b Giunta. c Rastrigin. d Griewank   56
Fig. 3.6   11 synthetic data spaces in 2D   59
Fig. 3.7   PSO clustering for 2D data spaces C1–C4 shown
            in Fig. 3.6   60
Fig. 3.8   Erroneous PSO clustering over data spaces C4, C5, C8
            and C9 shown in Fig. 3.6   60
Fig. 3.9   An example of a fully-connected feed-forward ANN   62
Fig. 3.10   Train (*top*) and test (*bottom*) error statistics vs. hash index
            plots from *shallow* BP- and PSO-training over the
            breast cancer dataset.   68
Fig. 3.11   Train (*top*) and test (*bottom*) error statistics vs. hash index
            plots from *deep* BP- and PSO-training over the
            breast cancer dataset.   69
Fig. 3.12   Train (*top*) and test (*bottom*) error statistics vs. hash index
            plots from *shallow* BP- and PSO-training over the
            heart disease dataset   70
Fig. 3.13   Train (*top*) and test (*bottom*) error statistics vs. hash index
            plots from *deep* BP- and PSO-training over the
            heart disease dataset   71
Fig. 3.14   Train (*top*) and test (*bottom*) error statistics vs. hash index
            plots from *shallow* BP- and PSO-training over the
            diabetes dataset   72
Fig. 3.15   Train (*top*) and test (*bottom*) error statistics vs. hash index
            plots from *deep* BP- and PSO-training over the *diabetes*
            dataset.   73
Fig. 3.16   Error statistics (for network configuration with the hash index
            d = 16) vs. training depth plots using BP and PSO over
            the *breast cancer* (*top*), *heart disease* (*middle*), and *diabetes*
            (*bottom*) datasets.   74
Fig. 4.1   2D synthetic data spaces carrying different
            clustering schemes   84

Fig. 4.2   An Illustrative MD PSO process during which particles
            7 and 9 have just moved 2D and 3D solution spaces at time t;
            whereas particle a is sent to 23rd dimension   86
Fig. 4.3   Sample MD PSO (*right*) vs. bPSO (*left*) particle structures.
            For MD PSO$\{D_{min} = 2, D_{max} = 10\}$ and at time t,
            $xd_a(t) = 2$ and $\tilde{xd}_a(t) = 3$ 88
Fig. 4.4   GUI of **PSOTestApp** with several MD PSO applications (*top*)
            and MD PSO Parameters dialog is activated when pressed
            "Run" button (*bottom*)   95
Fig. 5.1   A sample FGBF in 2D space   103
Fig. 5.2   Fitness score (*top* in log-scale) and dimension (*bottom*)
            plots vs. iteration number for MD PSO (*left*)
            and *bPSO* (*right*) operations both of which run over
            De Jong function   107
Fig. 5.3   Fitness score (*top* in log-scale) and dimension (*bottom*)
            plots vs. iteration number for a MD PSO run over *Sphere*
            function with (*left*) and without (*right*) FGBF   108
Fig. 5.4   Particle index plot for the MD PSO with FGBF operation
            shown in Fig. 5.3.   108
Fig. 5.5   Fitness score (*top* in log-scale) and dimension (*bottom*)
            plots vs. iteration number for a MD PSO run over *Schwefel*
            function with (*left*) and without (*right*) FGBF   109
Fig. 5.6   Fitness score (*top* in log scale) and dimension (*bottom*)
            plots vs. iteration number for a MD PSO run over *Giunta*
            function with (*left*) and without (*right*) FGBF   109
Fig. 5.7   MD PSO with FGBF operation over *Griewank* (*top*)
            and *Rastrigin* (*bottom*) functions with $d_0 = 20$ (*red*) and
            $d_0 = 80$ (*blue*) using the swarm size, $S = 80$ (*left*)
            and $S = 320$ (*right*).   110
Fig. 5.8   Current error at the beginning of a run   121
Fig. 5.9   Effect of multi-swarms on results   122
Fig. 5.10   Effect of FGBF on results   122
Fig. 5.11   Optimum dimension tracking in a MD PSO run   124
Fig. 5.12   Current error at the beginning of a MD PSO run   125
Fig. 5.13   Optimum dimension tracking without multi-swarms
            in a MD PSO run   126
Fig. 5.14   Effect of multi-swarms on the performance   126
Fig. 5.15   Optimum dimension tracking *without* FGBF
            in a MD PSO run   127
Fig. 5.16   Effect of FGBF on the performance   127
Fig. 6.1   The formation of the centroid subset in a sample clustering
            example. The black dots represent data points over 2D
            space and each *colored* ' + ' represents one centroid
            (dimension) of a swarm particle   154

Fig. 6.2   Typical clustering results via MD PSO with FGBF.
            Over-clustered samples are indicated with *.   156
Fig. 6.3   Fitness score (*top*) and dimension (*bottom*) plots vs. iteration
            number for a MD PSO with FGBF clustering operation
            over $C4$. 3 clustering snapshots at iterations 105, 1,050
            and 1,850, are presented below.   157
Fig. 6.4   Fitness score (*top*) and dimension (*bottom*) plots vs. iteration
            number for a MD PSO with FGBF clustering operation
            over $C9$. 3 clustering snapshots at iterations 40, 950
            and, 1,999, are presented below   158
Fig. 6.5   Particle index plot for the MD PSO with FGBF clustering
            operation shown in Fig. 6.4   158
Fig. 6.6   Fuzzy model for distance computation in HSV and HSL color
            domains (*best viewed in color*).   164
Fig. 6.7   Number of DC plot from three MPEG-7 DCDs
            with different parameter set over the sample database.   165
Fig. 6.8   The DC extraction results over 5 images from the sample
            database (*best viewed in color*).   166
Fig. 6.9   The DC extraction results over 5 images from the sample
            database (*best viewed in color*).   167
Fig. 6.10   DC number histograms of 2 sample images using
            3 parameter sets. Some typical back-projected images
            with their DC number pointed are shown within the
            histogram plots (*best viewed in color*).   169
Fig. 6.11   2D synthetic data spaces carrying different
            clustering schemes   171
Fig. 6.12   Some clustering runs with the corresponding
            fitness scores (*f*)   174
Fig. 6.13   The worst and the best clustering results using
            standalone (*left*) and SA-driven (*right*) MD PSO   175
Fig. 7.1   The function $y = \cos(x/2)\sin(8x)$ plot in interval
            $\{-\pi, \pi\}$ with 100 samples   195
Fig. 7.2   Error statistics from exhaustive BP training (*top*) and *dbest*
            histogram from 100 MD PSO evolutions (*bottom*)
            for $y = \cos(x/2)\sin(8x)$ function approximation.   196
Fig. 7.3   Training MSE (*top*) and dimension (bottom) plots vs. iteration
            number for 17th (*left*) and 93rd (*right*) MD PSO runs.   197
Fig. 7.4   MSE plots from the exhaustive BP training (*top*) and a single
            run of MD PSO (*bottom*).   198
Fig. 7.5   Error statistics from exhaustive BP training (*top*) and *dbest*
            histogram from 100 MD PSO evolutions (*bottom*)
            for 10-bit parity problem   198

Fig. 7.6   Error (MSE) statistics from exhaustive BP training (*top*)
            and *dbest* histogram from 100 MD PSO evolutions (*bottom*)
            for the *two-spirals* problem   199
Fig. 7.7   Error statistics from exhaustive BP training over
            Breast Cancer (*top*) and *Diabetes* (*bottom*) datasets.   202
Fig. 7.8   Error statistics from exhaustive BP training (*top*)
            and *dbest* histogram from 100 MD PSO evolutions (*bottom*)
            over the *Heart Disease* dataset   202
Fig. 7.9   Overview of the evolutionary RBF network classifier
            design for polarimetric SAR image   210
Fig. 7.10   Pauli image of 600×600 pixel subarea of San Francisco
            Bay (*left*) with the 5×5 refined Lee filter used. The training
            and testing areas for three classes are shown using red
            rectangles and circles respectively. The aerial photograph
            for this area (*right*) provided by the U.S. Geological Survey
            taken on Oct, 1993 can be used as ground-truth   211
Fig. 7.11   The classification results of the RBF-MDPSO classifier
            over the 600×600 sub-image of San Francisco Bay
            (*black* denotes *sea*, *gray* urban areas, *white*
            vegetated zones)   213
Fig. 7.12   The classification results of the RBF-MDPSO technique
            for the original (900×1024) San Francisco Bay image
            (*black* denotes sea, *gray* urban areas, *white*
            vegetated zones)   214
Fig. 7.13. Fitness score (*left top*) and dimension (*left bottom*) plots
            vs. iteration number for a typical MD PSO run. The resulting
            histogram plot (*right*) of cluster numbers which are
            determined by the MD PSO method.   215
Fig. 7.14   The classification results on the L-band AIRSAR data
            over Flevoland   217
Fig. 8.1   Patient-specific ECG classification system   232
Fig. 8.2   Sample beat waveforms, including Normal (N), PVC (V),
            and APC (S) AAMI heartbeat classes, selected from
            record 201 modified-lead II from the MIT/BIH arrhythmia
            database and corresponding TI DWT decompositions
            for the first five scales.   236
Fig. 8.3   Power spectrum of windowed ECG signal from record 201
            for Normal (N), PVC (V), and APC (S) AAMI heartbeat
            classes, and equivalent frequency responses of FIR digital
            filters for a quadratic spline wavelet at 360 Hz
            sampling rate   237
Fig. 8.4   Scatter plot of Normal (N), PVC (V), and APC (S) beats
            from record 201 in terms of the first and third principal
            components and RR_i time interval   239

Fig. 8.5   Error (MSE) statistics from exhaustive BP training (*top*)
           and *dbest* histogram from 100 MD PSO evolutions (*bottom*)
           for patient record 222   241
Fig. 8.6   The overview of the systematic approach for long-term
           ECG classification   247
Fig. 8.7   Sample beat waveforms, including Normal (N), PVC (V),
           and APC (S) AAMI [13] heartbeat classes from the
           MIT-BIH database. Heartbeat fiducial point intervals
           (RR-intervals) and ECG morphology features
           (samples of QRS complex and T-wave) are extracted   248
Fig. 8.8   Excerpt of raw ECG data from patient record 14,046
           in the MIT-BIH long-term database. The three key-beats,
           having morphological and RR-interval differences,
           are chosen by the systematic approach presented   251
Fig. 8.9   Excerpt of raw ECG data from patient record 14,172
           in the MIT-BIH long-term database. The key-beats extracted
           by the systematic approach are indicated.   251
Fig. 9.1   The overview of the CBIR framework   263
Fig. 9.2   Evolutionary update in a sample AS for MLP configuration
           arrays $R_{min} = \{15, 1, 2\}$ and $R_{max} = \{15, 4, 2\}$ where
           $N_R = 3$ and $N_C = 5$. The best runs for each configurations
           are highlighted and the best configuration in each run
           is tagged with '*'.   265
Fig. 9.3   Topology of the CNBC framework with C classes
           and N FVs.   266
Fig. 9.4   Illustration of the two-phase evolution session over
           BCs' architecture spaces in each NBC   268
Fig. 9.5   8 sample queries in *Corel_10* (qA and qB), and
           Corel_Caltech_30 (qC and qD) databases with
           and without CNBC. The *top-left* image is the query image . . .   279
Fig. 9.6   Two sample retrievals of sample queries qA and qB,
           performed at each stage from classes 2 and 6.
           The *top-left* is the query image.   280
Fig. 10.1 An illustrative EFS, which is applied to 2D feature vectors
           of a 3-class dataset   299
Fig. 10.2 Two sample feature synthesis performed on 2-D (FS-1)
           and 1-D (FS-2) feature spaces   300
Fig. 10.3 The block diagram of Feature eXtraction (FeX) and the
           EFS technique with R runs   301
Fig. 10.4 Encoding jth dimensional component of the particle a
           in dimension d for K-depth feature synthesis   302
Fig. 10.5 Four sample queries using original (*left*) and synthesized
           features with single (*middle*) and four (*right*) runs.
           Top-left is the query image   313

Chapter 1
Introduction

God always takes the simplest way

Albert Einstein

Optimization as a generic term is defined by the Merriam-Webster dictionary as: *an act, process, or methodology of making something (as a design, system, or decision) as fully perfect, functional, or effective as possible; specifically: the mathematical procedures (as finding the maximum of a function) involved in this.*

Dante Aligheiri, around the year 1300, elevated the simple principle of optimization to a virtue:

All that is superfluous displeases God and Nature
All that displeases God and Nature is evil.

A number of medieval philosophers and thinkers defended the principle that nature strives for the optimal or the best path. For instance, the famous French mathematician Maupertuis proclaimed: *If there occur some changes in nature, the amount of action necessary for this change must be as small as possible.* This was also indicated by William of Occam, the most influential philosopher of the fourteenth century, who quoted the principle of Economics as: *Entities are not to be multiplied beyond necessity.* In science this is best known as: *What can be done with fewer is done in vain with more.* Above all, optimization is founded and developed by a certain type of ingenious and creative people, the so-called Engineers. As a common misconception, English speakers tend to think that the word "engineering" is related to the word of "engine," thus engineers are people who work with engines. In fact, the word "engineer" comes from the French word "ingénieur" which derives from the same Latin roots as the words "ingenuity" and "genius". Therefore, "Optimization" is the very essence of engineering as engineers (at least the good ones) are not interested with *any* solution of a given problem, but the best possible or *as fully perfect, functional, or effective as possible* one. In short, engineering is the art of creating optimum solutions and the optimality *therein* can be defined by the conditions and constraints of the problem in hand.

The first step of the solution lies in the mathematical modeling of the problem and its constraints. A mathematical model is needed for the proper representation of the variables, features, and constraints. Once the model is formulated in terms of a so-called "objective function," then an efficient mathematical optimization technique can be developed to search for the extremum point of the function,

S. Kiranyaz et al., *Multidimensional Particle Swarm Optimization for Machine Learning and Pattern Recognition*, Adaptation, Learning, and Optimization 15, DOI: 10.1007/978-3-642-37846-1_1, © Springer-Verlag Berlin Heidelberg 2014

which corresponds to the optimum solution of the problem. In mathematical terms, let $f : S \rightarrow R$ be the objective function from a set S to the real numbers. An optimization technique searches for the extremum point x_0 in S such that either $f(x_0) \leq f(x)$ or $f(x_0) \geq f(x)$ for all x in S. In this way the original problem within which an optimal solution is sought, is transformed to an equivalent (or sometimes approximating) function optimization problem. The original problem can be a multi-objective problem where there is more than one objective present. For example, one might wish to design a video encoder with the lowest possible complexity and highest compression rate. There will be one design with the lowest complexity but possibly with an inferior compression rate and another one with the highest compression rate but possibly with a very high complexity. Obviously, there will be an infinite number of encoders with some compromise of complexity and compression efficiency. As these objectives—usually—conflict with each other, a trade-off will naturally occur. One way to tackle this kind of problem is to perform a *regularization* technique, which will properly blend the two or multiple objectives into a single objective function.

1.1 Optimization Era

The optimization era started with the early days of Newton, Lagrange, and Cauchy. Particularly, the development of the mathematical foundations such as differential calculus methods that are capable of moving toward an optimum of a function was possible thanks to the contributions of Newton, Gauss, and Leibnitz to calculus. Cauchy proposed the first steepest descent method to solve unconstrained optimization problems. Furthermore, Bernoulli, Euler, Lagrange, Fermat, and Weistrass developed the foundations of function minimization in calculus while Lagrange invented the method of optimization for constrained problems using the unknown multipliers called after him, i.e., *Lagrange* multipliers. After the second half of the twentieth century, with the invention of digital computers, massive number of new techniques and algorithms were developed to solve complex optimization problems and such ongoing efforts stimulated further research on different and entirely new areas in optimization era. A major breakthrough was "linear programming", which was invented by George Dantzig. To name few other milestones in this area:

- Kuhn and Tucker in 1951 carried out studies later leading to the research on *Nonlinear programming*, which is the general case where the objective function or the constraints or both contain nonlinear parts.
- Bellman in 1957 presented the principle of optimality for *Dynamic programming* with an optimization strategy based on splitting the problem into smaller sub-problems. The equation given by his name describes the relationship between these sub-problems.

- *Combinatorial optimization* is a generic term for a set of optimization methods encapsulating operations research, algorithm theory, and computational complexity theory. Methods in this domain search for a set of feasible solutions in discrete spaces with the goal of finding the optimum solution where exhaustive (or sequential) search is not feasible. It has important applications in several fields, including artificial intelligence, machine learning, mathematics, and software engineering. It is also applied to certain optimization problems that involve uncertainty. For example, many real-world problems almost invariably include some unknown parameters.
- Dantzig, Charnes, and Cooper developed *Stochastic programming*, which studies the case in which some of the constraints or parameters depend on random variables.
- *Robust programming* is, like stochastic programming, an attempt to capture uncertainty in the data underlying the optimization problem. This is not done through the use of random variables, but instead, the problem is solved taking into account inaccuracies in the input data.
- Simulated Annealing and the family of evolutionary algorithms (EAs) are sometimes called *Meta-heuristics*, which make few or no assumptions about the problem being optimized and can thus search for the global optimum over a large set of candidate solutions. However, there is no guarantee that the optimal solution will ever be found.

Among all these methods and many others hereby unmentioned, it is hard to classify the optimization methods as one method can be classified into many categories. A crude classification can divide them into *linear* and *nonlinear* programming. The former intends to seek an optimal solution to problems represented by a set of linear equations. The Minimum Spanning Tree [1] and Simplex methods [2] are examples of linear programming. An example of problems solved by linear programming is that of the "traveling salesman," seeking a minimal traveling distance between two connected graph vertices. Nonlinear programming (NLP) attempts to solve problems by nonlinear equations and can be divided further into two sub-categories: *deterministic* and *stochastic*. The former performs an iterative search within the solution (or error) space based on the gradient information. In each iteration, the search is carried out toward the direction in which the function is minimized. This is different in stochastic methods, which perform the search through the rules of probability. This gives them advantage to be applicable to *any* optimization problem as they do not require the gradient information. Due to their extensive number, in this book, we can only discuss some major and related optimization methods in detail among a wide variety of global optimization techniques while giving an extensive survey on the amazing history of optimization.

1.2 Key Issues

Optimization problems are often multi-modal, meaning that there exist some deceiving local optima, as seen in the two examples illustrated in Fig. 1.1. This is one of the most challenging type of optimization problems since the optimum solution can be hard to find (e.g., see the last example in the figure)—if not impossible. Recall that deterministic optimization methods are based on the calculation of derivatives or their approximations. They converge to the position where the function gradient is null following the direction of the gradient vector. When the solution space of the problem is uni-modal, they are reliable, robust, and fast in finding the global optimum; however, due to their iterative approach they usually fail to find the global optimum in multi-modal problems. Instead, they get trapped into a local minimum. Random initialization of multiple runs may help find better local optima; however, finding the global solution is never guaranteed. For multi-modal problems with a large number of local optima, random initialization may result in random convergence, therefore; the results are unrepeatable and sub-optimum. Furthermore, the assumptions for their applications may seldom hold in practice, i.e., the derivative of the function may not be defined.

Besides function minimization in calculus, deterministic optimization methods are commonly used in several important application areas. For example in artificial neural networks, the well-known training method, back propagation (BP) [3], is indeed a gradient descent learning algorithm. The parameters (weights and biases) of a feed-forward neural network are randomly initialized and thus each individual BP run performs a gradient descent in the solution (error) space to converge to a (new) set of parameters. Another typical example is the expectation-maximization (EM) method, which finds the maximum likelihood or maximum a posteriori (MAP) estimates of the parameters in some statistical models. The most common models are Gaussian mixture models (GMMs) that are parametric probability density functions represented as a weighted sum of Gaussian probability density functions. GMMs are commonly used as a parametric model of the probability distribution of the properties or features in several areas in signal processing, pattern recognition, and many other related engineering fields. In order to determine the parameters of a particular GMM, EM is used as an iterative method which alternates between performing an expectation (E) step, which computes the

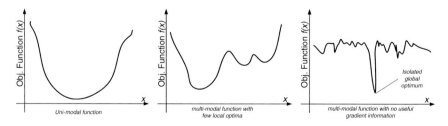

Fig. 1.1 Some sample uni- and multi-modal functions

expectation of the log-likelihood evaluated using the current estimate for the parameters, and a maximization (M) step, which computes the parameters maximizing the expected log-likelihood found in step E. These estimates are then used to determine the distribution of the latent variables in the next E step and so on. EM is a typical example of deterministic optimization methods, which performs greedy descent in the error space and if the step sizes are chosen properly, it can be easily shown that EM becomes identical to the gradient descent method. GMMs are especially useful as a data mining tool and frequently used to model data distributions and to cluster them. K-means [4] is another example of a deterministic optimization technique, and perhaps one of the most popular data clustering algorithm ever proposed. Similarly, when applied to complex clustering problems in high-dimensional data spaces, as a natural consequence of the highly multi-modal nature of the solution space, K-means cannot converge to the global optimum—meaning that the true clusters cannot be revealed and either over- or usually under-clustering occurs. See, for instance, the clustering examples in Fig. 4.1 where clusters are colored into red, green, and blue for a better visualization and the data points are represented by colored pixels with the cluster centroids shown by a white '+'. In the simulations, it took 4, 13, and 127 K-means runs, respectively, to extract the true clusters in the first three examples as shown in the figure. Even though we performed more than 10,000 runs, no individual run successfully clusters the last example with 42 clusters. This is an expected outcome considering the extremely multi-modal error space of the last example with a massive number of local optima.

Such a deficiency turned the attention toward stochastic optimization methods and particularly to evolutionary algorithms (EAs) such as genetic algorithm (GA) [5], genetic programming (GP) [6], evolution strategies (ES), [7] and evolutionary programming (EP), [8]. All EAs are population-based techniques which can often avoid being trapped in a local optimum; however, finding the optimum solutions is never guaranteed. On the other hand, another major drawback still remains unanswered for all, that is, the inability to find the true dimension of the solution space in which the global optimum resides. Many problems, such as data clustering, require this information in advance without which convergence to the global optimum is simply not possible, e.g., see the simple 2D data clustering examples in Fig. 1.2 where the true number of clusters (K) must be set in advance. In many clustering problems, especially complex ones with many clusters, this

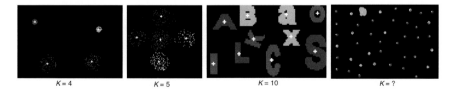

$K = 4$ $K = 5$ $K = 10$ $K = ?$

Fig. 1.2 Sample clustering operations in 2D data space using K-means method where the true K value must be set before

may not be feasible, if not impossible, to determine in advance, and thus the optimization method should find it along with the optimum solution in that dimension. This is also a major problem in many optimization methods mentioned earlier. For instance, BP can only train a feed-forward ANN without searching for the optimum configuration for the learning problem in hand. Therefore, what a typical BP run can indeed accomplish is the sub-optimum parameter setting of a sub-optimum ANN configuration.

All optimization methods so far mentioned and many more are applicable only to static problems. Many real-world problems are dynamic and thus require systematic re-optimizations due to system and/or environmental changes. Even though it is possible to handle such dynamic problems as a series of individual processes via restarting the optimization algorithm after each change, this may lead to a significant loss of useful information, especially when the change is not too drastic, but rather incremental in nature. Since most of such problems have a multi-modal nature, which further complicates the dynamic optimization problems, the need for powerful and efficient optimization techniques is imminent. Due to the reasons mentioned earlier, in the last decade the efforts have been focused on EAs and particularly on particle swarm optimization (PSO) [9–11], which has obvious ties with the EA family, lies somewhere between GA and EP. Yet unlike GA, PSO has no complicated evolutionary operators such as *crossover, selection,* and *mutation* and it is highly dependent on stochastic processes. However, PSO might exhibit some major problems and severe drawbacks such as parameter dependency [12] and loss of diversity [13]. Particularly, the latter phenomenon increases the probability of being trapped in local optima and it is the main source of premature convergence problem especially when the dimensionality of the search space is large [14] and the problem to be optimized is multi-modal [13, 15].

Low-level features (also called descriptors in some application domains) play a central role in many computer vision, pattern recognition, and signal processing applications. Features are various types of information extracted from the raw data and represent some of its characteristics or signatures. However, especially the (low-level) features, which can be extracted automatically, usually lack the discrimination power needed for accurate processing especially in the case of a large and varied media content data reserves. Especially in content-based image indexing and retrieval (CBIR) area, this is referred to as "Semantic Gap" problem, which defines a rather large *gap* between the low-level features and their limited ability to represent the "content." Therefore, it is crucial to optimize these features particularly for achieving a reasonable performance on multimedia classification, indexing, and retrieval applications, and perhaps for many other domains in various related fields. Since features are in high dimensions in general, the optimization method should naturally tackle with multi-modality and the phenomenon so-called "the curse of dimensionality." Furthermore, in such application domains, the features may not be static, rather dynamically changing (new features can be extracted or some features might be modified). This brings the *scalability* issue along with the instantaneous *adaptability* to whatever (incremental) change may occur in time. All in all, these issues are beyond the capability of the basic PSO or any other EA method alone and will thus be the major subject of this book.

1.3 Synopsis of the Book

This book, first of all, is *not* about PSO or any traditional optimization method proposed since there are many brilliant books and publications for them. As the key-issues are highlighted in the previous section, we shall basically start over where they left. In this book after a proper introduction to the general field and related work, we shall first present a novel optimization technique, the so-called Multi-dimensional particle swarm optimization (MD PSO), which re-forms the native structure of swarm particles in such a way that they can make inter-dimensional passes with a dedicated dimensional PSO process. Therefore, in a multi-dimensional search space where the optimum dimension is unknown, swarm particles can seek for both positional and dimensional optima. This eventually negates the necessity of setting a fixed dimension a priori, which is a common drawback for the family of swarm optimizers. Therefore, instead of operating at a fixed dimension N, the MD PSO algorithm is designed to seek both positional and dimensional optima within a dimension range, ($D_{min} \leq N \leq D_{max}$). Nevertheless, MD PSO is still susceptible to premature convergence as inherited by the basic PSO. To address this problem we shall then introduce an enhancement procedure, called Fractional global best formation (FGBF) technique, which basically collects all promising dimensional components and fractionally creates an *artificial global-best* particle (*aGB*) that has the potential to be a better "guide" than the PSO's native *gbest* particle. We shall further enrich this scope by introducing the application of stochastic approximation (SA) technique to further "guide the guide." As an alternative and generic approach to FGBF, SA-driven PSO, and its multi-dimensional extension, SA-driven also addresses the premature convergence problem in a more generic way. Both SA-driven and MD PSO with FGBF can then be applied to solve many practical problems in an efficient way. The first application is nonlinear function minimiza-tion. We shall demonstrate the capability of both techniques for accurate root finding in challenging high-dimensional nonlinear functions. The second application domain is dynamic data clustering, which presents highly complex and multi-modal error surface where traditional optimization or clustering techniques usually fails. As discussed earlier, the dynamic clustering problem requires the determination of the solution space dimension (i.e., number of clusters) and an effective mechanism to avoid local optima traps (both dimensionally and spatially) particularly in complex clustering schemes in high dimensions. The former requirement justifies the use of the MD PSO technique while the latter calls for FGBF (or SA-driven). We shall present that, if properly applied, the true number of clusters with accurate center localization can be achieved. Several practical applications shall be demonstrated based on this dynamic clustering technique, i.e., to start with, 2D synthetic data spaces with ground truth clusters are first examined to show the accuracy and efficiency of the method. In a specific application in the biomedical engineering, we shall then present a novel and personalized long-term ECG classification system, which addresses the problem within a long-term ECG signal, known as *Holter* register, recorded from an individual patient. Due to the massive amount of ECG data in a *Holter* register, visual inspection is quite difficult and cumbersome, if not infeasible. Therefore, the system helps professionals to quickly and accurately diagnose any latent heart disease by examining only the representative beats (the so-called master key-beats) each of which is automatically extracted from a cluster of homogeneous (similar) beats. Furthermore, dynamic clustering method shall be used to "evolve" radial basis function (RBF) neural networks so that evolutionary RBF networks can "adapt" or "search" for the optimum RBF architecture for the problem in hand. We shall show that both training and test classification results indicate a superior performance with respect to the traditional back propagation (BP) method over real datasets (e.g., features from *Benthic* macroinvertebrate images).

Another novel technique we shall present is the automatic design of artificial neural networks (ANNs) for a given problem by evolving the network configuration toward the optimal one within a given architecture space. This is entirely based on multi-dimensional particle swarm optimization (MD PSO) technique. With the proper encoding of the network configurations and parameters into particles, MD PSO can then seek positional optimum in the error space and dimensional optimum in the architecture space. The optimum dimension converged at the end of a MD PSO process corresponds to a unique ANN configuration where the network parameters (connections, weights, and biases) can then be resolved from the positional optimum reached on that dimension. Furthermore, the proposed technique generates a ranked list of network configurations, from the best to the worst. This is indeed a crucial piece of information, indicating what potential configurations can be alternatives to the best one, and which configurations should not be used at all for a particular problem. In this book, the architecture space will be defined over feed-forward, fully-connected ANNs so as to use the conventional techniques for comparison such as BP and some other evolutionary methods in this field. This technique is then applied over the most challenging synthetic problems (e.g., 10-bit parity problem, highly dynamic function interpolation, *two-spirals* problem, etc.) to test its performance on evolving networks. Additional optimization tests are performed over several benchmark problems to test their generalization capability and compare them with several competing techniques. We shall show that MD PSO evolves to optimum or near-optimum networks in general and has a superior generalization capability. Furthermore, MD PSO naturally favors a low-dimension solution when it exhibits a competitive performance with a high dimension counterpart and such a native tendency eventually leads the evolution process to favor compact network configurations in the architecture space rather than the complex ones, as long as optimality prevails. The main application area of this elegant technique shall be generic and patient-specific classification system designed for robust and accurate detection of electrocardiogram (ECG) heartbeat patterns. We shall deliberately show and explain that this system can adapt to significant inter-patient variations in

ECG patterns by evolving the optimal network structure and thus achieves a high accuracy over large datasets. One major advantage of this system is that due to its parameter invariance, it is highly generic and thus applicable to any ECG dataset.

We then focus on the multimedia classification, indexing, and retrieval problem and provide detailed solutions within two chapters. In this particular area, the following key questions still remain unanswered. (1) how to select relevant features so as to achieve the highest discrimination over certain classes, (2) how to combine them in the most effective way, (3) which distance metric to apply, (4) how to find the optimal classifier configuration for the classification problem at hand, (5) how to scale/adapt the classifier if a large number of classes/features are present, and finally, (6) how to train the classifier efficiently to maximize the classification accuracy. Traditional classifiers, such as support vector machines (SVMs), random forest (RF), and artificial neural networks (ANNs), cannot cope up with such requirements since a single classifier, no matter how powerful and well-trained it can be, cannot discriminate efficiently a vast amount of classes, over an indefinitely large set of features, where both classes and features are not static, e.g., in the case of multimedia repositories. Therefore, in order to address these problems and hence to maximize the classification accuracy which will in turn boost the retrieval performance, we shall present a novel and generic-purpose framework design that embodies a collective networks of evolutionary classifiers. At a given time, this allows the creation and design of a dedicated classifier for discriminating a certain class from the others. Each evolution session will "learn" from the current best classifier and can improve it further, possibly as a result of the (incremental) optimization, which may find another configuration in the architecture space as the "optimal." Moreover, with each evolution, new classes or features can also be introduced which signals the collective classifier network to create new corresponding networks and classifiers within to *adapt* dynamically to the change. In this way, the collective classifier network will be able to dynamically *scale* itself to the indexing requirements of the multimedia database while striving for maximizing the classification and retrieval accuracies thanks to the dedicated classifiers within.

Finally, the multimedia indexing and retrieval problem further requires a decisive solution for the well-known "Semantic Gap" problem. Initially, the focus of the research in this field was on content analysis and retrieval techniques linked to a specific medium. More recently, researchers have started to combine features from various media. They have also started to study the benefit of knowledge discovery in accurate content descriptions, refining relevance feedback, more generally, improving indexing. All in all, narrowing the semantic gap basically requires advanced approaches that depend on a central element to describe a medium's content: the features. Features are the basis of a content-based indexing and retrieval system. They represent the information extracted from a medium in a suitable way, they are stored in an index, and used during the query processing. They basically characterize the medium's signature. However, especially the low-level features, which can be extracted automatically, currently lack the discrimination power needed for accurate retrievals in large multimedia collections. Therefore, we shall focus on the following major research objective: how to improve these features within an evolutionary feature synthesis framework that will be based upon the MD PSO. In that, the features extracted will just be the initial phase of the process while they will be subject to ongoing evolutionary processes that will improve their discrimination capabilities based on the users' (relevance) feedbacks/interactions with the retrieval system.

As a conclusion, the book shall mainly cover the theory and major applications of the MD PSO technique, with some additional state-of-the-art algorithms to significantly improve the PSO's global convergence performance. MD PSO is a recent optimization technique, which can be applied to many engineering problems if properly adapted. Therefore, the applications that will be presented in the book are guiding examples of "how to" perform MD PSO in an effective way. We have particularly chosen the aforementioned problem domains, i.e., dynamic data clustering, evolutionary ANNs, collective network of evolutionary classifiers, and evolutionary feature synthesis, for which MD PSO will serve as the backbone technique.

References

1. A. Kruger, "Median-cut color quantization," Dr. Dobb's J., 46–54 and 91–92, Sept. (1994)
2. S. Smale, On the average number of steps of the simplex method of linear programming. Math. Program., 241–262 (1983)
3. Y. Chauvin, D.E. Rumelhart, *Back Propagation: Theory, Architectures, and Applications* (Lawrence Erlbaum Associates Publishers, UK, 1995)
4. G. Hammerly, C. Elkan, Alternatives to the k-means algorithm that find better clusterings, in *Proceedings of the 11th ACM CIKM*(2002), 600–607
5. D. Goldberg, Genetic *Algorithms in Search, Optimization and Machine Learning* (Addison-Wesley, MA, 1989) pp. 1–25
6. S. Kirkpatrick, C.D. Gelatt, M.P. Vecchi, Optimization by simulated annealing. Science **220**, 671–680 (1983)
7. T. Back, F. Kursawe, Evolutionary algorithms for fuzzy logic: a brief overview, In *Fuzzy Logic and Soft Computing*, World Scientific (Singapore, 1995), pp. 3–10
8. U.M. Fayyad, G.P. Shapire, P. Smyth, R. Uthurusamy, *Advances in Knowledge Discovery and Data Mining* (MIT Press, Cambridge, 1996)
9. A. P. Engelbrecht, Fundamentals of Computational Swarm Intelligence (Wiley, New York, 2005)
10. J. Kennedy, R Eberhart, Particle swarm optimization, in *Proceedings of IEEE International Conference on Neural Networks*, vol. 4 (Perth, Australia, 1995), pp. 1942–1948
11. M.G. Omran, A. Salman, A.P. Engelbrecht, *Particle Swarm Optimization for Pattern Recognition and Image Processing* (Springer, Berlin, 2006)
12. M. Løvberg, T. Krink, Extending Particle Swarm Optimizers with Self-Organized Criticality. Proc. IEEE Congr. Evol. Comput. **2**, 1588–1593 (2002)
13. M. Riedmiller, H. Braun, "A Direct Adaptive Method for Faster Backpropagation Learning: The RPROP Algorithm," in *Proceedings of the IEEE International Conference on Neural Networks* (1993), pp. 586–591

14. G-J Qi, X-S Hua, Y. Rui, J. Tang, H.-J. Zhang, Image Classification With Kernelized Spatial-Context, *IEEE Transactions on Multimedia* 12(4), 278–287, June (2010). doi:10.1109/TMM.2010.2046270
15. K. Ersahin, B. Scheuchl, I. Cumming, Incorporating texture information into polarimetric radar classification using neural networks," in *Proceedings of the IEEE International Geoscience and Remote Sensing Symp* (Anchorage, USA, 2004), pp. 560–563

Chapter 2
Optimization Techniques: An Overview

Since the fabric of the universe is most perfect, and is the work of a most wise Creator, nothing whatsoever takes place in the universe in which some form of maximum or minimum does not appear.

Leonhard Euler

It is an undeniable fact that all of us are optimizers as we all make decisions for the sole purpose of maximizing our quality of life, productivity in time, as well as our welfare in some way or another. Since this is an ongoing struggle for creating the best possible among many inferior designs, optimization was, is, and will always be the core requirement of human life and this fact yields the development of a massive number of techniques in this area, starting from the early ages of civilization until now. The efforts and lives behind this aim dedicated by many brilliant philosophers, mathematicians, scientists, and engineers have brought the high level of civilization we enjoy today. Therefore, we find it imperative to get to know first those major optimization techniques along with the philosophy and long history behind them before going into the details of the method detailed in this book. This chapter begins with a detailed history of optimization, covering the major achievements in time along with the people behind them. The rest of the chapter then draws the focus on major optimization techniques, while briefly explaining the mathematical theory and foundations over some sample problems.

2.1 History of Optimization

In its most basic terms, *Optimization* is a mathematical discipline that concerns the finding of the extreme (minima and maxima) of numbers, functions, or systems. The great ancient philosophers and mathematicians created its foundations by defining the optimum (as an extreme, maximum, or minimum) over several fundamental domains such as numbers, geometrical shapes optics, physics, astronomy, the quality of human life and state government, and several others. This era started with *Pythagoras of Samos* (569 BC to 475 BC), a Greek philosopher who made important developments in mathematics, astronomy, and the theory of music. He is often described as the first pure mathematician. His most important philosophical foundation is [1]: "that at its deepest level, reality is mathematical in nature."

S. Kiranyaz et al., *Multidimensional Particle Swarm Optimization for Machine Learning and Pattern Recognition*, Adaptation, Learning, and Optimization 15, DOI: 10.1007/978-3-642-37846-1_2, © Springer-Verlag Berlin Heidelberg 2014

Zeno of Elea (490 BC to 425 BC) who was a Greek philosopher famous for posing so-called paradoxes was the first to conceptualize the notion of extremes in numbers, or infinitely small or large quantities. He took a controversial point of view in mathematical philosophy, arguing that any motion is impossible by performing infinite subdivisions described by *Zeno's Dichotomy*. Accordingly, one cannot even start moving at all. Probably, *Zeno* was enjoying the challenging concept of "infinity" with his contemporaries without the proper formulation of the limit theory and calculus at the time.

Later, *Plato* (427 BC to 347 BC) who is one of the most important Greek philosophers and mathematicians, gained from the disciples of *Pythagoras*, and formed his idea [2], … "that the reality which scientific thought is seeking must be expressible in mathematical terms, mathematics being the most precise and definite kind of thinking of which we are capable. The significance of this idea for the development of science from the first beginnings to the present day has been immense." About 75 years earlier, *Euclid* wrote *The Elements, Plato* wrote *The Republic* around 375 BC, where he was setting his ideas on education: In that, one must study the five mathematical disciplines, namely arithmetic, plane geometry, solid geometry, astronomy, and harmonics. After mastering mathematics, one can proceed to the study of philosophy. The following dialog is a part of the argument he made:

> "…But when it is combined with the perception of its opposite, and seems to involve the conception of plurality as much as unity, then thought begins to be aroused within us, and the soul perplexed and wanting to arrive at a decision asks "What is absolute unity?" This is the way in which the study of the one has a power of drawing and converting the mind to the contemplation of reality."
>
> "And surely," he said, "this characteristic occurs in the case of one; for we see the same thing to be both one and infinite in multitude?"
>
> "Yes," I said, "and this being true of one, it must be equally true of all number?"
>
> "Certainly"

Aristotle (384 BC to 322 BC), who was one of the most influential Greek philosophers and thinkers of all times, made important contributions by systematizing deductive logic. He is perhaps best described by the authors of [3] as, "Aristotle, more than any other thinker, determined the orientation and the content of Western intellectual history. He was the author of a philosophical and scientific system that through the centuries became the support and vehicle for both medieval Christian and Islamic scholastic thought: until the end of the seventeenth century, Western culture was Aristotelian. And, even after the intellectual revolutions of centuries to follow, Aristotelian concepts and ideas remained embedded in Western thinking. "He introduced the well-known principle," The whole is more than the sum of its parts." Both the Greek philosophers, *Plato* and *Aristotle*, used their "powers of reasoning" to determine the best style of human life. Their goal was to develop the systematic knowledge of how the behavior of both individuals and society could be optimized. They particularly focused on questions of ethics (for optimizing the lifestyle of an individual) and politics (for optimizing the functioning of the state). At the end, both Plato and Aristotle recognized that the knowledge of how the members of society could *optimize* their lives was crucial. Both believed that the proper development of an individual's character traits was the key to living an *optimal* lifestyle.

As a follower of Plato's philosophy, *Euclid* of Alexandria (325 BC to 265 BC) was the most prominent antique Greek mathematician best known for his work on geometry, *The Elements,* which not only makes him the leading mathematician of all times but also one who influenced the development of Western mathematics for more than 2,000 years [4]. It is probable that no results in *The Elements* were first proved by *Euclid* but the organization of the material and its exposition are certainly due to him. He solved some of the earliest optimization problems in Geometry, e.g., in the third book, there is a proof that the greatest and least straight lines can be drawn from a point to the circumference of a circle; in the sixth book it is proven that a square has the *maximum* area among all rectangles with given total length of the edges.

Archimedes of Syracuse (287 BC to 212 BC) is considered by most historians of mathematics as one of the greatest mathematicians of all times [5]. He was the inventor of the water pump, the so-called *Archimedes' screw* that consists of a pipe in the shape of a helix with its lower end dipped in the water. As the device is rotated the water rises up the pipe. This device is still in use in many places in the world. Although he achieved great fame due to his mechanical inventions, he believed that pure mathematics was the only worthy pursuit. His achievements in calculus were outstanding. He perfected a method of integration which allowed him to find areas, volumes, and surface areas of many bodies by using the method of exhaustion, i.e., one can calculate the area under a curve by approximating it by the areas of a sequence of polygons. In Heath [6], it is stated that "Archimedes gave birth to the calculus of the infinite conceived and brought to perfection by Kepler, Cavalieri, Fermat, Leibniz and Newton." Unlike *Zeno* and other Greek philosophers, he and *Euclid* were the first mathematicians who were not troubled by the apparent contradiction of the *infinite* concept. For instance, they contrived the method of exhaustion technique to find the area of a circle *without* knowing the exact value of π.

Heron of Alexandria (~ 10 AC to ~ 75 AC) who was an important geometer and worker in mechanics wrote several books on mathematics, mechanics, and even optics. He wrote the book, *Catoptrica,* which is attributed by some historians to *Ptolemy* although most now seem to believe that this was his genuine work indeed. In this book, *Heron* states that vision occurs as a result of light emissions by the eyes with infinite velocity. He has also shown that light travels between two points through the path of the *shortest* length.

Pappus of Alexandria (~ 290 AC to ~ 350 AC) is the last of the great Greek geometers and made substantial contributions on many geometrical optimization problems. He proved what is known as the "honeycomb conjecture" that the familiar honeycomb shape, which is a repeating hexagonal pattern (volumetric hexagonal-shaped cylinders, stacked one against the other in an endless array) was the *optimal* way of storing honey. *Pappus* introduces this problem with one of the most charming essays in the history of mathematics, one that has frequently been excerpted under the title: *On the Sagacity of Bees*. In that, he speaks poetically of the divine mission of bees to bring from heaven the wonderful nectar known as honey, and says that in keeping with this mission they must make their honeycombs without any cracks through which honey could be lost. Having also a divine sense of symmetry, the bees had to choose among the regular shapes that could fulfill this condition, (e.g. *triangles, squares,* and *hexagons*). At the end they naturally chose the hexagon because a hexagonal prism required the *minimum* amount of material to enclose a given volume. He collected these ideas in his Book V and states his aim that [7], "Bees, then, know just this fact which is useful to them, that the hexagon is greater than the square and the triangle and will hold more honey for the same expenditure of material in constructing each. But we, claiming a greater share in wisdom than the bees, will investigate a somewhat wider problem, namely that, of all equilateral and equiangular plane figures having an equal perimeter, that which has the greater number of angles is always the greater, and the greatest of then all is the circle having its perimeter equal to them."

Also in Book V, *Pappus* discusses the 13 semi-regular solids discovered by *Archimedes* and solves other *isoperimetric* problems which were apparently discussed by the Athenian mathematician *Zenodorus* (200 BC to 140 BC). He compares the areas of figures with equal perimeters and volumes of solids with equal surface areas, proving that the sphere has the *maximum* volume among regular solids with equal surface area. He also proves that, for two regular solids with equal surface area, the one with the greater number of faces has the greater volume. In Book VII, *Pappus* defines the two basic elements of analytical problem solving, the analysis and synthesis [7] as, … "in analysis we suppose that which is sought to be already done, and inquire what it is from which this comes about, and again what is the antecedent cause of the latter, and so on until, by retracing our steps, we light upon something already known or ranking as a first principle… But in synthesis, proceeding in the opposite way, we suppose to be already done that which was last reached in analysis, and arranging in their natural order as consequents what were formerly antecedents and linking them one with another, we finally arrive at the construction of what was sought…"

During the time of the ancient great Greek philosophers and thinkers, arithmetic and geometry were the two branches of mathematics. There were some early attempts to do algebra in those days; however, they lacked the formalization of algebra, namely the arithmetic operators that we take for granted today, such as "+, −, ×, ÷" and of course, "=". Much of the world, including Europe, also lacked an efficient numeric system like the one developed in the Hindu and Arabic cultures. *Al'Khwarizmi* (790–850) was a Muslim Persian mathematician who wrote on Hindu–Arabic numerals and was among the first to use the number *zero* as a place holder in positional base notation. Algebra as a branch of mathematics can be said to date to around the year 825 when *Al'Khwarizmi* wrote the earliest known algebra treatise, *Hisab al-jabr w'al-muqabala*. The word "algebra" comes from the Persian word *al'jabr* (that means "to restore") in the title. Moreover, the English term "algorithm," was derived from *Al'Khwarizmi* 's name as the way of a Latin translation and pronunciation: *Algoritmi*.

Ibn Sahl (940–1000) was a Persian mathematician, physicist, and optics engineer who was credited for first discovering the law of refraction, later called as the Snell's law. By means of this law, he computed the *optimum* shapes for lenses and curved mirrors. This was probably the first application of optimization in an engineering problem.

Further developments in algebra were made by the Arabic mathematician *Al-Karaji* (953–1029) in his treatise *Al-Fakhri*, where he extends the methodology to incorporate integer powers and integer roots of unknown quantities. Something close to a proof by mathematical induction appears in a book written by *Al-Karaji* who used it to prove the binomial theorem, Pascal's triangle, and the sum of integral cubes. The historian of mathematics, Woepcke in [8], credits him as *the first who introduced the theory of algebraic calculus*. This was truly one of the cornerstone developments for the area of optimization as it is one of the uses of calculus in the real world.

René Descartes (1596–1650) was a French mathematician and philosopher and his major work, *La Géométrie*, includes his linkage of algebra to geometry from which we now have the Cartesian geometry. He had a profound breakthrough when he realized he could describe any position on a 2D plane using a pair of numbers associated with a horizontal axis and a vertical axis—what we call today as "coordinates." By assigning the horizontal measurement with x's and the vertical measurement with y's, *Descartes* was the first to define any geometric object such as a line or circle in terms of algebraic equations. Scott in [9] praises his work for four crucial contributions:

1. He makes the first step toward a theory of invariants, which at later stages derelativises the system of reference and removes arbitrariness.
2. Algebra makes it possible to recognize the typical problems in geometry and to bring together problems which in geometrical dress would not appear to be related at all.
3. Algebra imports into geometry the most natural principles of division and the most natural hierarchy of method.
4. Not only can questions of solvability and geometrical possibility be decided elegantly, quickly, and fully from the parallel algebra, without it they cannot be decided at all.

The seminal construction of what we call *graphs* was obviously the cornerstone achievement without which any formulation of optimization would not be possible. In that, *Descartes* united the analytical power of algebra with the descriptive power of geometry into the new branch of mathematics, he named as *analytic geometry,* a term which is sometimes called as *Calculus with Analytic Geometry*. He was one of the first to solve the tangent line problem (i.e., the slope or the derivative) for certain functions. This was the first step toward finding the *maxima* or *minima* of any function or surface, the foundation of all analytical optimization solutions. On the other hand, when *Descartes* published his book, *La Géometrie* in 1637, his contemporary *Pierre de Fermat* (1601–1665) was already working on analytic geometry for about 6 years and he also solved the tangent line problem with a different approach, which is based on the approximation of the slope, converging to the exact value on the limit. This is the pioneer work for finding the derivative and henceforth calculating the optimum point when the slope is zero. Due to this fact, *Lagrange* stated clearly that he considers *Fermat* to be the inventor of the calculus. But at that time, such an approximation-based approach was perhaps why *Fermat* did not get the full credit for his work. There was an ongoing dispute between the two because *Descartes* thought that *Fermat's* work was reducing the importance of his own work *La Géometrie*. Fermat initiated the technique for solving $\mathrm{d}f(x)/\mathrm{d}x = 0$ to find the local optimum point of the function $f(x)$ and this is perhaps the basis in applied mathematics and its use in optimization. Another reason for the dispute between them might be that *Fermat* found a mistake in a book by *Descartes* and corrected it. *Descartes* attacked *Fermat's* method of maxima, minima, and tangents but in turn, *Fermat* proved correct and eventually *Descartes* accepted his mistake. Fermat's most famous work, called *Fermat's Last Theorem*, was the proof for the statement, $x^n + y^n = z^n$, has no integer solutions for $n > 2$. His proof remains a mystery till today since Fermat wrote it as, "I have discovered a truly remarkable proof which this margin is too small to contain." He also deduced the most fundamental optimization phenomenon in the optics, "the light always follows the shortest possible path", (or similarly, "the light follows the path which takes the shortest time").

Like most developments, the calculus too was the culmination of centuries of work. After these pioneers, the two most recognized discoverers of calculus are *Isaac Newton* of England (1643–1727) and a German, *Gottfried Wilhelm Leibniz* (1646–1716). Both deserve equal credit for independently coming up with calculus; however, at that time a similar rivalry and dispute occurred between the two as each accused the other of plagiarism for the rest of their lives. The mathematics community today has largely adopted *Leibniz's* calculus symbols but on the other hand, the calculus he discovered allowed *Newton* to establish the well-known physics laws which are sufficient in macro scale to explain many physical phenomena in nature to a remarkable accuracy. Their approach to calculus was also totally different, i.e., *Newton* considered functions changing in time, whereas *Leibniz* thought of variables x, y as ranging over sequences of infinitely close values, $\mathrm{d}x$ and $\mathrm{d}y$; however, he never thought of the derivative as a limit. In 1666, Newton found out the slope of a function by the derivative and solved the inverse problem by taking the integral, which he used to calculate the area under any function. Therefore, this work contains the first clear statement of the *Fundamental Theorem of Calculus*. As *Newton* did not publish his findings until 1687, unaware that he had discovered similar methods, Leibniz developed his calculus in Paris between 1673 and 1676. In November 1675, he wrote a manuscript using the common integral notation, $\int f(x)\,\mathrm{d}x$, for the first time. The following year, he discovered the power law of differentiation, $\mathrm{d}(x^n) = nx^{n-1}\mathrm{d}x$ for both integer and fractional n. He published the first account of differential calculus in 1684 and then published the explanation of integral calculus in 1686. There were rumors that *Leibniz* was following *Newton's* studies from their common colleagues and occasional discussion letters between the two, but despite of all his correspondences with *Newton*, he had already come to his own conclusions about calculus. In 1686 Leibniz published a paper, in *Acta Eruditorum*, dealing with the integral calculus with the first appearance of the integral notation in print. *Newton's* famous work, *Philosophiae Naturalis Principia Mathematica*, surely the greatest scientific book ever written, appeared in the following year. The notion that the Earth rotated around the Sun was already known by ancient Greek philosophers, but it was Newton who explained *why*, and henceforth, the great scientific revolution began with it.

During his last years, *Leibniz* published *Théodicée* claiming that the universe is in the *best* possible form but imperfect; otherwise, it would not be distinct from God. He invented more mathematical terms than anyone else, including "function," "analysis situ," "variable," "abscissa," "parameter," and "coordinate." His childhood IQ has been estimated as the second-highest in all of history, behind only Goethe [5]. Descriptions that have been applied to *Leibniz* include "one of the two greatest universal geniuses" (*da Vinci* was the other) and the "Father of the Applied Science." On the other hand, *Newton* is the genius who began revolutionary advances on calculus, optics, dynamics, thermodynamics, acoustics, and physics; it is easy to overlook that he too was one of the greatest geometers for he calculated the *optimum* shape of the bullet earlier than his invention of calculus. Among many brilliant works in mathematics and especially in calculus, he also discovered the Binomial Theorem, the polar coordinates, and power series for exponential and trigonometric functions. For instance, his equation, $e^x = \sum x^k / k!$, has been called as "the most important series in mathematics." Another optimization problem he solved is the *brachistochrone*, which is the curve of *fastest* descent between two points by a point-like body with a zero velocity while the gravity is the only force (with no friction). This problem had defeated the best mathematicians in Europe but it took Newton only a few hours to solve it. He published the solution anonymously, yet upon seeing the solution, *Jacob Bernoulli* immediately stated "I recognize the lion by his footprint."

After the era of *Newton* and *Leibniz,* the development of the calculus was continued by the Swiss mathematicians, *Bernoulli* brothers, *Jacob Bernoulli* (1654–1705), and *Johann Bernoulli* (1667–1748). Jacob was the first mathematician who applied separation of variables in the solution of a first-order nonlinear differential equation. His paper of 1690 was indeed a milestone in the history of calculus since the term integral appears for the first time with its *integration* meaning. *Jacob* liked to pose and solve physical *optimization* problems such as the catenary (which is the curve that an idealized hanging chain assumes under its own weight when supported only at its ends) problem. He was a pioneer in the field of calculus of variations, and particularly differential equations, with which he developed new techniques to many optimization problems. In 1697, he posed and partially solved the isoperimetric problem, which is a class of problems of the calculus of variations. The simplest of them is the following: among all curves of given length, find the curve for which some quantity (e.g., area) dependent on the curve reaches a minimum or maximum. Other optimization problems he solved include isochronous curves and curves of *fastest* descent. *Johann Bernoulli* on the other hand, followed a similar path like his brother. He mostly learned from him and also from *Leibniz*, and later on he alone supported *Leibniz* for the Newton–Leibniz controversy. He showed *Leibniz*'s calculus can solve certain problems where *Newton* had failed. He also became the principle teacher of *Leonhard Euler*. He developed the exponential calculus and together with his brother *Jacob*, founded the calculus of variations. Although their work of line was pretty similar, there were no common papers published, because after a while a bitter jealousy led to another famous rivalry, this time between the *Bernoulli* brothers, who—especially *Johann*—began claiming each other's work. Later, a similar jealousy arose between *Johann* and his son, *Daniel Bernoulli* (1700–1782), where this time *Johann* started to compete with him on his most important work, *Hydrodynamica* in 1734 which *Daniel* published in 1738 at about the same time as *Johann* published a similar version of it, *Hydraulica*. However, he discovered *L'Hôpital*'s Rule 50 years before *Guillaume de L'Hôpital* (1661–1704) did, and according to some mathematics historians, he solved the catenary problem before his brother did, although he used ideas that *Jacob* had given when he posed the problem. He attained great fame in his life and made outstanding contributions in calculus and physics for solving many real optimization problems, e.g., about vibrations, elastic bodies, optics, tides, and ship sails.

Calculus of variations is an area of calculus that deals with the *optimization* of *functionals*, which are mappings from a set of functions to real numbers and are often expressed as definite integrals involving functions and their derivatives. A physical system can be modeled by functionals, with which their variables can be optimized considering the constraints. Calculus of variations and the use of differential equations for the general solution of many optimization problems may not be possible without the Swiss mathematician, *Leonhard Euler* (1701–1783) who is probably the most influential mathematician ever lived. He is the father of mathematical analysis and his work in mathematics is so vast that we shall only name the few crucial developments herein for calculus and optimization. He took marvelous advantage of the analysis of *Fermat, Newton, Leibniz,* and the *Bernoulli* family members, extending their work to marvelous heights. To start with, many fundamental calculus and mathematical foundations that are used today were created by *Euler*, i.e., in 1734, he proposed the notation $f(x)$ for a function, along with three of the most important constant symbols in mathematics: e for the base of natural logs (in 1727), i for the square root of -1 (in 1777), π for pi, and several mathematical notations such as "$\sum$" for summation (in 1755), finite differences Δx, $\Delta^2 x$, and many others. In 1748, he published his major work, *Introductio in analysin infinitorum* in which he presented that mathematical analysis was the study of functions. This was the pioneer work, which bases the foundations of calculus on the theory of elementary functions rather than on geometric curves, as had been done earlier. Also in this work, he unifies the trigonometric and exponential functions by his famous formula: $e^{ix} = \cos x + i\sin x$. (The particular setting of $x = \pi$ yields $e^{i\pi} + 1 = 0$ that fits the three most important constants in a single equation.)

Euler was first partially and later totally blind during a long period of his life. From the French Academy in 1735, he received a problem in celestial mechanics, which had required several months to solve by other mathematicians. *Euler*, using his improved methods solved it in 3 days (later, with his superior methods, *Gauss* solved the same problem within an hour!). However, the strain of the effort induced a fever that caused him the loss of sight in his right eye. Stoically accepting the misfortune, he said, "Now I will have less distraction." For a period of 17 years, he was almost totally blind after a cataract developed in his left eye in 1766. Yet he possessed a phenomenal memory, which served him well during his blind years as he calculated long and difficult problems on the blackboard of his mind, sometimes carrying out arithmetical operations to over 50 decimal places. The *calculus of variations* was created and named by *Euler* and in that he made several fundamental discoveries. His first work in 1740, *Methodus inveniendi lineas curvas* initiated the first studies in the calculus of variations. However, his contributions already began in 1733, and his treaty, *Elementa Calculi Variationum* in 1766 gave its name. The idea was already born with the *brachistochrone curve* problem raised by *Johann Bernoulli* in 1696. This problem basically deals with the following: Let a point particle of mass m on a string whose endpoints are at $a = (0,0)$ and $b = (x,y)$, where $y < 0$. If gravity acts on the particle with force $F =$ mg, what path of string *minimizes* its travel time from a to b, assuming no friction? The solution of this problem was one of the first accomplishments of the calculus of variations using which many optimization problems can be solved. Besides *Euler*, by the end of 1754, *Joseph-Louis Lagrange* (1736–1813) had also made crucial discoveries on the *tautochrone* that is the curve on which a weighted particle will always arrive at a fixed point in a fixed amount of time independent of its initial position. This problem too contributed substantially to the calculus of variations. *Lagrange* sent *Euler* his results on the *tautochrone* containing his method of maxima and minima in a letter dated 12 August 1755, and *Euler* replied on 6 September saying how impressed he was with *Lagrange's* new ideas (he was 19 years old at the time). In 1756, *Lagrange* sent *Euler* results that he had obtained on applying the calculus of variations to mechanics. These results generalized results which *Euler* had himself obtained. This work led to the famous *Euler–Lagrange* equation, the solution of which is applied on many optimization problems to date. For example using this equation, one can easily show that the closed curve of a given perimeter for which the area is a *maximum*, is a circle, the *shortest* distance between two fixed points is a line, etc. Moreover, *Lagrange* considered optimizing a functional with an added constraint and he turned the problem using the method of Lagrange multipliers to a single optimization equation that can then be solved by the *Euler–Lagrange* equation.

Euler made substantial contributions to differential geometry, investigating the theory of surfaces and curvature of surfaces. Many unpublished results by *Euler* in this area were rediscovered by *Carl Friedrich Gauss* (1777–1855). In 1737, *Euler* wrote the book, *Mechanica* which provided a major advance in mechanics. He analyzed the motion of a point mass both in a vacuum, in a resisting medium, under a central force, and also on a surface. In this latter topic, he solved various problems of differential geometry and geodesics. He then addressed the problem of ship propulsion using both theoretical and applied mechanics. He discovered the *optimal* ship design and first established the principles of hydrostatics.

Based on these pioneers' works, during the nineteenth century, the first optimization algorithms were developed. During this era, many brilliant mathematicians including *Jakob Steiner* (1796–1863), *Karl Theodor Wilhelm Weierstrass* (1815–1897), *William Rowan Hamilton* (1805–1865), and *Carl Gustav Jacob Jacobi* (1804–1851) made significant contributions to the field of calculus of variations. The first iterative optimization technique that is known as Newton's method or Newton–Raphson method was indeed developed by four mathematicians: *Isaac Newton, Joseph Raphson* (1648–1715), *Thomas Simpson* (1710–1761), and *Jean-Baptiste-Joseph Fourier* (1768–1830). *Newton* in 1664 found a non-iterative algebraic method of root finding of a polynomial and in 1687, he described an application of his procedure to a non-polynomial equation in his treaty *Principia Mathematica* where this was originated from *Kepler's* equation. Note that this was a purely algebraic and non-iterative method. In 1690, *Raphson* turned *Newton's* method into an iterative one, applying it to the solution of polynomial equations of degree up to 10. However, the method is still not based on calculus; rather explicit polynomial expressions are used in function form, $f(x)$ and its derivative, $f'(x)$. Simpson in 1740 was the first to formulate the *Newton–Raphson* method on the basis of calculus, extending it to an iterative solver for the multivariate minimization. *Fourier* in 1831 brought the method as we know of it today and published in his famous book, *Analyse des équations déterminées*. The method finds the root of a scalar function, $f(x) = 0$, iteratively by the following equation using only the first-order derivative, $f'(x)$.

$$x_{k+1} = x_k - f(x_k)/f'(x_k), k = 0, 1, 2, \ldots \qquad (2.1)$$

where the initial guess, x_0, is usually chosen randomly. Similarly, finding the root of f', which is equivalent to the optimum (or stationary) point of the function, $f(x)$, can similarly be expressed as,

$$x_{k+1} = x_k - f'(x_k)/f''(x_k), \quad k = 0, 1, 2, \ldots \qquad (2.2)$$

Augustin-Louis Cauchy (1789–1857) did important work on differential equations and applications to mathematical physics. The four-volume series, *Exercices d'analyse et de physique mathématique* published during 1840–1847 was a major work in this area in which he proposed the method of the *steepest descent* (also called as *gradient descent*) in 1847. This method is perhaps one of the most fundamental and basic derivative-based iterative procedures for unconstrained minimization of a differentiable function. Given a differentiable function in N-D, $f(\bar{x})$, the gradient method in each step moves along the direction that minimizes ∇f, that is, the direction of *steepest* descent and thus perpendicular to the slope of the curve at that point. The method stops when it reaches to a local minimum (maximum) where $\nabla f = 0$ and thus no move is possible. Therefore, the update equation is as follows:

$$\bar{x}_{k+1} = \bar{x}_k - \lambda_k \nabla f(\bar{x}_k), \quad k = 0, 1, 2, \ldots \tag{2.3}$$

where $\bar{x}_0$ is the initial starting point in the N-D space. An advantage of *gradient descent* compared to *Newton–Raphson* method is that it only utilizes first-order derivative information about the function when determining the direction of movement. However, it is usually slower than *Newton–Raphson* to converge and it tends to suffer from very slow convergence especially as a stationary point is approached.

A crucial optimization application is the *least-square approximation*, which finds the approximate solution of sets of equations in which there are more equations than unknowns. At the age of 18, *Carl Friedrich Gauss* who was widely agreed to be the most brilliant and productive mathematician ever lived, invented a solution to this problem in 1795, although it was first published by *Lagrange* in 1806. This method basically minimizes the sum of the squares of the residual errors, that is, the overall solution minimizes the sum of the squares of the errors made in the results of every single equation. The most important application is in data fitting and the first powerful demonstration of it was made by *Gauss*, at the age of 24 when he used it to predict the future location of the newly discovered asteroid, *Ceres*. In June 1801, *Zach*, an astronomer whom *Gauss* had come to know two or three years earlier, published the orbital positions of *Ceres*, which was discovered by an Italian astronomer *Giuseppe Piazzi* in January, 1801. *Zach* was able to track its path for 40 days before it was lost in the glare of the sun. Based on this data, astronomers attempted to determine the location of *Ceres* after it emerged from behind the sun without solving the complicated *Kepler's* non-linear equations of planetary motion. *Zach* published several predictions of its position, including one by Gauss which differed greatly from the others. When *Ceres* was rediscovered by Zach on 7 December 1801, it was almost exactly where *Gauss* had predicted using the *least-squares method*, which was not published at the time.

The twentieth century brought the proliferation of several optimization techniques. Calculus of variations was further developed by several mathematicians including *Oskar Bolza* (1857–1942) and *Gilbert Bliss* (1876–1951). *Harris Hancock* (1867–1944) in 1917 published the first book on optimization, *Theory of Maxima and Minima*. One of the crucial techniques of the optimization, *Linear Programming* (LP), was developed in 1939 by the Russian mathematician, *Leonid Vitaliyevich Kantorovich* (1912–1986); however, the method was kept secret until the time the American scientist, *George Bernard Dantzig* (1914–2005) published the *Simplex* method in 1947. LP, sometimes called linear optimization, is a mathematical method for determining a way to achieve the optimal outcome in a given mathematical model according to a list of requirements that are predefined by some linear relationships. More formally, LP is a technique for the optimization of a linear objective function, subject to constraints expressed by linear (in)equalities. In an LP, the variables are continuous while the objective function and constraints must be linear expressions. An expression is linear if it can be expressed in the form, $c_1x_1 + c_2x_2 + \ldots + c_nx_n$ for some constants $c_1, c_2,\ldots,c_n$. The solution space corresponds to a convex polyhedron, which is a set defined as the intersection of finitely many half spaces, each of which is defined by a linear inequality. Its objective function that is to be optimized (under the given constraints) is a real-valued affine function defined on this polyhedron. In short, the LP method finds an optimum point in the polyhedron—if it exists. The *Simplex* method, on the other hand, is an efficient method for finding the optimal solution in one of the corners of the N-D polyhedron where N is the number of linear (in)equalities each intersecting to yield a corner. The *Simplex* method iteratively searches each corner to find the optimum one, which corresponds to the optimum solution. Finding each of the N corners is a matter of solving a system of N equations that can be done by Gaussian elimination method.

John Von Neumann (1903–1957) developed the theory of duality as an LP solution, and applied it in the field of game theory. If an LP exists in the maximization linear form, which is called the *primal* LP, its dual is formed by having one variable for each constraint of the primal (not counting the non-negativity constraints of the primal variables), and having one constraint for each variable of the primal (plus the non-negative constraints of the dual variables); then the maximization can be switched to minimization, the coefficients of the objective function are also switched with the right-hand sides of the inequalities, and the matrix of coefficients of the left-hand side of the inequalities are transposed. In 1928, *Von Neumann* proved the *minimax* theorem in the game theory, which indicates that there exists a pair of strategies for both players that allows each one to minimize his maximum losses. Each player examines every possible strategy and must consider all the possible responses of his adversary. He then plays out the strategy which will result in the minimization of his maximum loss. Such a strategy, which minimizes the maximum loss for each player is called the optimal *minmax* solution. Alternatively, the theorem can also be thought of as *maximizing* the minimum gain (*maximin*).

In 1939, *Nonlinear Programming* (NLP or *Nonlinear Optimization*) was first developed by a graduate student *William Karush* (1917–1997), who was also the first to publish the necessary conditions for the inequality constrained problem in his Master's thesis, *Minima of Functions of Several Variables with Inequalities as Side Constraints*. The optimal solution by the NLP was only widely recognized after a seminal conference paper in 1951 by *Harold William Kuhn* (born in 1925) and *Albert William Tucker* (1905–1995). Thus the theory behind NLP was called the *Karush–Kuhn–Tucker* (KKT) *Theory*, which provided necessary and sufficient conditions for the existence of an optimal solution to an NLP. NLP has a particular importance in optimal control theory and applications since optimal control problems are optimization problems in (infinite-dimensional) functional spaces, while NLP deals with the optimization problems in Euclidean spaces; optimal control can indeed be seen as a generalization of NLP.

In 1952, *Richard Bellman* (1920–1984) made the first publication on *dynamic programming* (DP), which is a commonly used method of optimally solving complex problems by breaking them down into simpler problems. DP is basically an algorithmic technique, which uses a recurrent formula along with one or more starting states. A subsolution of the problem is constructed from those previously found. DP solutions have a polynomial complexity, which assures a much faster running time than other techniques such as backtracking and brute-force. The problem is first divided into "states," each of which represents a sub-solution of the problem. The state variables chosen at any given point in time are often called the "control" variables. Finally, the *optimal* decision rule is the one that achieves the best possible value from the objective function, which is written as a function of the state, called the "value" function. *Bellman* showed that a DP problem in discrete time can be stated in a recursive form by writing down the relationship between the value function in one period and the value in the next period. The relationship between these two value functions is called the *Bellman* equation. In other words, the *Bellman* equation, also known as a DP equation, is a necessary condition for optimality. The Bellman equation can be solved by *backwards induction*, either analytically in a few special cases, or numerically on a computer. Numerical *backwards induction* is applicable to a wide variety of problems, but may be infeasible when there are many state variables, due to the "Curse of Dimensionality," which is a term coined by Bellman to describe the problem caused by the exponential increase in volume associated with adding extra dimensions to the (search) space. One implication of the curse of dimensionality is that some methods for numerical solution of the Bellman equation require vastly more computer time when there are more state variables in the value function.

All of the optimization methods described till now have been developed for deterministic processes applied over known differentiable (and double differentiable) functions. Optimization by *Stochastic Approximation* (SA) aims at finding the minima or maxima of an unknown function with unknown derivatives, both of which can be confounded by random error. Therefore, SA methods belong to the family of iterative stochastic optimization algorithms, which converge to the optimum points of such functions that cannot be computed directly, but only estimated via noisy observations. SA is a part of the stochastic optimization (SO) methods that generate and use random variables, which appear in the formulation of the optimization problem itself, along with the random objective functions or random constraints. Stochastic approximation was introduced in 1951 by the American mathematicians *Herbert Ellis Robbins* (1915–2001) and his student, *Sutton Monro* (1919–1995) [10]. This algorithm is a root finder of the equation, $\theta(x) = 0$ which has a unique root at $x = \alpha$. It is assumed that one cannot observe directly the function, $\theta(x)$, rather we have the noisy measurements of it, $N(X)$, where $E(N(X)) = \theta(x)$ ($E(.)$ is the mathematical expectation operation). The algorithm then iterates toward the root in the form: $x_{k+1} = x_k + c_k(\alpha - N(x))$ where $c_1, c_2, \ldots$ is a sequence of positive step sizes. They suggested the form of $c_k = c/k$ and proved that under certain conditions, x_k converges to the root, α.

Motivated by the publication of the *Robbins-Monro* algorithm in 1951, the *Kiefer-Wolfowitz* algorithm [11] was introduced in 1952 by the Polish mathematician, *Jacob Wolfowitz* (1910–1981) and the American statistician, *Jack Kiefer* (1924–1981). This is the first stochastic optimization method which seeks the maximum point of a function. This method suffers from heavy function computations since it requires $2(d + 1)$ function computations for each gradient computation, where d is the dimension of the search space. This is a particularly a significant drawback in high dimensions. To address this drawback, *James Spall* in [12], proposed the use of simultaneous perturbations to estimate the gradient. This method would require only two function computations per iteration, regardless of the dimension. For any SA method, applied over unimodal functions, it can be shown that the method can converge to the local optimum point with probability one.

However, for multimodal functions, SA methods, as all other gradient-based deterministic algorithms may be stuck on a local optimum. The convergence to the global optimum point is a crucial issue, yet most likely infeasible by any of these gradient-based methods. This brought the era of probabilistic metaheuristics. The American physicist *Nicholas Metropolis* (1915–1999) in 1953 co-authored the paper, *Equation of State Calculations by Fast Computing Machines*, a technique that was going to lead to the first probabilistic metaheuristics method, now known as *simulated annealing*. After this landmark publication, *Keith Hastings* (born in 1930) extended it to a more general case in 1970, by developing the *Metropolis–Hastings* algorithm, which is a Markov chain Monte-Carlo method for creating a series of random numbers from a known probability density function. In 1983, the adaptation of this method led to the simulated annealing method [13], which is a generic probabilistic metaheuristics for the global optimization problem so as to converge to the global optimum of any function in a large search space. The name *annealing* basically mimics the process undergone by misplaced atoms in a metal when it is first heated and then slowly cooled. With a similar analogy, each step of the simulated annealing attempts to replace the current solution with a new solution chosen randomly according to a certain probability distribution. This new solution may then be accepted with a probability that depends both on the difference between the corresponding function values and also on a global parameter T (called the temperature) that is gradually decreased during the process (*annealing*). When T is large, the choice between the new and the previous solution becomes almost purely random and as T goes to zero; it consistently selects the best solution between the two, mimicking a steepest descent (or ascent) method. Therefore, especially when T is large (during the early stages of *simulated annealing*'s iterative algorithm), it prevents the early trappings to a local minima and then yields the convergence to the optimum point as T goes to zero. While this technique cannot guarantee finding the optimum solution, it can often find a suboptimum point in the close vicinity of it, even in the presence of noisy data.

The 1950s and early 1960s were the times when the use of the computers became popular for a wide range of optimization problems. During this era, *direct search methods* first appeared, whereas the name "direct search" was introduced in 1961 by *Robert Hooke* and *T. A. Jeeves*. This was a pattern search method, which is better than a random search due to its search directions by exploration in the search space. After this key accomplishment, in 1962 the first simplex-based direct search method was proposed by *W. Spendley, G. R. Hext,* and *F. R. Himsworth* in their paper, *Sequential Application of Simplex Designs in Optimisation and Evolutionary Operation*. Note that this is an entirely different algorithm than the *Simplex* method for LP as discussed earlier. It uses only two types of transformations to form a new simplex (e.g., vertices of a triangle in 2D) in each step: *reflection* away from the worst vertex (the one with the highest function value), or *shrinking* toward the best vertex (the one with the lowest function value). For each iteration, the angles between simplex edges remain constant during both operations, so the working simplex can change in size, but not in shape. In 1965, this method was modified by *John Ashworth Nelder* (1924–2010) and *Roger Mead* who added two more operators: *expansion* and *contraction* (in and out), which allow the simplex to change not only its size, but also its shape [14]. Their modified simplex method, known as *Nelder-Mead (or simplex) method,* became immediately famous due to its simplicity and low storage requirements, which makes it an ideal optimization technique especially for the primitive computers at that time. During the 1970s and 1980s, it was used by several software packages while its popularity grew even more. It is now a standard method in MATLAB© where it can be applied by the command: *fminsearch*. Nowadays, despite its long past history, the *simplex method,* is still one of the most popular heuristic optimization techniques in use.

During the 1950s and 1960s, the concept of artificial intelligence (AI) was also born. Along with the AI, a new family of metaheuristic optimization algorithms in stochastic nature was created: *evolutionary algorithms* (EAs). An EA uses mechanisms inspired by biological *evolution* such as reproduction, *mutation*, *recombination*, and *selection*. It is also a stochastic method as in simulated annealing; however, it is based on the collective behavior of a population. A potential solution of the optimization problem plays the role of a member in the population, and the fitness function determines the search space within which the solutions lie. The earliest instances of EAs appeared during the 1950s and early 1960s, simulated on computers by evolutionary biologists who were explicitly seeking to model aspects of natural evolution. At first, it did not occur to any of them that this approach might be generally applicable to optimization problems. The EAs were first used by a Norwegian-Italian mathematician; *Nils Aall Barricelli* (1912–1993) who applied to evolutionary simulations. By 1962, several researchers developed evolution-inspired algorithms for function optimization and machine learning, but at the time their work only attracted little attention. The first development in this field for optimization came in 1965, when the German scientist *Ingo Rechenberg* (born in 1934), developed a technique called *evolution strategy* (ES), which uses natural problem-dependent representations, and primarily mutation and selection, as search operators in a loop where each iteration is called *generation*. The sequence of generations is continued until a termination criterion is met.

The next EA member came in 1966, when an American aerospace engineer, *Lawrence Fogel* (1928–2007) developed *evolutionary programming* (EP) where a potential solution of a given problem in hand is represented by simple finite-state machines as predictors. Similar to evolution strategies, EP performs random *mutation* of a simulated machine and keeps the best one. However, both EAs still lack a crucial evolutionary operator, the *crossover*. As early as 1962, *John Holland* (born in 1929) performed the pioneer work on adaptive systems, which laid the foundation for a new EA, *genetic algorithms* (GAs). Holland was also the first to explicitly propose crossover and other recombination operators. In 1975 he wrote the ground-breaking book on GA, "Adaptation in Natural and Artificial Systems." Based on earlier work on EAs by himself and by colleagues at the University of Michigan, this book was the first to systematically and rigorously present the concept of adaptive digital systems using evolutionary operators such as *mutation*, *selection* and *crossover*, simulating processes of natural evolution. In a GA, a population of strings (called chromosomes), which encodes potential solutions (called individuals, creatures, or phenotypes) of an optimization problem, evolves toward better solutions using these operators in an iterative way. Traditionally, solutions are represented in binary strings of 0s and 1s, but other encodings are also possible. The evolution usually starts from a population of randomly generated individuals and in each generation, the fitness of every individual in the population is evaluated, multiple individuals are stochastically selected from the current population (based on their fitness), and modified (recombined and possibly randomly mutated) to form a new generation, and so on. The GA process is terminated either when a successful solution or a maximum number of generations is reached. These foundational works established more widespread interest in evolutionary computation. By the early to mid-1980s, GAs were being applied to a broad range of fields, from abstract mathematical problems to many engineering problems such as pipeline flow control, pattern recognition and classification, and structural optimization.

In 1995, differential evolution (DE) as the most recent EA was developed by *Rainer Storn* and *Kenneth Price*. Similar to GA and many other EAs, DE is a population-based technique, which performs evolutionary operators, *mutation, crossover,* and *selection* in a certain way and the candidate solutions are represented by agents based on floating point number arrays (or vectors). As any other EA, it is a generic optimization method that can be used on optimization problems that are noisy, dynamic, or not even continuous. In each generation, it creates new candidate solutions by combining existing ones according to its simple expression, and then keeping whichever candidate solution has the best score or fitness. This is a typical process of an EA, especially resembling GA; however, DE has a distinct property of interaction among individuals, that is, each individual (agent) is mutated with respect to three others. A similar concept of interaction became the basis and the key element in one of the latest and the most successful methods in the era of probabilistic metaheuristics, the *Particle Swarm Optimization* (PSO), which was proposed in 1995 by *Russell C. Eberhart* and *James Kennedy*. PSO was first intended as a simulation program for the social behavior and stylized representation of the movement of organisms in a bird flock or fish school. The algorithm was simplified and it was observed to be performing optimization. We shall cover the details and the philosophy behind the PSO in Chap. 3 and, therefore, in the forthcoming sections in this chapter, we shall now detail the major optimization methods prior to PSO.

2.2 Deterministic and Analytic Methods

Assume an unconstrained optimization problem, such as $\min\limits_{x \in \Re^n} f(x)$, where the objective function $f : \Re^n \to \Re$ is sufficiently smooth with continuous second derivative. It is well known from the theory of functions from Calculus that the necessary and sufficient conditions for x^* to be a local minimum are (1) *gradient* $f'(x^*) = 0$ and (2) *Hessian* $H(x^*)$ $(= \nabla^2 f(x))$ is positively definite. For some problems, the solution can be obtained analytically by determining the zeros of the gradient and verifying positive definiteness of the *Hessian* matrix at these points. One particularly interesting property of an objective function is convexity. If f is a convex function, satisfying $f(\alpha x + (1 - \alpha)y) \le \alpha f(x) + (1 - \alpha)f(y), \alpha \in [0, 1]$, then it has only one (global) minimum. There are effective methods for solving convex optimization problems [15].

For one-dimensional (and possibly multi-dimensional) unconstrained optimization problems, such as $\min\limits_{x \in \Re^n} f(x)$, search methods explore the parameter space *iteratively* by adjusting the search direction and the search range in every iteration in order to find lower values of the objective function. Search methods are generally classified into three groups based on their use of (1) objective function evaluations, (2) gradient of the objective function, and (3) Hessian of the objective function. There are several iterative search methods, usually called "line search methods" and designed to solve one-dimensional, *unimodal* unconstrained optimization problems [16]. Some of these methods can be analogously applied to multi-dimensional unconstrained problems. The *generic* pseudo-code for a line search method is given in Table 2.1, where the steps 2.1 and 2.2 will differ for a specific search method.

2.2.1 Gradient Descent Method

When the search direction is chosen as the gradient descent direction, $-\nabla f(x)$, the corresponding iterative search is called the method of gradient descent (also known as steepest descent or Cauchy's method). The direction of the negative gradient along which the objective function decreases fastest is the most natural choice. This simple algorithm for continuous optimization uses gradient of the objective function in addition to the function value itself, hence f must be a

Table 2.1 Pseudo-code for generic line search method

Generic Line Search Method (termination criteria: $\|dx\| = \|x(k+1) - x(k)\| = \|\alpha(k)d(k)\| < \varepsilon$)
1. **Pick** an initial starting point, $x(0) \in \Re^n$ at $k = 0$
2. **Repeat:**
2.1. **Calculate** a search direction $d(k)$ from $x(k)$ such that it is a descent direction for $f(x(k))$
2.2. **Calculate** a step size $\alpha(k)$ ensuring $f(x(k+1)) < f(x(k))$
2.3. $k \leftarrow k+1$
2.4. **Update** $x(k+1) = x(k) + \alpha(k)d(k)$
3. **Until** $x(k)$ converges to $x_{opt} \in \Re^n$ for which the 1st and 2nd order optimality conditions are satisfied.

differentiable function. The principle of the gradient descent algorithm can be obtained by setting the search direction as $d(k) = -\nabla f(x)$ in step 2.1 and the *optimum* step size as $\alpha(k) = \arg\min_{\alpha \in \Re^+} f(x(k) - \alpha(k-1)\Delta f(x))$, in step 2.2 of the generic line search method, resulting in the position update as

$$x(k+1) = x(k) - \left(\arg\min_{\alpha \in \Re^+} f(x(k) - \alpha\nabla f(x)) \right) \nabla f(x) \qquad (2.4)$$

By using the optimum $\alpha(k)$ the gradient descent technique is guaranteed to converge to a local minimum from any starting point $x(0)$. Additionally, for the *exact* line search version of the algorithm described, it can be shown that the next step will be taken in the direction of the negative gradient at this new point and the step size will be chosen such that the successive search directions are orthogonal. In practice, there are inexact line search methods that use different criteria to find a suitable step size avoiding too long or too short steps to improve efficiency. The termination criterion is usually of the form $\|\nabla f(x)\| \leq \eta$ where η is small ($1e - 6$) and positive. However, the gradient method requires a large number of iterations for convergence when the Hessian of f near minima has a large condition number (linear dependence). The plots of the gradient descent method with the fixed ($\alpha(k) = 0.001$) and the exact (optimal $\alpha(k)$) step size over Rosenbrock (banana) function,$f(x, y) = 100(y - x^2)^2 + (1 - x)^2$,are illustrated in Fig. 2.1. The total numbers of iterations for the corresponding plots are 26093 and 6423, respectively.

2.2.2 Newton–Raphson Method

Assuming the objective function $f(x)$ is a twice differentiable function, *Newton–Raphson* method is based on the second order Taylor series expansion of the function f around the point x:

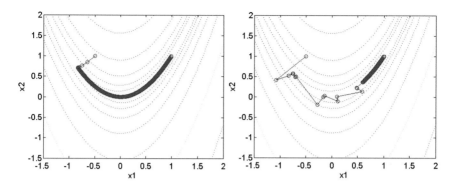

Fig. 2.1 Iterations of the fixed (*left*) and optimum $\alpha(k)$ (*right*) line search versions of gradient descent algorithm plotted over the Rosenbrock objective function, $x(0) = [-0.5, 1.0]$

$$f(x + \delta) \cong f(x) + \Delta f(x)\delta + \frac{1}{2}\delta^T H_f(x)\delta \qquad (2.5)$$

where the Hessian matrix $H_f(x) = \nabla^2 f(x)$ is assumed to be positive definite near local minimum x^*. Therefore, *Newton–Raphson* method utilizes both the first and the second partial derivatives of the objective function to find its minimum. Similar to the gradient descent method, it can be implemented as an iterative line search algorithm using $d(k) = -H_f(x)^{-1}\nabla f(x)$ as the search direction (Newton direction or the direction of the curvature) in 2.1 yielding the position update:

$$x(k+1) = x(k) - \alpha(k)H_f(x)^{-1}\nabla f(x) \qquad (2.6)$$

At each iteration, *Newton–Raphson* method approximates the objective function by a quadratic function around $x(k)$ and moves toward its minima. In the original algorithm, the step size, $\alpha(k)$, is fixed to 1. While the convergence of *Newton–Raphson* method is fast in general, being quadratic near x^*, the computation and the storage of the inverse Hessian is costly. Quasi-Newton methods, which compute the search direction (through inverse Hessian approximation) with less computation can be alternatively employed. In the left plot in Fig. 2.2, iterations of the Quasi-Newton method over the Rosenbrock function are shown. Note that the convergence to the optimum point, (1, 1) is impressively faster than of gradient descent as shown in Fig. 2.1 (33 iterations versus 6,423 iterations with the optimal $F(x, y) = 2y^2 + x^2$). In the right plot, iterations of both gradient descent and Quasi-Newton methods over a quadratic objective function, $F(x, y) = 2y^2 + x^2$, are shown. It took 6 iterations for Quasi-Newton and 13 iterations for gradient descent to converge ($\alpha(0) = 0.05$ for both methods).

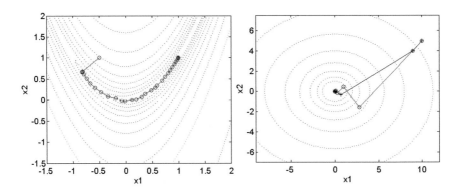

Fig. 2.2 (*Left*) Iterations of the Quasi-Newton method plotted over Rosenbrock function, $x_0 = [-0.5, 1.0]$. (*Right*) Iterations of the gradient descent (*red*) versus Quasi-Newton (*black*) methods plotted over a quadratic objective function, $x(0) = [10, 5]$

2.2.3 Nelder–Mead Search Method

Nelder–Mead or downhill simplex method [14], is a heuristic algorithm for multidimensional unconstrained optimization problems. The Nelder–Mead algorithm falls in the more general class of direct search algorithms which use only the function values, thus it depends on neither the first nor the second order gradients. This heuristic search method depends on the comparison of the objective function values at the $(n + 1)$ vertices of a general simplex, followed by replacement of the vertex with the highest function value by another point. It is, therefore, based on an iterative simplex search, keeping track of $n + 1$ points in n dimensions as vertices of a simplex (i.e., a triangle in 2 dimensions, a tetrahedron in 3 dimensions, and so on). It includes features, which enable the simplex to adapt to the local landscape of the cost function, i.e., at each iteration, the simplex moves toward the minimum by performing one of *reflection*, *expansion*, and *contraction* (in and out) operations. The stopping criterion is based on the standard deviation of the function value over the simplex. This is indeed a "greedy" method in the sense that the expansion point is kept if it improves the best function value in the current simplex. The convergence of a Nelder–Mead operation over the Rosenbrock's function is shown in the left plot of Fig. 2.3 and the right plot demonstrates the consecutive simplex operations during iterations 3–30 where the total number of iterations is 94. Note that the four operations are annotated in the plot.

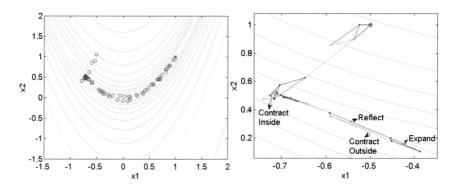

Fig. 2.3 *Left* Iterations of the Nelder–Mead method plotted over Rosenbrock function with $x(0) = [-0.5, 1.0]$. The vertices with the minimum function values are only plotted. *Right* consecutive simplex operations during the iterations 3–30

2.3 Stochastic Methods

2.3.1 Simulated Annealing

Metropolis in his groundbreaking paper, *Equation of State Calculations by Fast Computing Machines* in 1953 introduced the algorithm which simulates the evolution of a solid in a heat bath to thermal equilibrium. In physics, a thermal process for obtaining low energy states of a solid in a heat bath consists of the following two steps:

1. Increase the temperature of the heat bath to a maximum value at which the solid melts;
2. Slowly decrease the temperature until the particles arrange themselves in the ground state of the solid.

In the liquid phase, all particles are distributed randomly, whereas in the ground state of the solid the particles are arranged in a highly structured lattice, for which the corresponding energy is minimal. The ground state of the solid is obtained only if the maximum value of the temperature is sufficiently high and the cooling is performed slowly. Otherwise, the solid will be obtained in a meta-stable state rather than in the true ground state. This is the key for achieving the optimal ground state, which is the basis of the *annealing* as an optimization method. The *simulated annealing* method is a Monte Carlo-based technique and generates a sequence of states of the solid. Let i and j be the current and the subsequent state of the solid and ε_i and ε_j their energy levels. The state j is generated by applying a perturbation mechanism, which transforms the state i into j by a little distortion, such as a mere displacement of a particle. If $\varepsilon_j \leq \varepsilon_i$, the state j is accepted as the current state; otherwise, the state j may still be accepted as the current state with a probability

$$P(i \Rightarrow j) = \exp\left(\frac{\varepsilon_i - \varepsilon_j}{k_B T}\right) \tag{2.7}$$

where k_B is a physical constant called the Boltzmann constant; recall from the earlier discussion that T is the temperature of the state. This rule of acceptance is known as *Metropolis* criterion. According to this rule, the *Metropolis–Hastings* algorithm generates a sequence of solutions to an optimization problem by assuming: (1) solutions of the optimization problem are equivalent to the state of this physical system, and (2) the cost (fitness) of the solution is equivalent to the energy of a state. Recall from the earlier discussion that the temperature is used as the control parameter that is gradually (iteratively) decreased during the process (*annealing*). Simulated annealing can thus be viewed as an iterative *Metropolis–Hastings* algorithm, executed with the gradually decreasing values of T. With a given cost (fitness) function, f, let ψ be the continuously decreasing temperature function, T_0 is the initial temperature, N is the neighborhood function, which changes the state (candidate solution) with respect to the previous state in an appropriate way, ε_C is the minimum fitness score aimed, and x is the variable to be optimized in N-D search space. Accordingly, the pseudo-code of the simulated annealing algorithm is given in Table 2.2.

Note that a typical characteristic of the simulated annealing is that it accepts deteriorations to a limited extent. Initially, at large values of temperature, T, large deteriorations may be accepted; as T gradually decreases, the amount of deteriorations possibly accepted goes down and finally, when the temperature reaches absolute zero, deteriorations cannot happen at all—only improvements. This is why it mimics the family of steepest descent methods as T goes to zero. On the other hand, recall that simulated annealing and the family of Evolutionary

Table 2.2 Pseudo-code of the simulated annealing algorithm

Simulated Annealing (termination criteria:{*IterNo*, ε_C ,...}, ψ ,T_0, N)

1. **Randomize** *initial solution,* x_0, and **Let** $T = T_0$, $t = 0$
2. **Repeat:**
   2.1. **Generate** new solution: $x \leftarrow N(x_0)$
   2.2. **Let** $\Delta\varepsilon = f(x) - f(x_0)$
   2.3. **If** $\Delta\varepsilon \leq 0$ then $x_0 = x$
   2.4. **Else:**
       2.4.1. Generate a random number, $u \in [0,1]$
       2.4.2. If $u < \exp(-\dfrac{\Delta\varepsilon}{k_B T})$ then $x_0 = x$
   2.5. **Assign** new temperature: $T = \psi(T)$
   2.6. $t \leftarrow t+1$
3. **Until** $t > iterNo$ **OR** $f(x) < \varepsilon_C$

Fig. 2.4 The plot of 1,532 iterations of the simulated annealing method over Rosenbrock function with $x(0) = [-0.5, 1.0]$, $\varepsilon_C = 10^{-3}$, $T_0 = 1$, $\psi(T) = 0.95T$, $N(x_0) = x_0 + 0.01x_{range}r$ where $r \in N(0, 1)$, and x_{range} is the dimensional range, i.e., $x_{range} = 2 - (-2) = 4$

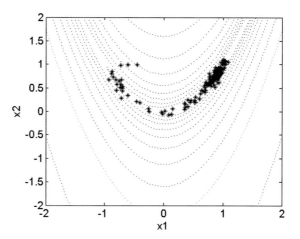

Algorithms (EAs) are sometimes called *meta-heuristics*, which make few or no assumptions about the problem being optimized and can thus search for the global optimum over a large set of candidate solutions. However, besides the population-based nature of EAs, this particular property is another major difference between them since EAs are all based on the "survival of the fittest" philosophy, whereas for the simulated annealing, worse solutions (generations in GA or particle positions in PSO) can still be "tolerated" for the sake of avoiding a local optimum.

Figure 2.4 shows the simulated annealing iterations plotted over the Rosenbrock function. The parameters and functions (for the temperature and neighborhood) used are: $\varepsilon_C = 10^{-3}$, $T_0 = 1$, $\psi(T) = 0.95T$, $N(x_0) = x_0 + 0.01x_{range}r$ where $r \in N(0, 1)$, and x_{range} is the dimensional range, i.e., $x_{range} = 2 - (-2) = 4$. Note that as $T \rightarrow 0$, it mimics the gradient descent method and hence took a longer time to converge to the global optimum.

2.3.2 Stochastic Approximation

Recall that the goal of deterministic optimization methods is to minimize a loss function $L : R^p \rightarrow R^1$, which is a differentiable function of θ and the minimum (or maximum) point θ^* corresponds to zero-gradient point, i.e.,

$$g(\theta) \equiv \left. \frac{\partial L(\theta)}{\partial \theta} \right|_{\theta = \theta^*} = 0 \tag{2.8}$$

As mentioned earlier, in cases where more than one point satisfies this equation (e.g., a multi-modal problem), then such algorithms may only converge to a local minimum. Moreover, in many practical problems, the exact gradient value, g, is not readily available. This makes the stochastic approximation (SA) algorithms quite popular. The general SA takes the following form:

$$\widehat{\theta}_{k+1} = \widehat{\theta}_k - a_k \widehat{g}_k\left(\widehat{\theta}_k\right) \qquad (2.9)$$

where $\widehat{g}_k\left(\widehat{\theta}_k\right)$ is the estimate of the gradient $g(\theta)$ at iteration k and a_k is a scalar gain sequence satisfying certain conditions. Unlike any steepest (gradient) descent method, SA assumes no direct knowledge of the gradient. To estimate the gradient, there are two common SA methods: finite difference stochastic approximation (FDSA) and simultaneous perturbation SA (SPSA) [17]. FDSA adopts the traditional Kiefer-Wolfowitz approach to approximate gradient vectors as a vector of p partial derivatives where p is the dimension of the loss (fitness) function. Accordingly, the estimate of the gradient can be expressed as follows:

$$\widehat{g}_k\left(\widehat{\theta}_k\right) = \begin{bmatrix} \dfrac{L\left(\widehat{\theta}_k + c_k\Delta_1\right) - L\left(\widehat{\theta}_k - c_k\Delta_1\right)}{2c_k} \\[2ex] \dfrac{L\left(\widehat{\theta}_k + c_k\Delta_2\right) - L\left(\widehat{\theta}_k - c_k\Delta_2\right)}{2c_k} \\[1ex] \vdots \\[1ex] \dfrac{L\left(\widehat{\theta}_k + c_k\Delta_p\right) - L\left(\widehat{\theta}_k - c_k\Delta_p\right)}{2c_k} \end{bmatrix} \qquad (2.10)$$

where Δ_k is the unit vector with a 1 in the kth place and c_k is a small positive number that gradually decreases with k. Note that separate estimates are computed for each component of the gradient, which means that a p-dimensional problem requires at least $2p$ evaluations of the loss function per iteration. The convergence theory for the FDSA algorithm is similar to that for the root-finding SA algorithm of Robbins and Monro. These are: $a_k > 0, c_k > 0, \lim_{k\to\infty} a_k = 0, \lim_{k\to\infty} c_k = 0,$ $\sum_{k=0}^{\infty} a_k < \infty$ and $\sum_{k=0}^{\infty} a_k/c_k < \infty$. The selection of these gain sequences is critical to the performance of the FDSA. The common choice is the following:

$$a_k = \frac{a}{(k+A+1)^{\upsilon}} \quad \text{and} \quad c_k = \frac{c}{(k+1)^{\tau}}, \qquad (2.11)$$

where a, c, υ and τ are strictly positive and $A \geq 0$. They are usually selected based on a combination of the theoretical restrictions above, trial-and-error numerical experimentation, and basic problem knowledge.

Figure 2.5 shows the FDSA iterations plotted over the Rosenbrock function with the following parameters: $a = 20, A = 250, c = 1, \upsilon = 1$ and $\tau = 0.75$. Note that during the early iterations, it performs a random search due to its stochastic nature with large a_k, c_k values but then it mimics the gradient descent algorithm. The large number of iterations needed for the convergence is another commonality with the gradient descent.

Fig. 2.5 The plot of 25,000 iterations of the FDSA method over Rosenbrock function with $x(0) = [-0.5, 1.0]$, $\varepsilon_C = 10^{-3}$, $a = 20$, $A = 250$, $c = 1$, $v = 1$, and $\tau = 0.75$

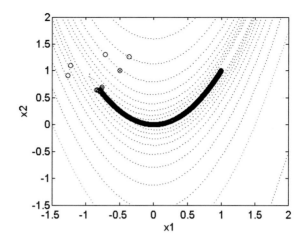

2.4 Evolutionary Algorithms

Among the family members of EAs, this section will particularly focus on GAs and DEs leaving out the details of both EP and ES while the next chapter will cover the details of PSO.

2.4.1 Genetic Algorithms

In nature, every living organism has a set of rules, a blueprint so to speak, describing how that organism is created (designed). The *genes* of an organism represent these rules and they are connected together into long strings called *chromosomes*. Each gene represents a specific property of the organism, such as eye or hair color and the collective set of gene settings are usually referred to as an organism's *genotype*. The physical expression of the genotype—the organism itself—is called the *phenotype*. The process of recombination occurs when two organisms mate and the genes are shared in the resultant offspring. In a rare occasion, a mutation occurs on a gene; however, this mutated gene will usually not affect the creation of the phenotype. Yet in rare cases, it will be expressed in the organism as a completely new trait. The ongoing cycle of natural selection, recombination, and mutation brought the evolution of the life on earth in addition to all such variations among the living organisms and of course their adaptation and survival instincts. The gene mutation plays a crucial role in the famous *Darwinian* rule of evolution, "the survival of the fittest." Genetic Algorithms (GAs) which are all inspired from the *Darwinian* evolution mimic all these natural evolutionary processes so as to search and find the optimum solution of the problem in hand.

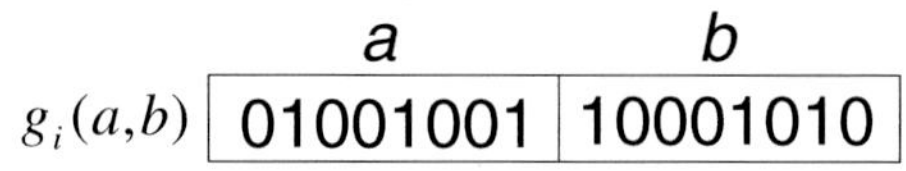

Fig. 2.6 A sample
chromosome representation
for the two problem variables,
a and b

$$g_i(a,b) \quad \boxed{\text{01001001}} \; \boxed{\text{10001010}}$$

(with column labels a above the first box and b above the second box)

In order to accomplish this, the first step is the encoding of the problem variables into genes in a way of strings. This can be a string of real numbers but more typically a binary bit string (series of 0s and 1s). This is the genetic representation of a potential solution. For instance, consider the problem with two variables, a and $b \ni 0 \leq a$, $b < 256$. A sample chromosome representation for the ith chromosome, $g_i(a, b)$, is shown in Fig. 2.6 where both a and b are encoded with 8-bits, therefore, the chromosome contains 16 bits. Note that examining the chromosome string alone yields no information about the optimization problem. It is only with the decoding of the chromosome into its phenotypic (real) values that any meaning can be extracted for the representation. In this case, as described below, the GA search process will operate on these bits (chromosomes), rather than the real-valued variables themselves, except, of course, where real-valued chromosomes are used.

The second requirement is a proper fitness function which calculates the fitness score of any potential solution (the one encoded in the chromosome). This is indeed the function to be optimized by finding the optimum set of parameters of the system or the problem in hand. The fitness function is always problem dependent. In nature, this corresponds to the organism's ability to operate and to survive in its present environment. Thus, the objective function establishes the basis for the proper selection of certain organism pairs for mating during the reproduction phase. In other words, the probability of selection is proportional to the chromosome's fitness. The GA process will then operate according to the following steps:

1. **Initialization**: The initial population is created while all chromosomes are (usually) randomly generated so as to yield an entire range of possible solutions (the search space). Occasionally, the solutions may be "seeded" in areas where optimal solutions are likely to be found. The population size depends on the nature of the problem, but typically contains several hundreds of potential solutions encoded into chromosomes.

2. **Selection**: For each successive generation, first the selection of a certain proportion of the existing population is performed to breed a new generation. As mentioned earlier, the selection process is random, however, favors the chromosomes with higher fitness scores. Certain selection methods rate the fitness of each solution and preferentially select the best solutions.

3. **Reproduction**: For each successive generation, the second step is to generate the next generation chromosomes from those selected through genetic operators such as *crossover* and *mutation*. These genetic operators ultimately result in the child (next generation) population of chromosomes that is different from the initial generation but typically shares many of the characteristics of its parents.

Fig. 2.7 A sample crossover operation over two chromosomes g_i and g_j after $L = 10$ bits. The resultant child chromosomes are c_i and c_j

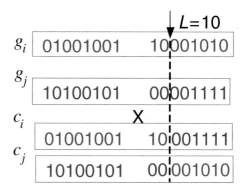

4. **Evaluation**: The child chromosomes are first decoded and then evaluated using the fitness function and they replace the least-fit individuals in the population so as to keep the population size unchanged. This is the only link between the GA process and the problem domain.
5. **Termination**: During the **Evaluation** step, if any of the chromosomes achieves the objective fitness score or the maximum number of generations is reached, then the GA process is terminated. Otherwise, steps 2 to 4 are repeated to produce the next generation.

To perform a crossover operation, an integer position, L, is selected uniformly at random between 1 and the chromosome string length minus one, and the genetic information exchanged between the individuals about this point, then two new offspring strings are produced. A sample operation for $L = 10$ is shown in Fig. 2.7. The crossover operation is applied with a probability, Px, over the pairs chosen for breeding. Crossover is a critical operator in GA due to two reasons: it greatly accelerates the search early in the evolution of a population and it leads to effective combination of subsolutions on different chromosomes. There is always a trade-off when setting its value, i.e., assigning a too high Px may lead to premature convergence to a local optimum and a too low value may deteriorate the rate of convergence.

The other genetic operator, *mutation*, is then applied to certain genes (bits) of the child chromosomes with a probability, Pm. In the binary string representation, a mutation will cause a single bit to change its state, i.e., $0 \Rightarrow 1$ or $1 \Rightarrow 0$. Assigning a very low Pm leads to genetic drift (which is non-ergodic in nature) and the opposite may lead to loss of good solutions unless there is an elitist selection. Without a proper Pm setting, GAs may converge toward local optima or even some arbitrary (non-optimum) points rather than the global optimum of the search space. This indicates that a sufficiently high Pm setting should be assigned to teach the algorithm how to "sacrifice" a short-term fitness in order to gain a longer term fitness. In contrast to the binary GA, the real-valued GA uses real values in chromosomes without any encoding and thus the fitness score of each chromosome can be computed without decoding. This is a more straightforward, faster, and efficient scheme than the binary counterpart. However, both crossover and mutation operations might

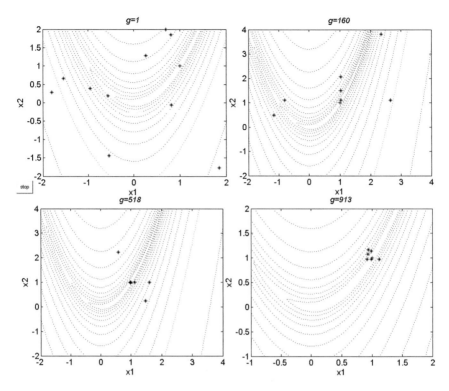

Fig. 2.8 The distributions of real-valued GA population for generations, $g = 1$, 160, 518, and 913 over Rosenbrock function with $S = 10$, $Px = 0.8$, and σ linearly decreases from x_{range} to 0

be different and there are numerous variants performing different approaches for each. A similar crossover operation as in binary GA can still be used (flipping with a probability, Px, over the pairs chosen for breeding). Another crossover operation common for real-valued GA is to exploit the idea of creating the child chromosome between parents via arithmetic recombination (linear interpolation), i.e., $z_i = \alpha$ $x_i + (1-\alpha) y_i$ where x_i, y_i and z_i are ith parent and child chromosomes, respectively, and $\alpha : 0 \leq \alpha \leq 1$. The parameter α can be a constant, or a variable changing according to some function or a random number. On the other hand, the most common mutation method is to shift by a random deviate, applied to each chromosome separately, taken from Gaussian distribution $N(0, \sigma)$ and then curtail it to the problem range. Note that the standard deviation, σ, controls the amount of shift. Figure 2.8 shows the distributions of a real-valued GA population for generations, $g = 1, 160, 518,$ and 913 over Rosenbrock function with the parameter settings as, $S = 10, Px = 0.8$, and an adaptive σ linearly decreasing from x_{range} to 0, where x_{range} is the dimensional range, i.e., $x_{\text{range}} = 2 - (-2) = 4$ for the problem shown in the figure. It took 160 generations for a GA chromosome to converge to the close vicinity of the optimum point, (1, 1).

2.4.2 Differential Evolution

Differential Evolution (DE) is an evolutionary algorithm, showing particular similarities to GA and hence can be called as a genetic-type method. DE has certain differences in that it is applicable to real-valued vectors, rather than bit-encoded strings. Accordingly, the ideas of mutation and crossover are substantially different. Particularly, the mutation operator is entirely different in a way that it is difficult to see why it is called mutation, except perhaps it serves the same purpose of avoiding early local trappings. DE has a notion of population similar to PSO rather than GA as its population members are called *agents* rather than *chromosomes*.

Suppose we optimize a real-valued (fitness) function in N-D, having N real variables. The ath agent in the population in the generation, g, represents the candidate solution of this function in the following array form: $x_a^g = \left[x_{a,1}^g, x_{a,2}^g, \ldots, x_{a,N}^g\right]$, $a \in [1, S]$ where S represents the size of the population. Then the DE process will follow the same path as GA except that the selection is performed after the reproduction, as follows:

1. **Initialization**: The initial population is created with $S > 3$. The range of each agent is defined for $g = 0$, i.e., $x_d^{\min} < x_{a,d}^0 < x_d^{\max}$ and agent vector elements are randomly initialized within this range, $\left[x_d^{\min}, x_d^{\max}\right]$.
2. **Reproduction**: For each successive generation, $g = 1, 2, \ldots$, first mutation and then the crossover operators are applied on each agent's vector. To perform the mutation over the ath agent vector, x_a^g, three distinct agents, b, c, and d, are first randomly chosen such that $a \neq b \neq c \neq d$. This is why $S > 3$. The so-called donor vector for agent a is formed as follows:

$$y_a^{g+1} = x_b^g - \text{Fr}\left(x_c^g - x_d^g\right) \qquad (2.12)$$

where $r \sim U(0, 1)$ is a random variable with a uniform distribution and F is a constant, usually assigned to 2. This is the mutation operation, which adds the weighted difference of the two of the vectors to the third, hence gives the name "differential" evolution. The following crossover operation then forms a trial vector from the elements of the agent vector, x_a^g, and the elements of the donor vector, y_a^{g+1}, each of which enters the trial vector with probability R.

$$u_{a,j}^{g+1} = \left\{ \begin{array}{ll} y_{a,j}^{g+1} & \text{if} \quad r \leq R \quad \text{or} \quad j = \delta \\ x_{a,j}^g & \text{if} \quad r > R \quad \text{and} \quad j \neq \delta \end{array} \right\} \qquad (2.13)$$

where $1 \leq \delta < N$ is a random integer ensuring that $u_{a,j}^{g+1} \neq x_{a,j}^g$. In other words, at least one element from the donor vector is ensured into the trial vector.

3. **Selection (with Evaluation)**: For each successive generation once the trial vector is generated, the agent vector, x_a^g, is compared with the trail vector, $u_{a,j}^{g+1}$, and the one with the better fitness is admitted to the next generation.

$$x_a^{g+1} = \left\{ \begin{array}{cc} u_a^{g+1} & \text{if } f\left(u_a^{g+1}\right) \leq f\left(x_a^g\right) \\ x_a^g & \text{else} \end{array} \right\} \qquad (2.14)$$

4. **Termination**: During the previous step, if any agent achieves the objective fitness score or the maximum number of generations is reached, then the DE process is terminated. Otherwise, steps 2 and 3 are repeated to produce the next generation.

Figure 2.9 illustrates the generation of the trial vector on a sample 2-D function. Note that the trial vector, u_a^{g+1}, gathers the first-dimensional element $\left(u_{a,1}^{g+1}\right)$ from the agent vector, $x_{a,1}^g$ and the second-dimensional element $\left(u_{a,2}^{g+1}\right)$ from the donor vector, $y_{a,2}^{g+1}$.

The choice of DE parameters F, S, and R can have a large impact on the optimization performance and how to select good parameters that yield good performance has therefore been subject to much research, e.g., see Price et al. [18] and Storn [19]. Figure 2.10 shows the distributions of DE population for generations, $g = 1, 20, 60$, and 86 over Rosenbrock function with the parameter settings as, $S = 10$, $F = 0.8$, and $R = 0.1$. Note that as early as 20th generations, a member of DE population already converged to the close vicinity of the optimum point, $(1, 1)$.

Fig. 2.9 A sample 2-D fitness function and the DE process forming the trial vector

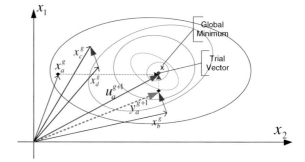

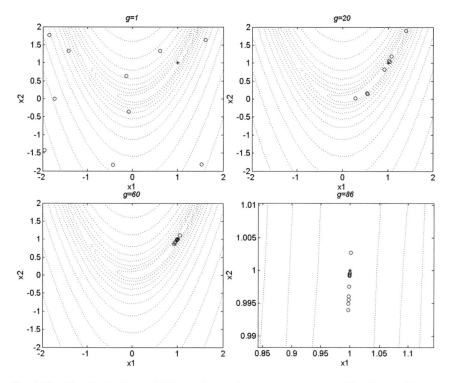

Fig. 2.10 The distributions of DE population for generations, $g = 1$, 20, 60, and 86 over Rosenbrock function with $S = 10$, $F = 0.8$ and $R = 0.1$

References

1. Biography in *Encyclopaedia Britannica.* http://www.britannica.com/eb/article-9062073/ Pythagoras
2. C Field, *The Philosophy of Plato* (Oxford, 1956)
3. Biography in *Encyclopaedia Britannica.* http://www.britannica.com/eb/article-9108312/ Aristotle
4. The MacTutor History of Mathematics archive. http://turnbull.dcs.st-and.ac.uk/history/
5. List of Greatest Mathematicians. http://fabpedigree.com/james/grmatm3.htm
6. T L Heath, *A History of Greek mathematics* II (Oxford, 1931)
7. I. Bulmer-Thomas, Selections illustrating the History of Greek mathematics II (London, 1941)
8. F. Woepcke, *Extrait du Fakhri, traité d'Algèbre par Abou Bekr Mohammed Ben Alhacan Alkarkhi* (1853)
9. J. F. Scott, *The Scientific Work of René Descartes* (1987)
10. H. Robbins, S. Monro, A stochastic approximation method. Ann. Math. Stat. **22**, 400–407 (1951)
11. J. Kiefer, J. Wolfowitz, Stochastic estimation of the maximum of a regression function. Ann. Math. Stat. **23**, 462–466 (1952)
12. J.C. Spall, Multivariate stochastic approximation using a simultaneous perturbation gradient approximation. IEEE Trans. Autom. Control **37**, 332–341 (1992)

13. S. Kirkpatrick, C.D. Gelatt, M.P. Vecchi, Optimization by simulated annealing. Science **220**, 671–680 (1983)
14. J.A. Nelder, R. Mead, A simplex method for function minimization. Comput. J. **7**, 308–313 (1965)
15. S. Boyd, L. Vandenberghe, *Convex Optimization* (Cambridge University Press, UK, 2004)
16. A. Antoniou, W.-S. Lu, *Practical Optimization, Algorithms and Engineering Applications* (Springer, USA, 2007)
17. R. Silipo, et al., ST-T segment change recognition using artificial neural networks and principal component analysis. Comput. Cardiol., 213–216, (1995)
18. K. Price, R. M. Storn, J. A. Lampinen, Differential Evolution: A Practical Approach to Global Optimization. Springer. ISBN 978-3-540-20950-8 (2005)
19. R. Storn, "On the usage of differential evolution for function optimization". Biennial Conference of the North American Fuzzy Information Processing Society (NAFIPS). pp. 519–523, (1996)

Chapter 3
Particle Swarm Optimization

*We converse as we live by repeating, by combining and
recombining a few elements over and over again just as nature
does when of elementary particles it builds a world.*

William Gass

The behavior of a single organism in a swarm is often insignificant, but their
collective and social behavior is of paramount importance. The particle swarm
optimization (PSO) was introduced by Kennedy and Eberhart [1] in 1995 as a
population-based stochastic search and optimization process. It is originated from
the computer simulation of the individuals (particles or living organisms) in a bird
flock or fish school [2], which basically show a natural behavior when they search
for some target (e.g., food). The goal is, therefore, to converge to the global optima
of some multidimensional and possibly nonlinear function or system. Henceforth,
PSO follows the same path of other evolutionary algorithms (EAs), [3] such as
genetic algorithm (GA) [4], genetic programming (GP) [5], evolution strategies
(ES) [6], and evolutionary programming (EP) [7]. Recall that the common point of
all is that EAs are in population-based nature and they can avoid being trapped in a
local optimum. Thus they can find the optimum solutions; however, this is never
guaranteed.

3.1 Introduction

PSO in most basic terms belongs to the swarm intelligence paradigm, which
studies the collective behavior and social characteristics of organized, decentral-
ized, and complex systems known as "swarms." A swarm is an apparently dis-
organized collection (population) of moving individuals that tend to cluster
together while each individual seems to be moving in a random direction. Each
individual in the swarm has the capability of interaction with the other individuals,
or the so-called "agents" (or "particles" in PSO), although the capabilities of each
agent are rather limited by certain set of rules. Therefore, the behavior of an agent
in a swarm is often insignificant, but their collective and social behavior is of
paramount importance, in that, the swarm intelligence comes both from the collective adaptation and stochastic nature of the swarm. The main motivation stems
directly from the organic swarms in nature such as bird flocks, fish schools, ant
colonies, and other animal herds and packs, which exhibit an amazing self-

S. Kiranyaz et al., *Multidimensional Particle Swarm Optimization for Machine Learning
and Pattern Recognition*, Adaptation, Learning, and Optimization 15,
DOI: 10.1007/978-3-642-37846-1_3, © Springer-Verlag Berlin Heidelberg 2014

organization and collective/social adaptation capabilities. This cannot be explained simply by the aggregated behavior of each individual member in the swarm but their collective adaptation to the environment, which in turn makes *the survival* in nature possible.

An intelligent swarm can, therefore, be defined as *a population of interacting individuals that optimizes a function or goal by collectively adapting to the local and/or global environment.* In the area of global optimization, the swarm intelligence first appeared with two methods: PSO in 1995 and ant colony optimization (ACO) in 1992. After their invention, there was an exponential growth in the number of scientific works related to swarm intelligence and the appearance of new journals devoted to the innovations in swarm intelligence.

In a PSO process, a swarm of particles, each of which represents a potential solution to the optimization problem in hand, navigate through the search space. The particles are initially distributed randomly over the search space, and the goal is to converge to the global optima of a function or a system. Each particle keeps track of its position in the search space and its best solution so far achieved. This is the personal best value (the so-called *pbest* [1]) and the PSO process also keeps track of the global best solution so far achieved by the swarm with its particle index (the so-called *gbest* [1]). So during their journey with discrete time iterations, the velocity of each agent in the next iteration is computed as a function of the best position of the swarm (position of the particle *gbest* as the *social* component), the best personal position of the particle (*pbest* as the *cognitive* component), and its previous velocity (the *memory* term). Both *social* and *cognitive* components contribute randomly to the position of the agent in the next iteration. This is illustrated in Fig. 3.1 where particle a has a new velocity update (at time $t + 1$), which can evade the nearby local optimum. This is of course an optimistic illustration and there is absolutely no guarantee that it will happen as such, since the cognitive and social components' contributions to the velocity update are all random; however, the tendency toward local and global best (similar to the "survival of the fittest" paradigm in the other EAs), and repeated trials with random scales *may* yield a convergence to the global optimum sooner or later. Note that its probability of success further rises due to the numerous number of particles in the swarm, since it does not matter if all fail to achieve this but one. This is the main philosophy behind the PSO and the next section is devoted to its detailed description and formulation.

3.2 Basic PSO Algorithm

In the basic PSO method (bPSO), a swarm of particles flies through an N-dimensional search space, where the position of each particle represents a potential solution to the optimization problem. Each particle a in the swarm, $\xi = \{x_1, .., x_a, .., x_S\}$, is represented by the following characteristics:

Fig. 3.1 Illustration of the basic velocity update mechanism in PSO

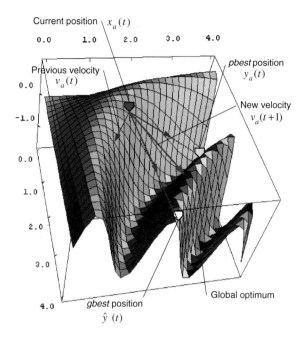

$x_{a,j}(t)$: jth dimensional component of the position of particle a, at time t

$v_{a,j}(t)$: jth dimensional component of the velocity of particle a, at time t

$y_{a,j}(t)$: jth dimensional component of the personal best (*pbest*) position of particle a, at time t

$\hat{y}_j(t)$: jth dimensional component of the global best position of swarm, at time t.

Let f denotes the fitness function to be optimized. Without loss of generality assume that the objective is to find the minimum of f in N-dimensional space. Then the personal best of particle a can be updated in iteration $t + 1$ as,

$$y_{a,j}(t+1) = \begin{cases} y_{a,j}(t) & \text{if } f(x_a(t+1)) > f(y_a(t)) \\ x_{a,j}(t+1) & else \end{cases} \forall j \in [1,N] \qquad (3.1)$$

Since *gbest* is the index of the GB particle, then $\hat{y}(t) = y_{gbest}(t) = \min(y_1(t), .., y_S(t))$. Then for each iteration in a PSO process, positional updates are performed for each particle, $a \in [1,S]$ and along each dimensional component, $j \in [1,N]$, as follows:

$$v_{a,j}(t+1) = w(t)\,v_{a,j}(t) + c_1 r_{1,j}(t)\big(y_{a,j}(t) - x_{a,j}(t)\big) + c_2 r_{2,j}(t)\big(\hat{y}_j(t) - x_{a,j}(t)\big)$$
$$x_{a,j}(t+1) = x_{a,j}(t) + v_{a,j}(t+1)$$

$$(3.2)$$

where w is the inertia weight [8], and c_1, c_2 are the acceleration constants. $r_{1,j} \sim U(0,1)$ and $r_{2,j} \sim U(0,1)$ are random variables with a uniform distribution. Recall from the earlier discussion that the first term in the summation is the *memory* term, which represents the contribution of previous velocity, the second term is the *cognitive* component, which represents the particle's own experience and the third term is the *social* component through which the particle is "guided" by the *gbest* particle toward the GB solution so far obtained. Note that the *gbest* particle is the common guide for all swarm particles since the third term exists in each velocity update equation. Although the use of inertia weight, w, was later added by Shi and Eberhart [8], into the velocity update equation, it is widely accepted as the basic form of PSO algorithm. Each PSO run updates the positions of the particles using Eq. (3.2). Depending on the problem to be optimized, PSO iterations can be repeated until a specified number of iterations, say *IterNo,* is exceeded, velocity updates become zero, or the desired fitness score is achieved (i.e.,$f < \varepsilon_C$, where f is the fitness function and ε_C is the cut-off error). Accordingly, the general pseudo-code of the *bPSO* is presented in Table 3.1.

Velocity clamping in step 3.4.1.2, also called as "dampening" with the user defined maximum range V_{max} (and $-V_{max}$ for the minimum) is one of the earliest attempts to control or prevent oscillations [9]. Figure 3.2 illustrates a typical

Table 3.1 Pseudo-code for the bPSO algorithm

---

bPSO (termination criteria:{IterNo, ε_C ,...}, V_{max})
1. **For** $\forall a \in [1, S]$ do:
   1.1. Randomize $x_a(1)$
   1.2. Let $y_a(0) = x_a(1)$
   1.3. Let $\hat{y}(0) = x_a(1)$
2. **End For.**
3. **For** $\forall t \in [1, IterNo]$ do:
   3.1. For $\forall a \in [1, S]$ do:
      3.1.1. **Compute** $y_a(t)$ using Eq. (15)
      3.1.2. **If** $\left(f(y_a(t)) < \min\left(f(\hat{y}(t-1), f(y_i(t))\atop 1 \le i < a \right) \right)$ then gbest = a and $\hat{y}(t) = y_a(t)$
   3.2. End For.
   3.3. If any *termination criterion* is met, then Stop.
   3.4. For $\forall a \in [1, S]$ do:
      3.4.1. For $\forall j \in [1, N]$ do:
         3.4.1.1. **Compute** $v_{a,j}(t+1)$ using Eq. (12)
         3.4.1.2. **If**$\left(\left| v_{a,j}(t+1) \right| > V_{max} \right)$ then clamp it to $\left| v_{a,j}(t+1) \right| = V_{max}$
         3.4.1.3. **Compute** $x_{a,j}(t+1)$ using Eq. (12)
      3.4.2. End For.
   3.5. End For.
4. **End For.**

Fig. 3.2 Particle velocity
tendency to explode without
velocity clamping

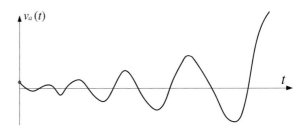

velocity explosion if the velocity clamping is not performed. Another way of
controlling the particle velocities is using the two acceleration constants, c_1, c_2,
which affect the trajectory of the particles. If they are set too small, the trajectory
of a particle falls and rises at a slow rate. As their value increased, the frequency of
the particle oscillations will be increased, so they are indeed the rate of velocity
change, or simply the accelerators. Eberhart and Shi suggested to use the inertia
weight which decreasing over time, typically from 0.9 to 0.4, with $c_1 = c_2 = 2$. It
has the effect of narrowing the search, gradually changing from an exploratory to
an exploitative mode.

Clerc and Kennedy [10] suggested a more generalized PSO, where a constriction coefficient, χ, is applied to both terms of the velocity formula. They show
that the constriction PSO can converge without using velocity clamping by V_{max},
and can be formulated as,

$$v_{a,j}(t + 1) = \chi\left(v_{a,j}(t) + c_1 r_{1,j}(t)\left(y_{a,j}(t) - x_{a,j}(t)\right) + c_2 r_{2,j}(t)\left(\hat{y}_j(t) - x_{a,j}(t)\right)\right)$$
$$x_{a,j}(t + 1) = x_{a,j}(t) + v_{a,j}(t + 1)$$

$$(3.3)$$

where usually $c_1 = c_2 = 2.05$, and the constriction factor χ is set to 0.7289. By
using the constriction coefficient, the amplitude of the particle's oscillation
gradually decreases, resulting in a convergence over time.

Some other important PSO variants and improvements will be covered in the
next section.

3.3 Some PSO Variants

The first set of improvements has been proposed for the problem-dependent performance of PSO due to its strong parameter dependency. There are mainly two
types of approaches: The first one is through self-adaptation, which has been
applied to PSO by Clerc [11], Yasuda et al. [12], Zhang et al. [13], and Shi and
Eberhart [14]. The other approach is via performing hybrid techniques, which are
employed along with PSO by Angeline [15], Reynolds et al. [16], Higashi and Iba
[17], Esquivel and Coello Coello [18], and many others. Finally, Van den Bergh

[19] showed that the following inequality should be satisfied in order to guarantee the convergence to (local) optima:

$$w > \frac{c_1 + c_2}{2} - 1 \qquad (3.4)$$

where w is the inertia weight, and c_1, c_2 are the acceleration constants used in Eq. (3.2). Mendes et al. [20] derived Fully Informed PSO (FIPS) from the constriction PSO and present its general form as,

$$v_{a,j}(t+1) = \chi \left(v_{a,j}(t) + \frac{1}{|N_a|} \sum_{n \in N_a} c_1 r_{1,j}(t) \left(y_{n,j}(t) - x_{a,j}(t) \right) \right)$$
$$x_{a,j}(t+1) = x_{a,j}(t) + v_{a,j}(t+1) \qquad (3.5)$$

where N_a defines a neighborhood of the particle a, and $|N_a|$ is the number of particles in it. In the FIPS, a particle is attracted by every other particle in its neighborhood. Therefore, the performance of FIPS is generally more dependent on the neighborhood topology (global best neighborhood topology is recommended).

The rest of the PSO variants presented in this section contains some improvements trying to avoid the premature convergence problem via introducing diversity to swarm particles. An earlier improvement to avoid premature convergence is the *craziness* operator, which has been first proposed by Kennedy and Eberhart [1]. At each iteration, a set of particles from the center of the swarm is selected and randomized within the search space. However, they concluded that it may not be a necessary operation since it does not contribute much to the performance of PSO.

Attractive and Repulsive PSO (ARPSO) proposed [21] alternates between attraction and repulsion phases. During attraction, ARPSO allows fast information flow between particles causing a low diversity but a better convergence to the solution. It is reported that 95 % fitness improvements can be obtained within this phase. In the repulsion phase, the particles are pushed away from the GB solution so far achieved to increase diversity. ARPSO exhibits a higher performance compared to both PSO and GA.

Note that according to the velocity update equation in Eq. (3.2), the velocity of the *gbest* particle will only depend on the memory term since $x_{gbest} = y_{gbest} = \hat{y}$. To address this problem Van den Bergh introduced a new PSO variant, the PSO with guaranteed convergence, (GCPSO) [22]. In GCPSO, a different velocity update equation is used for the *gbest* particle based on two threshold values that can be adaptively set during the process. It is claimed that GCPSO usually performs better than the *bPSO* when applied to unimodal functions and comparable for multimodal problems; however, due to its fast rate of convergence, GCPSO can be more likely to trap to a local optimum with a guaranteed convergence, whereas the *bPSO* may not. Based on GCPSO, Van den Bergh proposed the Multi-start PSO (MPSO) [22], which repeatedly runs GCPSO over randomized particles and stores the (local) optimum at each iteration. Yet, similar to *bPSO* and many of its variants, the performance still degrades significantly as the dimension of the search space increases [22].

Another attempt to improve the overall performance is to use multiple swarms instead of one. Lovberg et al. [23] proposed an approach, which divides the main swarm into several swarms where each swarm has its own *gbest* particle. The particles between different swarms can mate by using an arithmetic crossover operator with a certain probability, the so-called *breeding* operation. The results, however, show that this approach did not improve the overall performance since swarms with less particles do not have enough exploration power and no prevention is designed against such small-size swarms getting too similar to each other over time.

In another approach, Lovberg and Krink presented Self Organized Criticality (SOC) PSO [24, 25]. The criticality measures the proximity of particles, so that the particles that are too close to each other can be relocated in the search space to improve the diversity of the swarm. They propose two types of relocation: The first one is random initialization and the second one is random displacement of particles further in the search space. SOC PSO outperformed *bPSO* only in one out of four cases.

3.3.1 Tribes

In all PSO variants presented earlier including the basic (canonical) version, the description of the problem typically provides the following: the definition of the solution (or search) space; the fitness function to be optimized (the *objective function*) on each point of the search space; and finally, a stopping criterion (e.g., the maximum number of iterations or admissible error). The swarm, on the other hand, can be defined by the population size and other intrinsic parameters such as inertia factor w, acceleration constants c_1, c_2, or perhaps some other depending on the variant. The particular PSO variant, "Tribes", is a parameter-free PSO algorithm. Its major properties are:

- The swarm is divided into "tribes".
- At the beginning, the swarm is composed of only one particle.
- According to tribes' behaviors, particles are added or removed.
- According to the performances of the particles, their strategies of displacement are adapted.
- Adaptation of the swarm according to the performances of the particles.

In this process, some subgroups are defined in such a way that, inside each group, every particle informs all other particles, including itself. Therefore, these subgroups are called as *tribes*, a metaphor for different sized groups of particles moving about the search space, looking for the global solution of the problem in hand. In practice, this process is similar to *nesting* in GAs with the same purpose:

to explore several promising areas simultaneously, usually around local minima. When a particular tribe succeeds in finding a local minima, it informs others so that they may decide collectively where the global optimum resides. Therefore, there is an information network between tribes. Let A0 and B0 two particles in two tribes, A and B. This means that from any particle in A to any particle in B, there exists an information path, such as, "A0 informs A1, which informs A2…which informs B0,…". In short, each tribe can be defined as a dense network, and there are certain subnetworks among the tribes making the entire network fully connected. Such a networking structure must be automatically generated and updated by means of creation, evolution, and deletion of particles and tribes.

In brief, for tribes we can summarize the structural adaptations as follows:

- Definition of a status for each tribe: good, neutral, or bad.
- Definition of a status for each particle: good or neutral.
- Removal of a particle: remove the worst particle from a *good* tribe.
- Generation of a particle:improvement of performances of a *bad* tribe.

As in other canonical PSO types, each particle has current and personal best positions. A particle is said to be *good* if it has just improved its personal best performance, otherwise it is *neutral*. Note that this is a binary definition because improvement is not measured. We check only if it is strictly positive (real improvement) or null (no improvement). By definition, the best performance of a particle cannot deteriorate, and that is why there is no "bad" particle in the absolute, but only by comparison. The particle having the worst personal best position within a tribe is called as the *bad*. Similarly, the *best* particle is assigned relative to a tribe. Moreover, compared to canonical PSO, the particle memory is slightly improved, so that it remembers its last two performance variations, thus maintaining a short history of its moves. On the other hand, to measure the global performance of a tribe, two status assignments, *good* and *bad*, are used. It is determined by a simple rule: the higher the number of good particles in a tribe, the more the tribe is itself *good* and vice versa. More precisely, consider tribe T with size N, which can be assigned either *good* or *bad* according to:

$$T = \begin{pmatrix} \text{good} & \text{if } N_{\text{Good}} > rand(1,N) \\ \text{bad} & \text{else} \end{pmatrix} \tag{3.6}$$

where $N_{\text{Good}} < N$ is the number of good particles in T. Such a probabilistic approach will then lead to the construction of new tribes using the adaptation rules summarized earlier. For instance, only the *worst* particle in the *best* tribe is removed. Moreover, for each *bad* tribe, a free particle is created and initialized in such a way that the probability of a new region discovery can be higher.

The tribe creation process starts with the randomly generated initial particle, which also constitutes the initial tribe. It will then undergo to the same PSO velocity updates and if there is no improvement observed in the first iteration, then a second tribe is generated and initialized with its first particle, and so on. Therefore, the number of tribes along with their particles will be increased, in order to improve the search ability with the increasing population as long as no improvement is observed. As soon as a certain level of improvement is achieved, the excess population will be regulated by removing the worst particles in the best tribes.

3.3.2 Multiswarms

PSO was initially proposed as an optimization technique for static environments; however, many real problems are dynamic, meaning that the environment and the characteristics of the global optimum can change in time. Therefore, such problems require systematic re-optimizations due to system and/or environmental changes. Even though it is possible to handle such dynamic problems as a series of individual processes via restarting the optimization algorithm after each change, this may lead to a significant loss of useful information, especially when the change is not too drastic. The main problem of using the bPSO algorithm in a dynamic environment is that eventually the swarm will converge to a single peak—whether global or local. When another peak becomes the global maximum as a result of an environmental change, it is likely that the particles keep moving in the vicinity of the peak to which the swarm has converged earlier, and thus they cannot find the new global maximum. Blackwell and Branke have addressed this problem in [26] and [27] by introducing *multiswarms* that are actually separate PSO processes. Each particle is now a member of one of the swarms only and it is *unaware* of other swarms. This is one of the main differences compared to "Tribes", which otherwise is another good example of multiswarms. The main idea is that each swarm can converge to a separate peak. Swarms interact only by mutual repulsion that keeps them from converging to the same peak. For a single swarm, it is essential to maintain enough diversity, so that the swarm can track small location changes of the peak to which it is converging. For this purpose Blackwell and Branke introduced charged and quantum swarms, which are analogs to an atom having a nucleus and charged particles randomly orbiting it. The particles in the nucleus take care of the fine tuning of the result while the charged particles are responsible of detecting the position changes. However, it is clear that, instead of charged or quantum swarms, some other method can also be used to ensure sufficient diversity among particles of a single swarm, so that the peak can be tracked despite of small location changes.

As one might expect, the best results are achieved when the number of swarms is set equal to the number of peaks. However, it is then required that the number of peaks is known beforehand. In [28], Blackwell presents self-adapting multiswarms, which can be created or removed during the PSO process, and therefore it is not necessary to fix the number swarms beforehand. The repulsion between swarms is realized by simply reinitializing the worse of two swarms if they move within a certain range from each other. Using physical repulsion could lead to equilibrium, where swarm repulsion prevents both swarms from getting close to a

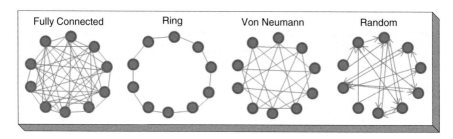

Fig. 3.3 Local PSO topologies

peak. A proper limit closer to which the swarms are not allowed to move, r_{rep} is attained by using the average radius of the peak basin, r_{bas}. If p peaks are evenly distributed in X^N, $r_{rep} = r_{bas} = X/p^{1/N}$.

Multiswarms, or the so-called subswarms also exists in local PSO topologies. Recall that in the (standard) global PSO, each particle is a neighbor of all other particles (i.e., = a fully connected topology). Therefore, the personal best position of the *gbest* particle guides the whole swarm as it affects all the velocity updates. Yet, if the current global optimum is not close to the global optimum solution, it may become hard, if not impossible, for the swarm to explore other areas of the search space. Generally speaking, global PSOs usually converge faster and may get trapped in a local optimum more easily. There are other local PSO variants where the particles are grouped within neighborhoods according to a certain strategy, to create subswarms. In this case, only the *gbest* particle in the subswarm can influence the velocity update of a given particle in that subswarm. Consequently, such local PSO variants (with certain subswarm topologies) converge slower than the global PSO, but they have higher chance of avoiding local minima due to greater population diversity [29]. Such a neighborhood approach actually models the social networks [30]. Four sample topologies are shown in Fig. 3.3, where the "Fully Connected" topology (also called as the "Star topology") corresponds to the global PSO. All the others are examples of local PSO topologies, where a local best (*lbest*) particle guides the subswarm with a neighborhood size, K. Henceforth, the same velocity update equation as given in Eq. (3.2) will be used while $\hat{y}_j(t)$ now represents the jth dimensional component of the *lbest* position of a subswarm, at time t.

Consider for instance the "Ring" topology; as the simplest example of local PSO, it connects each particle with the two immediate neighbors, e.g., $K = 2$ (left and right particles). The flow of information in this topology is drastically reduced compared to the global PSO (the star topology). Using the ring topology will slow down the convergence rate, because the best solution found has to propagate through several neighborhoods before affecting all particles in the swarm. This slow propagation will allow the particles to explore more areas in the search space and thus *may* decrease the chance of premature convergence. On the other hand, the choice of topology and thus the size of the neighborhood might be critical, and moreover it will induce new parameters to the PSO.

3.4 Applications

PSO as a generic optimization method can be applied to many problems in several areas, ranging from numerical optimization, to machine learning, and signal processing, just to name a few. For a given problem, that each particle represents a *potential* solution of the problem and the swarm particles move in the solution space seeking the optimum solution. Among many alternatives, we shall focus on three sample problem domains to demonstrate how PSO can be applied to search for the global solution.

3.4.1 Nonlinear Function Minimization

The first problem domain is nonlinear function minimization and several benchmark functions exist in high dimensions. Figure 3.4 presents six of these benchmark functions that are shown in 2D for illustration purposes. The general, unconstrained function minimization problem in d-dimensional space has the form:

$$x^* = \left[x_1^*, x_2^*, .., x_d^*\right] = \arg\min_{x \in R^d} f(x) \Leftrightarrow f(x^*) = \min_{x \in R^d} f(x) \qquad (3.7)$$

where $f(x)$ is the d-dimensional nonlinear function and suppose that it has a global minimum within a practical range of $\pm x_{\max}$, which corresponds to a d-dimensional cube representing the boundaries of the search space. All the benchmark functions in the figure have their global minimum at the origin, i.e., $x^* = [0, 0, .., 0]$. This is the most natural type of problem where the position of the PSO particle a can directly correspond to the points in the data space, i.e., the jth component of a d-dimensional point $(x_j, j \in [1, d])$ is stored in its positional component, $x_{a,j}(t)$. The PSO process can then start with a random initialization of each particle position within the search range, $\pm x_{\max}$ (i.e., as in step 1.1: **Randomize** $x_a(1)$).

Figure 3.5 presents the four plots of the personal best score of the *gbest* particle versus iteration (epoch) number obtained from the individual PSO runs for the four sample functions in dimensions $d = 20$ and $d = 80$. In all runs the following parameters are used: $c_1 = c_2 = 2$, $x_{\max} = 500$, $V_{\max} = x_{\max}/5 = 100$, $iterNo = 5000$, $\varepsilon_C = 10^{-4}$ and $S = 50$. The inertial factor, w, is linearly decreased from 0.9 to 0.4. Note that for unimodal functions in both dimensions, Sphere and Griewank, PSO runs successfully converged (within a vicinity of ε_C) to the global minimum within *iterNo* iterations. However, in the case of multimodal functions, i.e.,

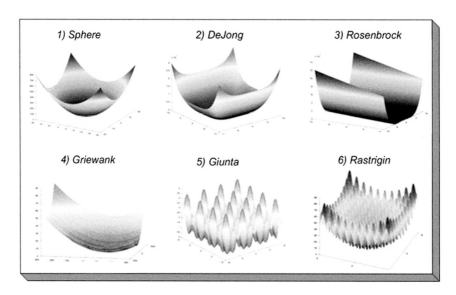

Fig. 3.4 Some benchmark functions in 2D

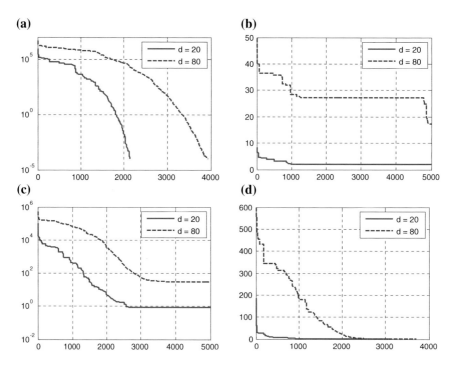

Fig. 3.5 The plots of the *gbest* particle's personal best scores for the four sample non-linear functions given in Fig. 3.4. **a** Sphere. **b** Giunta. **c** Rastrigin. **d** Griewank

Rastrigin and Giunta, a premature convergence to a local optimum is evident; nevertheless, the *gbest* scores indicate a good performance level particularly for the run on Rastrigin (a drastic drop from the level of 10^4 to 1.2). The effect of the dimension on the convergence is also quite visible. It is easy to observe that the higher the dimension it gets, the worse the convergence becomes. This observation is valid for uni- and multimodal functions.

3.4.2 Data Clustering

The second application domain is data clustering. As the process of identifying natural groupings in a multidimensional data based on some similarity measure (e.g., *Euclidean*), data clustering can be divided into two main categories: hierarchical and partitional [31]. Each category then has a wealth of subcategories and different algorithmic approaches for finding the natural groupings, or the so-called *clusters*. Clustering can also be performed in two different modes: hard (or crisp) and fuzzy. In the former mode, the clusters are disjoint, nonoverlapping, and each data point belongs to a single cluster, whereas in the latter case each data point can belong to all the clusters with some degree of membership [32]. *K*-means [33] is the well-known and widely used partitional clustering method, which first assigns each data point to one of the *K* cluster *centroids* which are then updated by the *mean* of the associated points. Starting from a random set of *K* centroids, this cycle is then iteratively performed until the convergence criteria, $\Delta_{K\text{means}} < \varepsilon$ is reached where the objective function, $\Delta_{K\text{means}}$ can be expressed as,

$$\Delta_{K\text{means}} = \sum_{k=1}^{K} \sum_{x_p \in c_k} \left\| c_k - x_p \right\|^2 \tag{3.8}$$

where c_k is the kth cluster center, x_p is the pth data point in cluster c_k and $\|.\|$ is the distance metric in the *Euclidean* space. As a hard clustering method, *K*-means suffers from the following drawbacks:

- The number of clusters K, needs to be set in advance.
- The performance of the method depends on the initial (random) centroid positions as the method converges to the closest local optima.
- The method is also dependent to the distribution of the data.

The fuzzy version of *K*-means, the so-called fuzzy C-means (FCM) (sometimes also called as fuzzy *K*-means) was proposed by Bezdek [34], and has become the most popular fuzzy clustering method so far. It is a fuzzy extension of the *K*-means with a similar objective function as follows:

$$\Delta_{\text{FCM}} = \sum_{k=1}^{K} \sum_{x_p \in c_k} u_{kp}^m \left\| c_k - z_p \right\|^2 \quad \text{with} \quad \sum_{k=1}^{K} u_{kp} = 1, \forall p \tag{3.9}$$

where $u_{kp} \leq 1$ is a positive membership value of the data point z_p to the cluster c_k and $m > 1$ is the fuzziness exponent. FCM usually achieves a better performance compared to the K-means [35] and is less data dependent; however, it still suffers from the same drawbacks, i.e., the number of clusters should be fixed a priori and unfortunately it may also converge to local optima [32]. Zhang and Hsu [36] proposed a novel fuzzy clustering technique, the so-called K-harmonic means (KHM), which is less sensitive to initial conditions and promises further improvements. The experimental results demonstrate that KHM outperforms both K-means and FCM [36, 37]. An extensive survey over various types of clustering techniques can be found in [32] and [38].

A hard clustering technique based on the bPSO was first introduced by Omran et al. in [39] and this work showed that the *bPSO* can outperform K-means, FCM, KHM, and some other state-of-the-art clustering methods in any (evaluation) criteria. This is indeed an expected outcome due to the PSO's aforementioned ability to cope up with the local optima by maintaining a guided random search operation through the swarm particles. In clustering, similar to other PSO applications, each particle represents a potential solution at a particular time t, i.e., the particle a in the swarm, $\xi = \{x_1, .., x_a, .., x_S\}$, is formed as $x_a(t) = \{c_{a,1}, .., c_{a,j}, .., c_{a,K}\} \Rightarrow x_{a,j}(t) = c_{a,j}$ where $c_{a,j}$ is the jth (potential) cluster centroid in N-dimensional data space and K is the number of clusters fixed in advance. Note that contrary to nonlinear function minimization in the earlier section, the data space dimension, N, is now different than the solution space dimension, K. Furthermore, the fitness function, f that is to be optimized, is formed with respect to two widely used criteria in clustering:

- *Compactness*: Data items in one cluster should be similar or close to each other in N-dimensional space and different or far away from the others belonging to other clusters.
- *Separation*: Clusters and their respective centroids should be distinct and well-separated from each other.

The fitness functions for clustering are then formed as a regularization function fusing both *Compactness* and *Separation* criteria and in this problem domain they are known as clustering validity indices. Omran et al. used the following validity index in their work [39]

$$f(x_a, Z) = w_1 \overline{d}_{\max}(x_a, Z) + w_2(Z_{\max} - d_{\min}(x_a)) + w_3 Q_e(x_a)$$

$$\text{where} \quad Q_e(x_a) = \frac{\sum_{j=1}^{K} \frac{\sum_{\forall z_p \in x_{a,j}} \|x_{a,j} - z_p\|}{\|x_{a,j}\|}}{K} \tag{3.10}$$

where Q_e is the quantization error (or the average intracluster distance), $\overline{d}_{\max}$ is the maximum average *Euclidean* distance of data points, $Z = \{z_p\}$, to their centroids, $x_a, \forall z_p \in x_{a,j}$, $Z_{\max}$ is a constant value for theoretical maximum intercluster distance, and $d_{\min}$ is the minimum centroid (intercluster) distance in the cluster centroid set x_a. The weights, w_1, w_2, w_3 are user defined regularization coefficients. So the minimization of the validity index $f(x_a, Z)$ will simultaneously try to minimize the intracluster distances (better *Compactness*) and maximize the intercluster distance (better *Separation*). In such a regularization approach, different priorities (weights) can be assigned to both objectives via proper setting of weight coefficients. Another traditional and well-known validity index is Dunn's index [40], which suffers from two drawbacks: It is computationally expensive and sensitive to noise [41]. Several variants of Dunn's index were proposed in [38], where robustness against noise is improved. There are many other validity indices, i.e., proposed by Turi [42], Davies and Bouldin [43], Halkidi and Vazirganis [44], etc. A throughout survey can be found in [41]. Most of them presented promising results; however, none of them can guarantee the "optimum" number of clusters in every clustering scheme. Especially for the aforementioned PSO-based clustering in [39], the clustering scheme further depends on weight coefficients and may therefore result in over- and under-clustering particularly in complex data distributions.

In order to test the clustering performance of the bPSO, we created 15 synthetic data spaces as shown in Fig. 3.6, and to make the evaluation independent from the choice of the parameters, we simply used Q_e in Eq. (3.10) as the clustering validity index function. For each clustering experiment, we manually set K as the true number of clusters existing in 2D synthetic data space. For illustration purposes each data space is formed in 2D; however, clusters are formed with different shapes, densities, sizes, and intercluster distances to test the robustness of

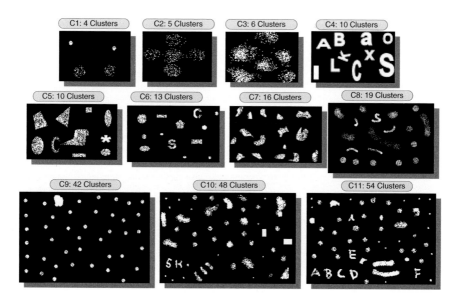

Fig. 3.6 11 synthetic data spaces in 2D

clustering methods against such variations. Furthermore, recall that the number of clusters determines the (true) dimension of the solution space in a PSO application, and hence data spaces with different numbers of clusters are used to test the convergence accuracy to the true (solution space) dimension. As a result, significantly varying complexity levels are achieved among the 11 data spaces to establish a general purpose evaluation of each technique. We set: $iterNo = 2,000$; however, the use of cut-off error as a termination criterion is avoided since it is not feasible to set a unique ε_C value for all clustering schemes. Therefore, each PSO run is executed with 2,000 iterations. The swarm size, $S = 80$, and the rest of the PSO parameters are set as the default values given earlier are also used in all experiments except the positional range, $\pm x_{max}$, since it can now be set simply as the natural boundaries of the 2D data space.

The first set of clustering operations is performed over the simple data spaces where they can yield accurate results, e.g., the results of clustering over the four data spaces at the top row in Fig. 3.6 are shown in Fig. 3.7 where each cluster is represented in one of the three color codes (red, green, and blue) for illustration purposes and each cluster centroid (each dimensional component of the *gbest* particle) is shown with a white '+'.

The convergence accuary of PSO tends to degrade with the increasing dimensionality and complexity due to the well-known "curse of dimensionality" phenomenon from which no optimization technique can be entirely immune. Figure 3.8 presents typical clustering results for $K \geq 10$ and while running each PSO operation till iteration number reaches to 20,000 (i.e., stagnation). $K = 10$ is indeed not a too high dimension for PSO but it particularly suffers from the highly complex clustering schemes as in C4 and C5 (i.e., varying sizes, shapes, and

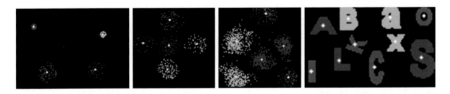

Fig. 3.7 PSO clustering for 2D data spaces C1-C4 shown in Fig. 3.6

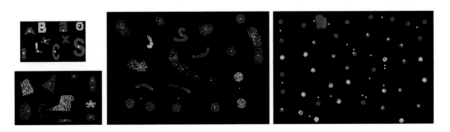

Fig. 3.8 Erroneous PSO clustering over data spaces C4, C5, C8, and C9 shown in Fig. 3.6

structures among clusters). Over a simpler data space, e.g., C6 with 13 clusters, we noticed that PSO occasionally yields accurate clustering but for those data spaces with 20–25 clusters or above, clustering errors become inevitable regardless of the level of complexity and errors tend to increase significantly in higher dimensions as a natural consequence of earlier local traps. A typical example is C9, which has 42 clusters in the simplest form (uniform size, shape, and density) and the clustering result presents many over- and under-clustering schemes with many occasional miss-located centroids. Much worse performance can be expected from their applications over C10 and C11.

As a result, although PSO-based clustering outperforms many well-known clustering methods, it still suffers from two major drawbacks. The number of clusters, K, (being the solution space dimension as well) should still be specified in advance and similar to other PSO applications, the method obviously tends to trap in local optima particularly when the complexity of the clustering scheme increases. This also involves the dimension of the solution space, i.e., convergence to "optimum" number of "true" clusters can only be guaranteed for low dimensions. Recall from the earlier discussion that this is also true for dynamic clustering schemes, DCPSO [45] and MEPSO [46], both of which eventually present results only in low dimensions ($K \leq 10$ in [45] and $K \leq 6$ in [46]) and for simple data distributions. The degradation is likely to be more severe particularly for DCPSO, since it entirely relies on K-means for actual clustering.

3.4.3 Artificial Neural Networks

3.4.3.1 An Overview

Another application domain is the training of artificial neural networks (ANNs) for supervised data classification. After the introduction of simplified neurons by McCulloch and Pitts in 1943 [20], ANNs have been widely applied to many application areas, most of which used feedforward ANNs with the back propagation (BP) training algorithm. An ANN consists of a set of connected processing units, usually called neurons or nodes. ANNs can be described as directed graphs, where each node performs some activation function on its inputs and then passes the result forward to be the input of some other neurons, until the output neurons are reached.

ANNs can be divided into feed forward and recurrent networks according to their connectivity. In a recurrent ANN there can be backward loops in the network structure, while in feedforward ANNs they are no loops. Furthermore, feedforward ANNs are usually organized into layers of parallel neurons and only connections between adjacent layers are possible. All layers besides the input and the output layers are called *hidden* layers. Commonly, the input layer is just a passive layer, where no computations are carried out and it is thus not counted in the total number of layers. The active neurons perform an activation function f of the form,

Fig. 3.9 An example of a
fully connected feed-forward
ANN

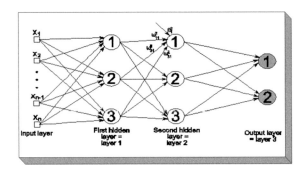

$$y_k^{p,l} = f\left(\sum_{j=1}^{N^{l-1}} w_{jk}^l y_j^{p,l-1} - \theta_k^l\right), \tag{3.11}$$

where $y_k^{p,l}$ is the output of neuron k at layer l, when pattern p is fed to the ANN,
N^{l-1} is the total number of neurons in layer $l\text{-}1$, w_{jk}^l is the connection weight
between neuron j at layer $l\text{-}1$ and neuron k at layer l, and θ_k^l is the bias of neuron
k. For the first processing layer (the layer after the input layer), $y_j^{p,l-1} = y_j^{p,0}$ is
naturally the jth dimension of the input x_j^p. The number of input neurons N_i and the
number of output neurons N_o for ANNs are defined by the problem, while the
number of hidden layers and the number of neurons in each hidden layer is
somehow decided usually by expert rule of thumb and with respect to the problem.
A sample feedforward ANN is illustrated in Fig. 3.9. It has three layers (two
hidden layers and the output layer). Figure 3.9 also shows the connection weights
w_{j1}^2 and the bias θ_1^2 for the first neuron in layer 2.

As an example illustrated in Fig. 3.9, the most common ANN type is the
multilayer perceptron (MLP) [47]. It is a feedforward network, which contains one
or more hidden layers, each with a given number of neurons. The degree of neuron
connectivity is usually high and the neurons have smooth nonlinear activation
functions. The use of nonlinear activation functions is essential, because otherwise
the MLP could always be reduced to a single-layer perceptron (SLP) without
changing its capabilities. Another popular type of feed-forward ANN is the radial
basis function (RBF) network [48], which has always two layers in addition to the
passive input layer: a hidden layer of RBF units and a linear output layer. Only the
output layer has connection weights and biases. The activation function of the kth
RBF unit is defined as,

$$y_k = \varphi\left(\frac{\|X - \mu_k\|}{\sigma_k^2}\right), \tag{3.12}$$

where φ is a radial basis function or, in other words, a strictly positive radially
symmetric function, which has a unique maximum at N-dimensional center μ_k and
whose value drops rapidly close to zero away from the center. σ_k is the width of

the peak around center μ_k. The activation function gets noteworthy values only when the distance between the N-dimensional input X and the center μ_k, $\|X - \mu_k\|$, is smaller than the width σ_k. The most commonly used activation function in RBF networks is the *Gaussian* basis function defined as,

$$y_k = \exp\left(-\frac{\|X - \mu_k\|^2}{2\sigma_k^2}\right), \tag{3.13}$$

where μ_k and σ_k are the mean and standard deviation of the Gaussian function and ‖.‖ is the *Euclidean* norm. While MLPs construct global approximations to non-linear input–output mappings, RBF is built from local approximations centered on clusters of input training samples, and both have a universal function approximation capability. More detailed information about MLPs and RBF networks the reader may consult [47].

Back propagation is the most commonly used training technique for feedforward ANNs. BP has the advantage of a directed search, where the weights are updated in such a way as to minimize the error. However, there are several aspects, which make the algorithm not guaranteed to be universally useful. Most troublesome is its strict dependency on the learning rate parameter, which, if not set properly, can either lead to oscillation or indefinitely long training time. Network paralysis might also occur, i.e., as the ANN trains, the weights tend to assume quite large values and the training process can come to a virtual standstill. Furthermore, BP eventually slows down by an order of magnitude for every extra (hidden) layer added to the ANN. After all, BP is simply a gradient descent algorithm over the error space, which can be complicated and may contain many deceiving local minima (multimodal). Therefore, BP most likely gets trapped into a local minimum, making it entirely dependent on the initial (weight) settings.

Let N_h^l be the number of hidden neurons in layer l of a MLP with input and output layer sizes N_I and N_O, respectively. The input neurons are merely fan out units, since no processing takes place there. Let F be the activation function applied over the weighted inputs plus a bias, as follows:

$$y_k^{p,l} = F(s_k^{p,l}) \quad \text{where} \quad s_k^{p,l} = \sum_j w_{jk}^{l-1} y_j^{p,l-1} + \theta_k^l \tag{3.14}$$

where $y_k^{p,l}$ is the output of the kth neuron of the lth hidden/output layer when pattern p is fed at the input, w_{jk}^{l-1} is the weight from the jth neuron in layer l-1 to the kth neuron in layer l, and θ_k^l is the bias value of the kth neuron of the lth hidden/output layer, respectively. The training mean square error, *MSE*, at the output layer is formulated as,

$$MSE = \frac{1}{2PN_O} \sum_{p \in T} \sum_{k=1}^{N_O} \left(t_k^p - y_k^{p,O}\right)^2 \tag{3.15}$$

where t_k^p is the target (desired) output and $y_k^{p,O}$ is the actual output from the kth neuron in the output layer, $l = O$, for pattern p in the training set T with size P, respectively.

The BP algorithm can be summarized as follows:

1. Initialize the weights w_{jk}^l and biases θ_k^l randomly. For RBF networks initialize also peak centers μ_k and sigmas o_k.
2. Feed pattern p to the network. Compute the output $y_k^{p,l}$ of each neuron.
3. Calculate the error between the computed output $y_k^{p,o}$ of each output neuron and the desired output t_k^p as $e_k^{p,o} = t_k^p - y_k^{p,o}$.
4. For each neuron k, calculate the local gradients $\frac{\partial E^p}{\partial h_k^l}$, where E^p is the total error energy defined as $E^p = \frac{1}{2}\sum_{k \in o} (e_k^{p,o})^2$ and h_k^l is a uniform symbol for each parameter w_{jk}^l, θ_k^l, μ_k ,and σ_k. The name BP algorithm comes from the fact that local gradients are computed starting from the output layer and then iteratively proceeding backwards toward the input layer. The formulas for calculating the local gradients for MLP and RBF can be found in [47] and [49], respectively.
5. Update the parameters as follows:

$$h_k^l(t + 1) = h_k^k(t) + \Delta h_k^k(t)$$
$$\Delta h_k^k(t) = -\eta \frac{\partial E^p}{\partial h_k^k} + \alpha \Delta h_k^k(t - 1) \qquad (3.16)$$

where, η is the learning rate parameter and α is the momentum constant.

One complete run over the training dataset is called an *epoch*. Usually many *epochs* are required to obtain the best training results; on the other hand, too many training *epochs* can lead to over fitting. In the above realization of the BP algorithm, the network parameters are updated after every training sample. This is called the *online* or *sequential* mode. The other possibility is the *batch* mode, where all the training samples are first presented to the network and then the parameters are adjusted, so that the total training error is minimized. The *sequential* mode is often favored over the *batch* mode as it requires less storage space. Moreover, the *sequential* mode is less likely to get trapped in a local minimum as updates at every training sample make the search stochastic in nature. Hence, *sequential* BP mode is used for MLP training.

PSO has been successfully applied for training feedforward [50–53] and recurrent ANNs [54, 55] and several works on this field have shown that it can achieve a superior learning ability to the traditional BP method in terms of accuracy and speed. At time t, suppose that particle a in the swarm, $\xi = \{x_1, \ldots, x_a, \ldots, x_S\}$, has the positional component:

$$x_a(t) = \left\{ \{w_{jk}^0\}, \{w_{jk}^1\}, \{\theta_k^1\}, \{w_{jk}^2\}, \{\theta_k^2\}, \ldots, \{w_{jk}^{O-1}\}, \{\theta_k^{O-1}\}, \{\theta_k^O\} \right\} \qquad \text{where}$$

$\{w_{jk}^l\}$ and $\{\theta_k^l\}$ represent the sets of weights and biases of layer l. Note that the input layer ($l = 0$) contains only weights, whereas the output layer ($l = O$) has only biases. With such a direct encoding scheme, the particle a represents all potential network parameters of the MLP architecture. In the next section, we shall present a direct comparison of PSO versus BP training of a collection of MLPs over supervised classification over several medical datasets.

3.4.3.2 BP versus PSO: Comparative Performance Evaluation Over Medical Datasets

An architecture space (AS) can be defined over a certain range of configurations, i.e., say from a SLP to complex MLPs with many hidden layers. Suppose, for the sake of simplicity, that a range is defined for the number of layers, $[L_{\min}, L_{\max}]$ and another for the number of neurons for each hidden layer l, $\left[N_{\min}^l, N_{\max}^l\right]$. Without loss of generality, assume that the size of both input and output layers is determined by the problem and hence fixed. Consequently, the architecture space can now be defined by only two (MLP) configuration sets, $R_{\min} = \left\{N_I, N_{\min}^1, \ldots, N_{\min}^{L_{\max}-1}, N_O\right\}$ and $R_{\max} = \left\{N_I, N_{\max}^1, \ldots, N_{\max}^{L_{\max}-1}, N_O\right\}$, one for the minimum and the other for the maximum number of neurons allowed for each layer of a MLP. The size of both arrays is naturally $L_{\max} + 1$, where the corresponding entries define the range of neurons possible on the lth hidden layer for all those MLPs, which can have an lth hidden layer. The size of input and output layers, $\{N_I, N_O\}$, is fixed and remains the same for all configurations in the architecture space within which any l-layer MLP can be defined providing that $L_{\min} \leq l \leq L_{\max}$. $L_{\min} \geq 1$ and $L_{\max}$ can be set to any reasonable value for the problem at hand. In this way, all network configurations in the architecture space are enumerated into a hash table with a proper hash function, which ranks the networks with respect to their complexity, i.e., associates higher hash indices to networks with higher complexity. The hash function then enumerates all potential MLP configurations into hash indices, starting from the simplest MLP with $L_{\min} - 1$ hidden layers, each of which has a minimum number of neurons given in $R_{\min}$, to the most complex network with $L_{\max} - 1$ hidden layers, each of which has the maximum number of neurons given in $R_{\max}$. Take, for instance, the following configuration sets, $R_{\min} = \{9, 1, 1, 2\}$ and $R_{\max} = \{9, 8, 4, 2\}$, which indicate that $L_{\max} = 3$. If $L_{\min} = 1$ then the hash function enumerates all MLP configurations in the architecture space as shown in Table 3.2 Note that in this example, the input and output layer sizes are nine and two, which are eventually fixed for all MLP configurations. The hash function associates the first index (d = 0) with the simplest possible architecture, i.e., a SLP with only input and output layers (i.e., 9×2). From indices 1 to 8, all configurations belong to 2-layers MLP with a single hidden layer containing a varying number of neurons between 1 and 8 (as specified in the 2nd entries of arrays $R_{\min}$ and $R_{\max}$). Similarly, for indices 9 and up, 3-layer MLPs are enumerated in which the number of neurons in the 1st and 2nd hidden layer sizes are varied according to the corresponding entries in $R_{\min}$

Table 3.2 A sample
architecture space for MLP
configuration sets $R_{min} = \{9, 1, 1, 2\}$ and
$R_{max} = \{9, 8, 4, 2\}$

Index	Configuration	Index	Configuration
0	9×2	21	$9 \times 5 \times 2 \times 2$
1	$9 \times 1 \times 2$	22	$9 \times 6 \times 2 \times 2$
2	$9 \times 2 \times 2$	23	$9 \times 7 \times 2 \times 2$
3	$9 \times 3 \times 2$	24	$9 \times 8 \times 2 \times 2$
4	$9 \times 4 \times 2$	25	$9 \times 1 \times 3 \times 2$
5	$9 \times 5 \times 2$	26	$9 \times 2 \times 3 \times 2$
6	$9 \times 6 \times 2$	27	$9 \times 3 \times 3 \times 2$
7	$9 \times 7 \times 2$	28	$9 \times 4 \times 3 \times 2$
8	$9 \times 8 \times 2$	29	$9 \times 5 \times 3 \times 2$
9	$9 \times 1 \times 1 \times 2$	30	$9 \times 6 \times 3 \times 2$
10	$9 \times 2 \times 1 \times 2$	31	$9 \times 7 \times 3 \times 2$
11	$9 \times 3 \times 1 \times 2$	32	$9 \times 8 \times 3 \times 2$
12	$9 \times 4 \times 1 \times 2$	33	$9 \times 1 \times 4 \times 2$
13	$9 \times 5 \times 1 \times 2$	34	$9 \times 2 \times 4 \times 2$
14	$9 \times 6 \times 1 \times 2$	35	$9 \times 3 \times 4 \times 2$
15	$9 \times 7 \times 1 \times 2$	36	$9 \times 4 \times 4 \times 2$
16	$9 \times 8 \times 1 \times 2$	37	$9 \times 5 \times 4 \times 2$
17	$9 \times 1 \times 2 \times 2$	38	$9 \times 6 \times 4 \times 2$
18	$9 \times 2 \times 2 \times 2$	39	$9 \times 7 \times 4 \times 2$
19	$9 \times 3 \times 2 \times 2$	40	$9 \times 8 \times 4 \times 2$
20	$9 \times 4 \times 2 \times 2$		

and R_{max}. Finally, the most complex MLP with the largest number of possible layers and the highest number of neurons is associated with the highest index, $d = 40$. Therefore, all 41 entries in the hash table span the architecture space with respect to the configuration complexity.

The comparative evaluations of both training algorithms were performed using a medical diagnosis benchmark dataset from the UCI Machine Learning repository [56], which is partitioned into three sets: training, validation, and testing. There are several techniques [57] to use training and validation sets individually to prevent overfitting and thus to improve the classification performance in the test data. However, there is no universally effective technique and there are several research articles reporting against the use of the cross validation technique in the design and training of MLP networks [57], [58]. In this book, for simplicity and to obtain an unbiased performance measure under equal training conditions, the validation and training sets are simply combined to be used for training. From *Proben1* repository [56], three benchmark classification problems, *breast cancer, heart disease*, and *diabetes*, are selected, which were commonly used by previous studies. These are medical diagnosis problems, which present the following attributes:

- All of them are real-world problems based on medical data from human patients.
- The input and output attributes are similar to those used by a medical doctor.
- Since medical samples and data are expensive to obtain, the training sets are quite limited.

We now briefly describe each classification problem next.

1. Breast Cancer

The objective of this dataset is to classify breast lumps as either benign or malignant according to microscopic examination of cells that are collected by needle aspiration. There are 699 exemplars of which 458 are benign and 241 are malignant and they are originally partitioned as 350 for training, 175 for validation, and 174 for testing. The dataset consists of 9 input attributes and 2 output attributes, i.e., each input pattern is described by 9-dimensional vector and there are two possible outcomes of the classifier. It is created at University of Wisconsin Madison by Dr. William Wolberg.

2. Heart Disease

The initial dataset consists of 920 exemplars with 35 input attributes, some of which are severely missing. Hence, a second dataset is composed using the cleanest part of this set, which was created at Cleveland Clinic Foundation by Dr. Robert Detrano. The Cleveland dataset is called "heartc" in Proben1 repository and contains 303 exemplars but 6 of them still contain missing data and are hence discarded. The remaining exemplars are partitioned as follows: 149 for training, 74 for validation, and 74 for testing. There are 13 input and 2 output attributes. The purpose is to predict the presence of a heart disease according to the input attributes.

3. Diabetes

This dataset is used to predict diabetes diagnosis among Pima Indians. The data is collected from female patients, aged 21 years or older. There are total of 768 exemplars of which 500 are classified as diabetes negative and 268 as diabetes positive. The dataset is originally partitioned as 384 for training, 192 for validation, and 192 for testing. It consists of eight input attributes and two output attributes.

The input attributes of all datasets are scaled within the range [0,1] by a linear function. Note that their output attributes are encoded using a 1-of-c representation for $c = 2$ classes. The winner-takes-all methodology is applied so that the output of the highest activation designates the class. Overall, the experimental setup becomes identical to those used in the previous studies and thus fair comparative evaluations can now be made over the classification error rate of the test data. In all experiments in this section, we use the sample architecture space given in Table 3.2, which has the generalized form as, $R_{\min} = \{N_I, 1, 1, N_O\}$ and $R_{\max} = \{N_I, 8, 4, N_O\}$ containing the compact 1-, 2-, or 3-layer MLPs where N_I and N_O, are determined by the number of input and output attributes of the classification problem. For BP, all networks were trained with 500 (shallow training) and with 5,000 (deep training) iterations with a low learning rate of 0.02 to prevent oscillations. For PSO training, in addition to default settings for the standard algorithm parameters as defined in Sect. 2.3, the number of particles was set to 40 ($S = 40$)

and the number of training iterations was set to 200 for the shallow and 2,000 for the deep training cases. For all experiments in this section, unless stated otherwise, 100 independent runs are performed for each configuration to compute the error statistics plots for each dataset. We mainly consider two major criteria for the performance assessment: (1) training MSE, which indicates the error minimization achieved by each method; (2) test CE, which is the primary objective of the classifier as it shows the classification accuracy level achieved as well as the generalization capability of each method. Using the corresponding error statistics plots, both criteria shall then be statistically evaluated by considering on the average (i.e., mean MSE and CE) and the best (i.e., minimum MSE and CE) performances achieved by each method, BP and PSO.

In order to perform a comprehensive and a systematic assessment of the performance of ANN classifiers in medical diagnosis, we apply exhaustive BP and PSO training for each network configuration in the architecture space, which is defined over MLPs with sigmoid activation functions. In this way we can escape from the bias or possible effect of a particular network over the performance, which was the case of many of the aforementioned studies that were mostly performed using only one or few fixed network architecture(s). Furthermore, to assess the effect of the training depth on both BP and PSO, both shallow and deep training will be applied over every network configuration in the architecture space by setting the number of iterations appropriately.

Figure 3.10 presents the corresponding error statistics plots from the shallow training over the breast cancer dataset. BP in general achieves the lowest average

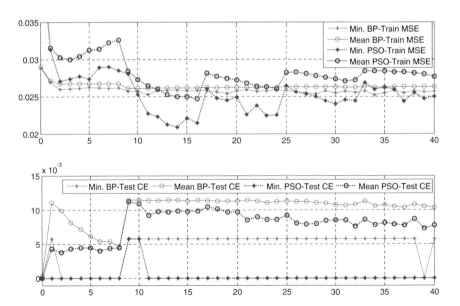

Fig. 3.10 Train (*top*) and test (*bottom*) error statistics vs. hash index plots from *shallow* BP- and PSO-training over the *breast cancer* dataset

training MSEs within a narrow variance except few network configurations with the corresponding indices, $d \in [13, 16]$ where PSO slightly surpasses BP. On the other hand, PSO achieves the best training performances (i.e., minimum MSEs) over the majority of network configurations except for the compact ones ($d \in [0, 9]$), where BP is consistently more successful. The lowest overall training MSE (both average and minimum) is too achieved by PSO using the configuration with the hash indices $d = 14$ and $d = 16$ (MLPs: $9 \times 6 \times 1 \times 2$ and $9 \times 8 \times 1 \times 2$), respectively. In terms of the classification performances over the test set, the results are consistently in favor of PSO, which performs better than BP with respect to both performance criteria. Particularly, PSO achieved the optimal 0 % CE (i.e.,100 % classification accuracy) as its best performance among all networks except for two networks ($d = 9$ and 10), whereas BP managed to achieve this only over the compact networks (i.e., $d = 0$ *and* $d \in [2, 8]$), plus the complex MLP with $d = 39$. Overall, PSO usually demonstrates a better classification performance for the breast cancer dataset with the shallow training.

The error statistics plots obtained from deep training of all networks in the architecture space by both methods are shown in Fig. 3.11. In this case, both BP and PSO achieve lower training MSEs as a natural consequence of the deep or overtraining, and BP in general achieves the lowest average training MSEs, particularly on complex networks with two hidden layers but it also surpasses PSO in terms of the minimum training MSEs except for only few networks. Due to such overtraining, the classification performance of both methods is expected to degrade, which is the case as shown by the bottom plots of Fig. 3.11. However, the

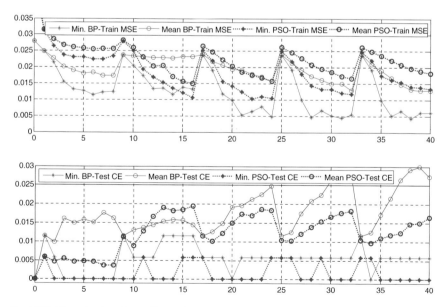

Fig. 3.11 Train (*top*) and test (*bottom*) error statistics vs. hash index plots from *deep* BP- and PSO-training over the *breast cancer* dataset

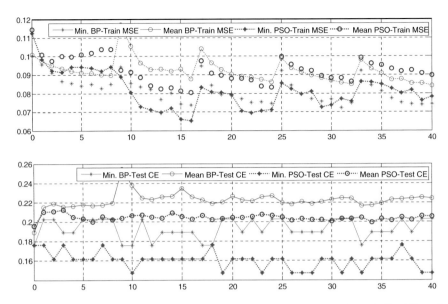

Fig. 3.12 Train (*top*) and test (*bottom*) error statistics vs. hash index plots from *shallow* BP- and PSO-training over the *heart disease* dataset

degradation on PSO's performance is not as severe as that of BP. Furthermore, note that there is almost no performance loss in the case of compact networks, due to PSO's global search ability. Particularly for complex networks, a significant performance gap in terms of average test CE occurs between the two methods in favor of PSO, since BP, as a deterministic local search method, is drastically affected by the overfitting of the training data and thus exhibits significantly worse average classification performance. This is also true but at a lesser extent for PSO; especially its minimum CE cannot anymore guarantee 0 % CE level for all network configurations.

Figure 3.12 presents the corresponding error statistics plots from the shallow training over the heart disease dataset. Both average and minimum training MSE statistics of both methods vary significantly with respect to the network configuration, making either method better than the other for a given network and error criterion. This is a good example that clearly shows the effect of different network configurations over the performance of each method. Therefore, using only one or few architectures for comparative evaluations may favor either method and hence such a static and limited approach is not sufficient to draw reliable conclusions. The lowest training MSE (both average and minimum) is achieved by PSO using the network with the hash index d = 16 (MLP: 13 × 8 × 1 × 2). As in the previous dataset, about the classification performances over the test set, the results are consistently in favor of PSO, which performs better than BP for all networks in the architecture space with respect to both performance criteria.

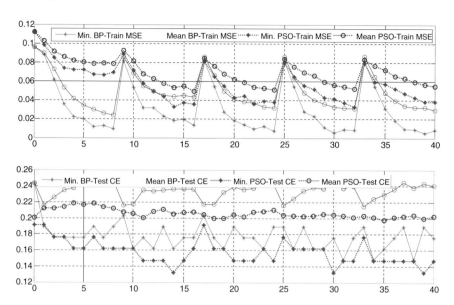

Fig. 3.13 Train (*top*) and test (*bottom*) error statistics vs. hash index plots from *deep* BP- and PSO-training over the *heart disease* dataset

Contrary to shallow training results, the top plot in Fig. 3.13 indicates that whenever deep training is performed over this dataset, BP surpasses PSO with respect to the training MSEs (both average and minimum) for all networks. Hence due to the overfitting of the training data, the classification performance of BP over the test set is quite degraded while no significant performance degradation occurs for PSO. PSO, once again, exhibits its relative immunity against overtraining due to its global search ability and yields the best classification performance over the test set (i.e., well generalization) regardless of the training depth. This is true for both average and the best performance criteria considered (see red and blue curves at the bottom plot in Fig. 3.13). PSO achieves the overall best classification performance, ~ 13 % CE, from the three different networks with the corresponding hash indices, $d = 14$, 30, and 39 although no network configuration makes too much difference when the average classification performance is concerned.

Figure 3.14 presents the corresponding error statistics plots from the shallow training over the diabetes dataset. Similar comments can be made about the training performance of PSO and BP as in the shallow training experiments over the heart disease dataset. That is, although BP is consistently better than PSO for compact networks, their training performances (minimum and average MSEs) are quite comparable and varying along with the network configuration. In terms of the classification performance over the test set, PSO usually achieves slightly lower CEs but the results are again quite comparable. When minimum CEs are concerned, from the network with the hash index $d = 16$ (MLP: $8 \times 8 \times 1 \times 2$) PSO achieved a minimum of 17.1 % CE that is slightly lower than the 18.8 %

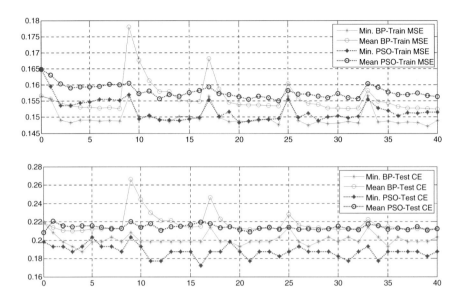

Fig. 3.14 Train (*top*) and test (*bottom*) error statistics vs. hash index plots from *shallow* BP- and PSO-training over the *diabetes* dataset

minimum CE achieved by BP from the network with the hash index d = 4 (MLP: $8 \times 4 \times 2$). Finally, according to the error statistics plots in Fig. 3.15 obtained by deep training over the same dataset, similar conclusions can be drawn about both training (MSE) and generalization (test CE) performances of PSO and BP as in the deep training experiments over the heart disease dataset, i.e., PSO yields almost the same average training MSE levels and BP significantly reduces both average and minimum MSE levels, as expected. One observation worth mentioning here is that the training and test performances of BP in both deep and shallow training exhibits a large variation with respect to the network configuration used (e.g., compare for instance the mean BP training MSE or test CE for $d = 8$ and $d = 9$), whereas the corresponding performance levels of PSO are more stable and usually with a smaller variance, regardless of the network configuration.

The overall test CE statistics of both training techniques (BP and PSO) computed over all configurations in selected MLP architecture spaces for each dataset are enlisted in Table 3.3. We used the following three statistics: minimum (*min*), mean (μ), and standard deviation (σ), respectively, which are computed per training depth (deep and shallow). The results in the table clearly indicate that PSO training on average achieves better classification performance than BP over the three benchmark medical diagnosis problems.

Finally, in order to accomplish the comparative performance evaluations of each method with respect to the variations in the training depth, we have selected a particular network configuration with the hash index $d = 16$, which is a relatively compact and one of the best performing classifier configuration within the sample

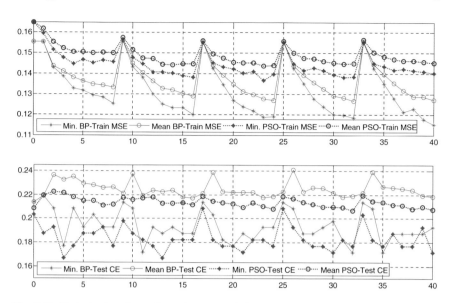

Fig. 3.15 Train (*top*) and test (*bottom*) error statistics vs. hash index plots from *deep* BP- and PSO-training over the *diabetes* dataset

Table 3.3 The Overall Test CE Statistics

DataSet	Training Method	Training Depth	Min. Test CE Statistics			Mean Test CE Statistics		
			min	μ	σ	min	μ	σ
Cancer	BP	Shallow	0	0.0045	0.0024	0.0003	0.0101	0.0024
	PSO		0	**0.0003**	0.0013	0	**0.0078**	0.0024
	BP	Deep	0	0.0055	0.0036	0	0.0176	0.0064
	PSO		0	0.0015	0.0026	0	0.0121	0.0052
Diabetes	BP	Shallow	0.1875	0.2002	0.0063	0.2101	0.2175	0.0112
	PSO		0.1719	0.1878	0.0067	0.2079	0.2137	0.0028
	BP	Deep	0.1719	0.1941	0.0129	0.2135	0.2246	0.0068
	PSO		0.1667	**0.1836**	0.0101	0.2069	**0.2135**	0.0040
Heart	BP	Shallow	0.1757	0.1928	0.0100	0.1893	0.2222	0.0087
	PSO		0.1471	0.1603	0.0092	0.1957	**0.2043**	0.0031
	BP	Deep	0.1486	0.1773	0.0171	0.2150	0.2340	0.0096
	PSO		0.1324	**0.1578**	0.0151	0.1976	0.2060	0.0054

architecture space, and we have performed exhaustive training (with 100 runs) for each of the 10 intermediate (training) depths, between the corresponding shallow and deep training, i.e., [500, 5,000] for BP and [200, 2,000] for PSO. Figure 3.16 shows the training MSE and test CE plots versus the training depth for all three datasets. From the figure, it is clear that both methods reduce the training MSE with increasing training depths, as a natural consequence of the overfitting of the training data. On the other hand, PSO achieves lower average and minimum

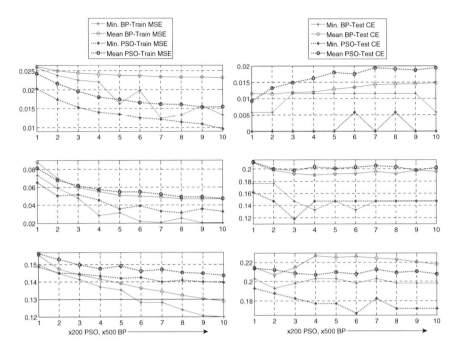

Fig. 3.16 Error statistics (for network configuration with the hash index d = 16) vs. training depth plots using BP- and PSO-over the *breast cancer* (*top*), *heart disease* (*middle*), and *diabetes* (*bottom*) datasets

training MSEs for the breast cancer, higher for the diabetes, and quite similar for the heart disease datasets, respectively. The classification performance of PSO shows a strong immunity against variations in the training depth and it generally achieves the lowest minimum CEs. For this particular network, either BP or PSO can achieve a better average classification performance depending on the training depth. Hence, this clearly draws the conclusion that the training depth too should be considered while comparing and/or analyzing individual performance of each method.

3.5 Programming Remarks and Software Packages

This is the first chapter from which we start to explain briefly the software packages we supply along with this book. All software programs are developed using C and C++ languages under Microsoft © Visual Studio 6.5 (VS6.5: version 6 with Service Pack 5). There are several applications developed and in this chapter, we shall start with describing the PSO test-bed application for nonlinear function minimization, namely **PSO_MDlib**. It is a simple console application which is a VS6.5 workspace with three projects in it: **MTL, PSO_MDlib**, and **PSO_MDmain**. The first one, **MTL**, stands for MUVIS **T**emplate **L**ibrary, which contains basic data structures such as link lists (queues), registers, threads, etc. (we shall later return to MUVIS). **PSO_MDlib** is a static library, where the basic (canonical and global) PSO has been implemented along with its multidimensional extension, MD PSO (to be discussed in Chap. 4). Finally, **PSO_MDmain.cpp** is the main console application, which offers three different entry point functions, **main()**, two of which are enabled with a compiler flag: **BATCH_RUN** or **MOVING_PEAKS_BENCHMARK**. The latter enables a special MD PSO application over a benchmark dynamic environment, which will be presented as the application for Sect. 5.2 in Chap. 5. The compiler flag BATCH_RUN can be enabled so as to test both PSO and/or MD PSO over all the functions, for both PSO types (BASIC vs. FGBF), over all test functions, and with several runs (FGBF will also be discussed in Chap. 5).

All intrinsic PSO (and MD PSO) parameters are stored in **PSOparam** structure (see the header file: **PSOparam.h**) (Table 3.4).

The first three parameters, S (**_noAgent**), *iterNo* (**_maxNoIter**), and ε_C(**_e-CutOff**) are common for PSO and MD PSO. The next three parameters are specific for MD PSO and will be covered in the Chap. 4. Finally, we shall perform bPSO application; therefore, **_mode** is set to BASIC. Since MD PSO is simply the multidimensional extension of the PSO, when **_dMin=_dMax-1,** then the MD PSO will be reduced to a regular PSO process at the solution space dimension **_dMin.**

PSO and its extension MD PSO are jointly implemented in an object-oriented, template-based, and morphological structure with four class implementations:

1. template <**class T, class X**> class **CPSO_MD** {…}
2. template <**class T, class X**> class **CParticle** {…}
3. template <**class X**> class **CSolSpace** {…}
4. class **COneClass** {…}

The main class: Template <**class T, class X**> class **CPSO_MD** is a template based class with two template class implementations, **class T** and **class X**. Such a

Table 3.4 The data structure, **PSOparam**

```
struct PSOparam
{
        int      _noAgent;      // No. of Agents..
        int      _maxNoIter;    // Max. No. of iterations..
        float    _eCutOff;      // CutOff percentage..

        int      _dMin, _dMax;  // Min & Max (Range of) dimension..
        int      _vdMin, _vdMax; // Min & Max (Range of) velocity for dim. agent..
        int      _xvMin, _xvMax; // Min & Max (Range of) velocity..
        modeMDPSO_mode;   // BASIC or FGBF..
};
```

template-based class structure makes MD PSO applicable to any problem as long as the potential solution of the problem (**class T**) along with the data space (**class X**) can be implemented accordingly. **CPSO_MD** contains an array of **CParticle** object pointers (**CParticle<T,X>** m_pPA**) with the total number of **m_noP.** Each particle has their regular PSO elements such as positions, velocities, and personal best positions, stored in the following array of pointers,

T**   m_pXX; // Current position of this particle in all dimensions.
T**   m_pVX; // Current velocity of this particle in all dimensions.
T**   m_pXY; // Personal best position of this particle in all dimensions.

The template class, **class T** represents the potential solution of the problem and we assign **class T** to the following template class: template **<class X>** *class* **CSolSpace <X>**. It simply contains the position of the solution, its dimension, and the boundaries, i.e.,

X*   m_pPos; // Current Position in N-dimensional solution space.
int   m_nDim; // Dimension of the solution space.
X   m_min, m_max; // Minimum and Maximum ranges (boundaries of the solution).

Finally, the template class, **class X,** represents the data space, where the real values of the PSO particle elements are stored. For instance, it can simply be set to standard data structures such as **float** or **double** for nonlinear function minimization, yet for a generic usage, we shall assign it to class **COneClass,** which contains nothing but a single floating point data element along with its individual score, i.e.,

float   m_x; // data per dimension.
float   m_bScore; // and the individual score of each dimension.

and the standard arithmetic operators (=, +, -,/, <, >, etc.) are implemented (for **m_x**) accordingly. We shall clarify the use of the member variable **m_bScore** in Chap. 5.

All nonlinear functions are implemented within MYFUN.cpp source file. To perform a PSO operation for nonlinear function minimization, a **CPSO_MD** object should be created with proper template classes: **<class T, class X>,** and initialized with: (1) the default PSO parameters stored in **PSOparam _psoDef**, and (2) the fitness function → any nonlinear function within MYFUN.cpp. In short, the entire MD PSO initialization can be summarized as:

  Create: CPSO_MD < CSolSpace < COneClass > ,COneClass > *pPSO = **new CPSO_MD < CSolSpace < COneClass > ,COneClass>**

  (_psoDef._noAgent,   _psoDef._maxNoIter,   _psoDef._eCutOff, _psoDef._mode);

  1) pPSO- > Init(_psoDef._dMin, _psoDef._dMax, _psoDef._vdMin, _psoDef._vdMax, -500, 500, _psoDef._xvMin, _psoDef._xvMax);/*** Initialize the object ***/

Table 3.5 Initialization of the PSO swarm

```
m_pPA = new CParticle<T,X>* [ m_noP ];
for(int a=0; a<m_noP; a++) // Create & Randomize all particles in all dimensions..
        m_pPA[a]  =  new  CParticle<T,X>  (m_xdMin, m_xdMax, m_vdMin, m_vdMax,
m_xxMin, m_xxMax, m_xvMin, m_xvMax);
```

Table 3.6 The main loop for (MD) PSO in **Perform**() function

```
for(int iter = 0; 1 ; ++iter)
{
        weight_up = (WMAX-WMIN) * (m_noIter - iter) /m_noIter + WMIN; //time variant weight..
        ...
```

2: pPSO- > SetFitnessFn(Sphere);/* PUT Fn here ***/**

Step 1 (the **Init(..)** call) basically creates and (randomly) initializes the MD PSO swarm, as shown in Table 3.5 where each particle, **m_pPA[a]**, has been randomized within the positional and dimensional ranges specified within the **_psoDef**.

After the initialization, the function call: **pPSO->Perform**(); will execute a PSO run over the function specified. In order to avoid the dimensional bias, note that *pDim (the target dimension for MD PSO) should be equal to _dMin=_d-Max-1. Each PSO run can be observed in real time (via DbgOut program UI) or offline using the two MATLAB files generated during the runtime. The first MATLAB file with the function name (*sphere50_d50_mode0.m*) contains the statistics for each run, such as mean and standard deviation of the best scores achieved. The second one stores the personal best scores (fitness) achieved by the *gbest* (GB) particles and the last particle that becomes *gbest* during the run. The call, **pPSO->Perform**(), has basically an iteration loop within which the (MD) PSO is executed, as given in Table 3.6.

In the main PSO loop, step 3 of the PSO pseudo-code given in Table 3.1 is performed. Step 3.1 where personal best positions of each particle (step 3.1.1) as well as the assignment of the *gbest* particle (step 3.1.2) are performed, as given in the Table 3.7.

Recall that the fitness function (sphere) has already been given to the PSO object and assigned to the member function pointer, **m_fpGetScore**, which computes the fitness of the current (or personal best) position of any particle in the swarm, **m_pPA[a]->GetPos**(), at dimension, **m_pPA[a]->GetDim**(). Once the fitness score has been computed (**minval**), it is compared with the past personal best score (**m_pPA[a]->GetPBScore**()), and if it is surpassed, the (new) personal best score and its new location are updated. This new personal best score is also compared with the *gbest* particle's personal best score and if it surpasses the old *gbest*, it will then become the new *gbest* particle. Since this is a bPSO operation

Table 3.7 Implementation of Step 3.1 of the PSO pseudo-code given in Table 3.1

```
for (int a=0;a<m_noP;a++)
{ // Finding local - global bests for all particles..
        doubleminval = m_fpGetScore(m_pPA[a]->GetPos(), m_pPA[a]->GetDim());
        m_pPA[a]->SetCurScore(minval); // Set the current score for Adaptive In. Weight setting..
        if(iter==0) m_pPA[a]->SetPBScore(minval); // At the first iteration, all scores are PB..

        If (mInval <= m_pPA[a]->GetPBScore())
        {
                m_pPA[a]->SetPBScore(minval); // the best score for particle a for its current dim..
                m_pPA[a]->StorePBPos(); // best local pos. for particle a in its currrent dim..
                int cur_dim_a = m_pPA[a]->GetDim(); // The current dim. of a.th particle..

                if (minval < m_pPA[m_gbest[cur_dim_a - m_xdMin]]->GetPBScore(cur_dim_a))
                        m_gbest[cur_dim_a - m_xdMin] = a; // The best particle for this dim. = cur_dim_a..

                if (minval < m_pPA[a]->GetPBScore(m_pPA[a]->GetPBDim()))
                        m_pPA[a]->SetPBDim(cur_dim_a); // the best local dim. for this particle, a..

                if (minval < m_pPA[m_gbest[m_dbest - m_xdMin]]->GetPBScore(m_dbest))
                        m_dbest = cur_dim_a; // the best global dimension for all particles - so far..
        } // if..
        ...
} // for all members..
```

where **cur_dim_a=m_xdMin,** all swarm particles reside in the current dimension: **cur_dim_a**.

For the moment, we shall skip another PSO mode, which enables the FGBF operation within the **if(m_mode==FGBF)** statement as it will be explained in Chap. 5.

The termination of the PSO process (by either *IterNo* or ε_C) is verified by the following code in Table 3.8. Note that the current dimension is equal to **m_dbest=m_xdMin,** since this is a PSO process over a single dimension. So the index of the *gbest* particle is stored in **m_gbest[m_dbest - m_xdMin]]=m_gbest[0].**

If neither of the termination criterion is reached, then step 3.4 is executed to update the position (and the dimension if it is a MD PSO run, as will be explained in Chap. 4) of each swarm particle as given in Table 3.9. Within the loop, each particle's cognitive and social components are computed and the velocity update is

Table 3.8 The termination of a PSO run

```
...
        //Check loop-end conditions before updates..
        if(m_pPA[m_gbest[m_dbest - m_xdMin]]->GetPBScore(m_dbest) <= m_eCutOff)
                break; // Global Minima is reached..
        if(iter == m_nolter-1) break; // iter. limit is reached..
...
```

Table 3.9 Implementation of Step 3.4 of the PSO pseudo-code given in Table 3.1

```
...
for (a=0;a<m_noP;a++)
{ // Update loop..
        // 1. Particle speed & location update in the prev. dimension..
        int cur_dim_a = m_pPA[a]->GetDim(); // The current dim. of a.th particle..
        T* vx = m_pPA[a]->GetVel(); // Current velocity of the particle a in cur. dim..
        T* xx = m_pPA[a]->GetPos(); // Current position of the particle a in cur. dim..
        T* xy = m_pPA[a]->GetPBPos(); // The personal best position of the particle a in cur. dim..

        // The best swarm pos. in current dim..
        T* xyb = m_pPA[m_gbest[cur_dim_a - m_xdMin]]->GetPBPos(cur_dim_a);
        temp->SetDim(cur_dim_a); // Set the temp particle to the same dim. as m_pPA[a]..
        T* cogn = temp->GetPos(); // The Cognitive Component part..
        T* social = temp->GetPBPos(); // The Social Component part..

        *vx = (*vx*weight_up) ;// The First term..
        cogn = *(*(cogn->Subtract2(*xy, *xx)) * c1) * (RandomizeX(cur_dim_a));
        *vx += cogn; // The Second term..
        social = *(*(social->Subtract2(*xyb, *xx)) * c2) * (RandomizeX(cur_dim_a));
        *vx += social; // The Third term..

        vx->CheckLimit();
        *xx += vx; // update..
        xx->CheckLimit();

        // 2. MD PSO DIMENSION UPDATE for the NEXT iteration..
        ... // TO BE COVERED IN CHAPTER 4..
} // for all members.....
...
```

composed over the previous velocity weighted with the inertia factor, **weight_up**. The velocity is clamped as in step 3.4.1.2 of the PSO pseudo-code (**vx->Check-Limit();**) and finally, the particle's current position is updated to a new position with the composed and clamped velocity term (i.e. ***xx +=vx**). As the last step, this new position is also verified whether it falls within the problem boundaries specified in advance ($\pm x_{max}$). Note that this practical verification step is omitted in the pseudo-code of the PSO given in Table 3.1.

As mentioned earlier, the multidimensional extension of the bPSO, the MD PSO will be explained in the Chap. 4 and the FGBF technique will be explained in Chap. 5. Accordingly, the programming details that have so far been skipped in this section shall be discussed at the end of those chapters.

References

1. J. Kennedy, R. Eberhart, Particle swarm optimization. In *Proceedings of IEEE International Conference on Neural Networks*, vol. 4, pp. 1942–1948, Perth, Australia, 1995
2. J.J. van Zyl, Unsupervised classification of scattering mechanisms using radar polarimetry data. IEEE Trans. Geosci. Remote Sens. **27**, 36–45 (1989)
3. A. Antoniou, W.-S. Lu, *Practical Optimization, Algorithms and Engineering Applications* (Springer, USA, 2007)
4. D. Goldberg, *Genetic Algorithms in Search, Optimization and Machine Learning* (Addison-Wesley, Reading MA, 1989), pp. 1–25
5. S. Kirkpatrick, C.D. Gelatt, M.P. Vecchi, Optimization by simulated annealing. Science **220**, 671–680 (1983)
6. T. Back, F. Kursawe, *Evolutionary Algorithms for Fuzzy Logic: A Brief Overview. In Fuzzy Logic and Soft Computing* (World Scientific, Singapore, 1995), pp. 3–10
7. U.M. Fayyad, G.P. Shapire, P. Smyth, R. Uthurusamy, *Advances in Knowledge Discovery and Data Mining* (MIT Press, Cambridge, MA, 1996)
8. Y. Shi, R.C. Eberhart, A modified particle swarm optimizer. In *Proceedings of the IEEE Congress on Evolutionary Computation*, pp. 69–73
9. R. Eberhart, P. Simpson, R. Dobbins, *Computational Intelligence. PC Tools* (Academic Press Inc., Boston, MA, USA, 1996)
10. M. Clerc, J. Kennedy, The particle swarm-explosion, stability and convergence in a multidimensional complex space. IEEE Trans. Evol. Comput. **6**(2), 58–73 (2002)
11. L.-Y. Chuang, H.W. Chang, C.J. Tu, C.H. Yang, Improved binary PSO for feature selection using gene expression data. Comput. Biol. Chem. **32**(1), 29–38 (2008)
12. K. Yasuda, A. Ide, N. Iwasaki, "Adaptive Particle Swarm Optimization. Proc. IEEE Int. Conf. Sys. Man Cybern. **2**, 1554–1559 (2003)
13. W.-J. Zhang, Y. Liu, M. Clerc, An adaptive PSO algorithm for reactive power optimization. Adv. Power Syst. Control Oper. Manage. (APSCOM) **1**, 302–307 (2003). (Hong Kong)
14. Y. Shi, R.C. Eberhart, Fuzzy adaptive particle swarm optimization. In *Proceedings of the IEEE Congress on Evolutionary Computation*. IEEE Press, vol. 1, pp. 101–106, 2001
15. P.J. Angeline, *Using selection to improve particle swarm optimization. In Proceedings of the IEEE Congress on Evolutionary Computation*, pp. 84–89. IEEE Press, 1998
16. R.G. Reynolds, B. Peng, J. Brewster, Cultural swarms II: Virtual algorithm emergence. In *Proceedings of IEEE Congress on Evolutionary Computation 2003 (CEC 2003)*, pp. 1972–1979, Australia, 2003
17. H. Higashi, H. Iba, Particle swarm optimization with Gaussian mutation. In *Proceedings of the IEEE Swarm Intelligence Symposium*, pp. 72–79, 2003
18. K. Ersahin, B. Scheuchl, I. Cumming, Incorporating texture information into polarimetric radar classification using neural networks. In *Proceedings of the IEEE International Geoscience and Remote Sensing Symposium*, pp. 560–563, Anchorage, USA, Sep 2004
19. F. Van den Bergh, A.P. Engelbrecht, A new locally convergent particle swarm optimizer. In *Proceedings of the IEEE International Conference on Systems, Man, and Cybernetics*, pp. 96–101, 2002
20. W. McCulloch, W. Pitts, A logical calculus of the ideas immanent in nervous activity. Bull. Math. Biophys. **7**, 115–133 (1943)
21. M. Riedmiller, H. Braun, A direct adaptive method for faster backpropagation learning: The RPROP algorithm. In *Proceedings of the IEEE International Conference on Neural Networks*, pp. 586–591, 1993
22. G.-J. Qi, X.-S. Hua, Y. Rui, J. Tang, H.-J. Zhang, Image classification with kernelized spatial-context. IEEE Trans on Multimedia **12**(4), 278–287 (2010). doi:10.1109/TMM.2010.2046270

23. M. Løvberg, T.K. Rasmussen, T. Krin, Hybrid particle swarm optimiser with breeding and sub-populations. In *Proceedings of GECCO2001—Genetic and Evolutionary Computation Conference*, p. 409, CA, USA. July 7–11, 2001

24. Y. Lin, B. Bhanu, Evolutionary feature synthesis for object recognition. IEEE Trans. Man Cybern. C **35**(2), 156–171 (2005)

25. M. Løvberg, T. Krink, Extending particle swarm optimizers with self-organized criticality. Proc. IEEE Congr. Evol. Comput. **2**, 1588–1593 (2002)

26. T.M. Blackwell, J. Branke, Multi-swarm optimization in dynamic environments. Appl. Evol. Comput. **3005**, 489–500 (2004). (Springer)

27. T.M. Blackwell, J. Branke, Multiswarms, exclusion, and anti-convergence in dynamic environments. IEEE Trans. Evol. Comput. **10/4**, 51–58 (2004)

28. T.M. Blackwell, Particle swarm optimization in dynamic environments. Evol. Comput. Dyn. Uncertain Environ. Stud. Comput. Intell. **51**, 29–49 (2007). (Springer)

29. R. Mendes, Population topologies and their influence in particle swarm performance. PhD thesis, Universidade do Minho, 2004

30. J. Kennedy, Small worlds and mega-minds: effects of neighborhood topology on particle swarm performance. In *Proceedings of the 1999 Congress on Evolutionary Computation*, vol. 3. doi:10.1109/CEC.1999.785509, 1999

31. H. Frigui, R. Krishnapuram, Clustering by competitive agglomeration. Pattern Recogn. **30**, 1109–1119 (1997)

32. A.K. Jain, M.N. Murthy, P.J. Flynn, Data clustering: A review. *ACM Computing Reviews*, Nov 1999

33. C.P. Tan, K.S. Lim, H.T. Ewe, Image processing in polarimetric SAR images using a hybrid entropy decomposition and maximum likelihood (EDML). In *Proceedings of International Symposium on Image and Signal Processing and Analysis (ISPA)*, pp. 418–422, Sep 2007

34. B. Bhanu, J. Yu, X. Tan, Y.Lin, Feature synthesis using genetic programming for face expression recognition. *Genetic and Evolutionary Computation (GECCO 2004), Lecture Notes in Computer Science*, vol. 3103, pp. 896–907, 2004

35. G. Hammerly, Learning structure and concepts in data through data clustering. PhD thesis, June 26, 2003

36. B. Zhang, M. Hsu, K-harmonic means—a data clustering algorithm. *Hewlett-Packard Labs Technical Report HPL-1999-124*, 1999

37. G. Hammerly, C. Elkan, Alternatives to the k-means algorithm that find better clusterings. In *Proceedings of the 11th ACM CIKM*, pp. 600–607, 2002

38. N.R. Pal, J. Biswas, Cluster validation using graph theoretic concepts. Pattern Recogn. **30**(6), 847–857 (1997)

39. T. Ojala, M. Pietikainen, D. Harwood, A comparative study of texture measures with classification based on feature distributions. Pattern Recogn. **29**, 51–59 (1996)

40. T.G. Dietterich, G. Bakiri, Solving multiclass learning problems via error-correcting output codes. J. Artif. Intell. Res. **2**, 263–286 (1995)

41. M. Halkidi, Y. Batistakis, M. Vazirgiannis, On cluster validation techniques. J. Intell. Inf. Syst. **17**(2, 3), 107–145 (2001)

42. T.N. Tran, R. Wehrens, D.H. Hoekman, L.M.C. Buydens, Initialization of Markov random field clustering of large remote sensing images. IEEE Trans. Geosci. Remote Sens. **43**(8), 1912–1919 (2005)

43. S.R. Cloude, E. Pottier, An entropy based classification scheme for land applications of polarimetric SAR. IEEE Trans. Geosci. Remote Sens. **35**, 68–78 (1997)

44. M. Halkidi, M. Vazirgiannis, Clustering validity assessment: finding the optimal partitioning of a dataset. In *Proceedings of First IEEE International Conference on Data Mining (ICDM'01)*, pp. 187–194, 2001

45. M.G. Omran, A. Salman, A.P. Engelbrecht, Dynamic clustering using particle swarm optimization with application in image segmentation. Patt. Anal. Appl. **8**, 332–344 (2006)

46. A. Abraham, S. Das, S. Roy, Swarm intelligence algorithms for data clustering. In *Soft Computing for Knowledge Discovery and Data Mining Book*, Part IV, pp. 279–313, Oct 25, 2007
47. S. Haykin, *Neural Networks: a Comprehensive Foundation* (Prentice hall, USA, 1998). June
48. S. Pittner, S.V. Kamarthi, Feature extraction from wavelet coefficients for pattern recognition tasks. IEEE Trans. Pattern Anal. Machine Intell. **21**, 83–88 (1999)
49. J.A. Nelder, R. Mead, A simplex method for function minimization. Comput. J. **7**, 308–313 (1965)
50. M. Carvalho, T.B. Ludermir, Particle swarm optimization of neural network architectures and weights. In *Proceedings of the 7th International Conference on Hybrid intelligent Systems*, pp. 336–339, Washington DC, 17–19 Sep 2007
51. M. Meissner, M. Schmuker, G. Schneider, Optimized particle swarm optimization (OPSO) and its application to artificial neural network training. BMC Bioinf **7**, 125 (2006)
52. Z. Ye, C.-C. Lu, Wavelet-based unsupervised SAR image segmentation using hidden markov tree models. In *Proceedings of the 16th International Conference on Pattern Recognition (ICPR'02)*, vol. 2, pp. 20729, 2002
53. C. Zhang, H. Shao, An ANN's evolved by a new evolutionary system and its application. In *Proceedings of the 39th IEEE Conference on Decision and Control*, vol. 4, pp. 3562–3563, 2000
54. H. Robbins, S. Monro, A stochastic approximation method. Ann. Math. Stat. **22**, 400–407 (1951)
55. J.F. Scott, The Scientific Work of René Descartes, 1987
56. L. Prechelt, Proben1—A set of neural network benchmark problems and benchmark rules. *Technical Report 21/94, Fakultät für Informatik*, Universität Karlsruhe, Germany, September, 1994
57. S. Amari, N. Murata, K.R. Muller, M. Finke, H.H. Yang, Asymptotic statistical theory of overtraining and cross-validation. IEEE Trans. Neural Networks **8**(5), 985–996 (1997)
58. T.L. Ainsworth, J.P. Kelly, J.-S. Lee, Classification comparisons between dual-pol, compact polarimetric and quad-pol SAR imagery. ISPRS J. Photogram. Remote Sens. **64**, 464–471 (2009)

Chapter 4
Multi-dimensional Particle Swarm Optimization

> *If you want your children to be intelligent, read them fairy tales. If you want them to be more intelligent, read them more fairy tales.*
>
> Albert Einstein

Imagine now that each PSO particle can also change its dimension, which means that they have the ability to jump to another (solution space) dimension as they see fit. In that dimension they simply do regular PSO moves but in any iteration they can still jump to any other dimension. In this chapter we shall show how the design of PSO particles is extended into Multi-dimensional PSO (MD PSO) particles so as to perform interdimensional jumps without altering or breaking the natural PSO concept.

4.1 The Need for Multi-dimensionality

The major drawback of the basic PSO algorithm and many PSO variants is that they can only be applied to a search space with fixed dimensions. However, in many of the optimization problems (e.g., clustering, spatial segmentation, optimization of the multi-dimensional functions, evolutionary artificial neural network design, etc.), the optimum dimension where the optimum solution lies is also unknown and should thus be determined within the PSO process. Take for instance, the optimization problem of multi-dimensional functions. Let us start with the d-dimensional sphere function,

$$F(\bar{x}) = \sum_{i=1}^{d} x_i^2 \qquad (4.1)$$

This is a unimodal function with a single minimum point at the origin. When d is fixed and thus known a priori, there are powerful optimization techniques, including PSO, which can easily find the exact minimum or converge to ε-neighborhood of the minimum. We can easily extend this to a family of functions, i.e.,

S. Kiranyaz et al., *Multidimensional Particle Swarm Optimization for Machine Learning and Pattern Recognition*, Adaptation, Learning, and Optimization 15, DOI: 10.1007/978-3-642-37846-1_4, © Springer-Verlag Berlin Heidelberg 2014

$$F(\bar{x}, d) = \sum_{i=1}^{d} x_i^2 \quad \forall d \in \{D_{\min}, D_{\max}\} \tag{4.2}$$

At any particular dimension, d, $F(\bar{x}, d)$ has the minimum at the origin in d-dimensional space, i.e., $F(\bar{x}, d) = 0$ for $\bar{x} = \{0, 0, \ldots, 0\}$. Now consider the following multi-dimensional form of $F(\bar{x}, d)$ with a dimensional bias term, $\Psi(d)$.

$$F(\bar{x}, d) = \sum_{i=1}^{d} x_i^2 + \Psi(d) \quad \forall d \in \{D_{\min}, D_{\max}\} \quad \text{where}$$

$$\Psi(d) = \left\{ \begin{array}{ll} 0 & \text{for} \quad d = d_0 \\ > 0 & \text{else} \end{array} \right\} \tag{4.3}$$

In this case, it is obvious that there is only one true optimum point at the origin of dimension d_0. In other words, at all other dimensions, the function $F(\bar{x}, d)$ has only suboptimal points at the origin in each dimension. One straightforward alternative for the optimization in this type of multi-dimensional functions is to run the method distinctively for every dimension in the range. However, this might be too costly—if not infeasible for many problems especially depending on the dimensional range.

Another typical example of multi-dimensional optimization problems is data clustering where the true number of clusters is usually unknown. Some typical 2D synthetic data spaces with ground truth clusters were shown in Fig. 3.17 in the previous chapter, some of which are also shown in Fig. 4.1. Recall that for illustration purposes each data space is formed in 2D; however, clusters are formed with different shapes, densities, sizes, and inter-cluster distances. Such a clustering complexity will make their error surfaces highly multimodal and the optimization method for clustering now has to find out the true number of clusters as well as the accurate cluster centroids around which clusters are formed. Only few PSO studies have so far focused on this problem, i.e., [1] and [2]. In Ref. [2], Omran et al.

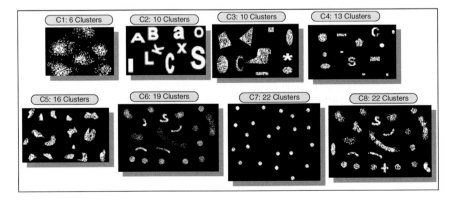

Fig. 4.1 2D synthetic data spaces carrying different clustering schemes

presented Dynamic Clustering PSO (DCPSO), which is in fact a hybrid clustering algorithm where binary PSO is used (only) to determine the number of clusters (and hence the dimension of the solution space) along with the selection of initial cluster centers whilst the traditional *K-means* method performs the clustering operation in that dimension (over the initial cluster centers). In Ref. [1], Abraham et al. presented the Multi-Elitist PSO (MEPSO), another variant of the basic PSO algorithm to address the premature convergence problem. In addition to being nongeneric PSO variants that are applicable only to clustering problems, both [1] and [2] do not clarify whether or not they can cope with higher dimensions of the solution space since the maximum number of clusters used in their experiments are only 6 and 10, respectively. This is also true for most of the static (fixed dimensional) PSO variants due to the aforementioned fact that the probability of getting trapped into a local optimum significantly increases in higher search space dimensions [3]. An extended survey about PSO and its variants can be found in Ref. [4].

In order to address this problem in an effective way, in this chapter we present a multi-dimensional PSO (MD PSO) technique, which negates the need of fixing the dimension of the solution space in advance.

4.2   The Basic Idea

In the simplest form, MD PSO reforms the native structure of swarm particles in such a way that they can make interdimensional passes with a dedicated dimensional PSO process. Therefore, in a multi-dimensional search space where the optimum dimension is unknown, swarm particles can seek for both positional and dimensional optima. This eventually negates the necessity of setting a fixed dimension a priori, which is a common drawback for the family of swarm optimizers. Therefore, instead of operating at a fixed dimension d, the MD PSO algorithm is designed to seek both positional and dimensional optima within a dimension range, $(D_{min} \leq d \leq D_{max}.)$

In order to accomplish this, each particle has two sets of components, each of which has been subjected to two independent and consecutive processes. The first one is a regular positional PSO, i.e., the traditional velocity updates and due positional shifts in N-dimensional search (solution) space. The second one is a dimensional PSO, which allows the particle to navigate through dimensions. Accordingly, each particle keeps track of its last position, velocity, and personal best position (*pbest*) in a particular dimension, so that when it revisits that the same dimension at a later time, it can perform its regular "positional" fly using this information. The dimensional PSO process of each particle may then move the particle to another dimension where it will remember its positional status and keep "flying" within the positional PSO process in this dimension, and so on. The swarm, on the other hand, keeps track of the *gbest* particles in all dimensions, each of which respectively indicates the best (global) position so far achieved and can

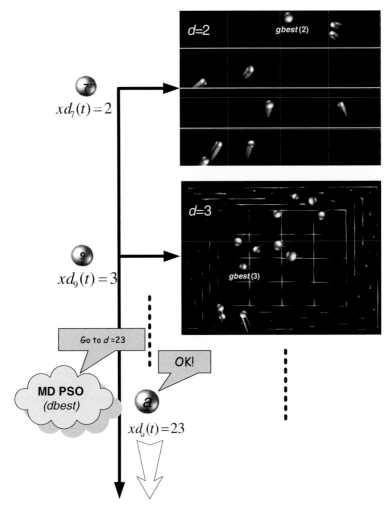

Fig. 4.2 An Illustrative MD PSO process during which particles 7 and 9 have just moved 2D and 3D solution spaces at time *t;* whereas particle *a* is sent to 23rd dimension

thus be used in the regular velocity update equation for that dimension. Similarly, the dimensional PSO process of each particle uses its personal best dimension in which the personal best fitness score has so far been achieved. Finally, the swarm keeps track of the global best dimension, *dbest*, among all the personal best dimensions.

Figure 4.2 illustrates a typical MD PSO operation at a time instance *t* where three particles have just been moved to the dimensions: $d = 2$ (particle 7), $d = 3$ (particle 9), and finally $d = 23$ (particle *a*), respectively by the dimensional PSO process with the guidance of *dbest*. The figure also shows illustrative 2D and 3D solution (search) spaces in which the particles (including 7 and 9) currently in these dimensions are making regular positional PSO moves by the guidance of the *gbest* particles (*gbest(2)* and *gbest(3)*). Afterwards each particle in these dimensions will have the freedom to leave to another dimension and similarly, new particles may come and perform the positional PSO operation and so on.

4.3 The MD PSO Algorithm

In a MD PSO process at time (iteration) t, each particle a in the swarm, $\xi = \{x_1, \ldots, x_a, \ldots, x_S\}$, encapsulates the following positional/dimensional PSO parameters:

$xx_{a,j}^{xd_a(t)}(t)$ jth component of the position vector of particle a, in dimension $xd_a(t)$

$vx_{a,j}^{xd_a(t)}(t)$ jth component of the velocity vector of particle a, in dimension $xd_a(t)$

$xy_{a,j}^{xd_a(t)}(t)$ jth component of the personal best (*pbest*) position vector of particle a, in dimension $xd_a(t)$

$gbest(d)$ Global best particle index in dimension d

$x\hat{y}_j^d(t)$ jth component of the global best position vector of swarm, in dimension d

$xd_a(t)$ Dimension of particle a

$vd_a(t)$ Velocity of dimension of particle a

$\tilde{xd}_a(t)$ Personal best dimension of particle a

$dbest$ Global best dimension ever achieved

Note that a simple two-letter naming convention is applied for these parameters. Each parameter has two letters (i.e., *xx, vx,* etc.). The first letter is either "*x*" or "*v*", representing either the positional or the velocity member of a particle. Since there are two interleaved PSO processes involved, the first letter will then indicate the position in the former and dimension in the latter process. The second character indicates the type of the PSO process or the type of the parameter (either current or personal best position for instance). Thus all parameters in *dimensional* PSO have the second character as either "*d*" or "$\tilde{d}$". Similarly, all parameters in the *positional* PSO have the second character as either "*x*" or "*y*". There are only two exceptions in this naming convention, *gbest(d)* and *dbest*, as we intend to keep the native PSO naming convention for *gbest* particle along with the fact that now there are distinct *gbest* particles for each dimension, thus yielding to *gbest(d)*. The same analogy also applies to the parameter *dbest* being as the global best dimension.

Let f denotes the dimensional fitness function that is to be optimized within a certain dimension range, $\{D_{min}, D_{max}\}$. Without loss of generality assume that the objective is to find the minimum (position) of f at the optimum dimension within a multi-dimensional search space. Assume that the particle a visits (back) the same dimension after T iterations (i.e., $xd_a(t) = xd_a(t + T)$), then the personal best position can be updated in iteration $t + T$ as follows:

$$
xy_{a,j}^{xd_a(t+T)}(t+T) = \begin{cases} xy_{a,j}^{xd_a(t)}(t) & \text{if} \quad f(xx_a^{xd_a(t+T)}(t+T)) > f(xy_a^{xd_a(t)}(t)) \\ xx_{a,j}^{xd_a(t+T)}(t+T) & \text{else} \end{cases}
$$
$$
j = 1, 2, \ldots, xd_a(t)
$$

$$(4.4)$$

Furthermore, the personal best dimension of particle a can be updated at iteration $t + 1$ as follows:

$$
\tilde{xd}_a(t+1) = \begin{cases} \tilde{xd}_a(t) & \text{if} \quad f(xx_a^{xd_a(t+1)}(t+1)) > f(xy_a^{\tilde{xd}_a(t)}(t)) \\ xd_a(t+1) & \text{else} \end{cases} \quad (4.5)
$$

Note that both Eqs. (4.4) and (4.5) are analogous to Eq. (3.1), which update the personal best position/dimension in the basic (global) PSO if a better current position/dimension is reached.

Figure 4.3 shows sample MD PSO and *bPSO* particles with indices a. A *bPSO* particle that is at a (fixed) dimension, $N = 5$, contains only positional components, whereas MD PSO particle contains both positional and dimensional components, respectively. In the figure the dimension range for the MD PSO is given in between 2 and 10; therefore, the particle contains nine sets of positional components. In this example as indicated by the arrows the current dimension the particle a resides is 2 ($xd_a(t) = 2$) whereas its personal best dimension is 3 ($\tilde{xd}_a(t) = 3$.) Therefore, at time t a positional PSO update is first performed over the positional elements, $xx_a^2(t)$ and then the particle may move to another dimension with respect

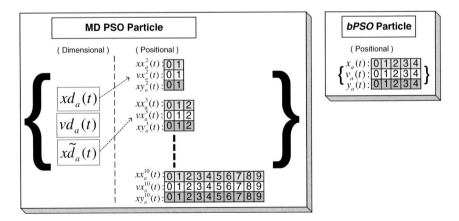

Fig. 4.3 Sample MD PSO (*right*) vs. bPSO (*left*) particle structures. For MD PSO$\{D_{\min} = 2$, $D_{\max} = 10\}$ and at time t, $xd_a(t) = 2$ and $\tilde{xd}_a(t) = 3$

to the dimensional PSO. Recall that each positional element, $xx_{a,j}^2(t), j \in \{0,1\}$, represents a potential solution in the data space of the problem.

Recall that $gbest(d)$ is the index of the global best particle at dimension d and let $S(d)$ be the total number of particles in dimension d, then $x\hat{y}^{dbest}(t) = xy_{gbest(dbest)}^{dbest}(t) = \arg \min_{\forall i \in [1,S]} (f(xy_i^{dbest}(t)))$. For a particular iteration t, and for a particle $a \in \{1, S\}$, first the positional components are updated in its current dimension, $xd_a(t)$ and then the dimensional update is performed to determine its next $(t + 1$st) dimension, $xd_a(t + 1)$. The positional update is performed for each dimensional component, $j \in \{1, xd_a(t)\}$,, as follows:

$$vx_{a,j}^{xd_a(t)}(t+1) = w(t)vx_{a,j}^{xdd_a(t)}(t) + c_1r_{1,j}(t)\left(xy_{a,j}^{xd_a(t)}(t) - xx_{a,j}^{xd_a(t)}(t)\right) + c_2r_{2,j}(t)\left(x\hat{y}_j^{xd_a(t)}(t) - xx_{a,j}^{xd_a(t)}(t)\right)$$
$$xx_{a,j}^{xd_a(t)}(t+1) = xx_{a,j}^{xd_a(t)}(t) + C_{vx}\left[vx_{a,j}^{xd_a(t)}(t+1), \{V_{min}, V_{max}\}\right]$$
$$xx_{a,j}^{xd_a(t)}(t+1) \leftarrow C_{xx}\left[xx_{a,j}^{xd_a(t)}(t+1), \{X_{min}, X_{max}\}\right]$$

$$(4.6)$$

where $C_{xx}[.,.] \equiv C_{vx}[.,.]$are the clamping operators applied over each positional component, $xx_{a,j}^d$ and $vx_{a,j}^d$. $C_{xx}[.,.]$ may or may not be applied depending on the optimization problem but $C_{vx}[.,.]$ is needed to avoid exploding. Each operator can be applied in two different ways,

$$C_{xx}[xx_{a,j}^d(t), \{X_{min}, X_{max}\}] = \begin{cases} xx_{a,j}^d(t) & \text{if } X_{min} \leq xx_{a,j}^d(t) \leq X_{max} \\ X_{min} & \text{if } xx_{a,j}^d(t) < X_{min} \\ X_{max} & \text{if } xx_{a,j}^d(t) > X_{max} \end{cases} \quad (a)$$

$$C_{xx}[xx_{a,j}^d(t), \{X_{min}, X_{max}\}] = \begin{cases} xx_{a,j}^d(t) & \text{if } X_{min} \leq xx_{a,j}^d(t) \leq X_{max} \\ U(X_{min}, X_{max}) & \text{else} \end{cases} \quad (b)$$

$$(4.7)$$

where the option (a) is a simple thresholding to the range limits and (b) reinitializes randomly the positional component in the jth dimension $(j < d)$.

Note that the particle's new position, $xx_a^{xd_a(t)}(t+1)$, will still be in the same dimension, $xd_a(t)$; however, the particle may jump to another dimension afterwards with the following dimensional update equations:

$$vd_a(t+1) = \lfloor vd_a(t) + c_1r_1(t)(x\tilde{d}_a(t) - xd_a(t)) + c_2r_2(t)(dbest - xd_a(t)) \rfloor$$
$$xd_a(t+1) = xd_a(t) + C_{vd}[vd_a(t+1), \{VD_{min}, VD_{max}\}]$$
$$xd_a(t+1) \leftarrow C_{xd}[xd_a(t+1), \{D_{min}, D_{max}\}]$$

$$(4.8)$$

where $\lfloor . \rfloor$ is the *floor* operator, $C_{xd}[.,.]$ and $C_{vd}[.,.]$ are the clamping operators applied over dimensional components, $xd_a(t)$ and $vd_a(t)$, respectively. Though we employed the inertia weight for positional velocity update in Eq. (4.5), we have witnessed no benefit of using it for dimensional PSO, and hence we left it out of Eq. (4.8) for the sake of simplicity. Note that both velocity update Eqs. (4.6) and (4.8) are similar to those for the basic PSO given in Eq. (3.2). $C_{vd}[.,.]$ is similar to $C_{vx}[.,.]$, which is used to avoid exploding. This is accomplished by basic thresholding expressed as follows:

$$C_{vd}[vd_a(t),\{VD_{\min}, VD_{\max}\}] = \begin{cases} vd_a(t) & \textit{if } VD_{\min} \leq vd_a(t) \leq VD_{\max} \\ VD_{\min} & \textit{if } vd_a(t) < VD_{\min} \\ VD_{\max} & \textit{if } vd_a(t) > VD_{\max} \end{cases} \quad (4.9)$$

$C_{xd}[.,.]$, on the other hand, is a mandatory clamping operator, which keeps the dimensional jumps within the dimension range of the problem, $\{D_{\min}, D_{\max}\}$. Furthermore within $C_{xd}[.,.]$, an optional in-flow buffering mechanism can also be implemented. This can be a desired property, which avoids the excess number of particles on a certain dimension. Particularly, *dbest* and dimensions within its close vicinity have a natural attraction and without such buffering mechanism, the majority of swarm particles may be hosted within this local neighborhood, and hence other dimensions might encounter a severe depletion. To prevent this, the buffering mechanism should control the in-flow of the particles (by the dimensional velocity updates) to a particular dimension. On some early *bPSO* implementations over problems with low (and fixed) dimensions, 15–20 particles were usually sufficient for a successful operation. However, in high dimensions this may not be so since more particles are usually needed as the dimension increases. Therefore, we empirically set the number of particles to be proportional to the solution space dimension and not less than 15. At time t, let $P_d(t)$ be the number of particles in dimension d. $C_{xd}[.,.]$ can then be expressed with the (optional) buffering mechanism as follows:

$$C_{xd}[xd_a(t),\{D_{\min}, D_{\max}\}] = \begin{cases} xd_a(t-1) & f\ P_d(t) \geq \max(15, xd_a(t)) \\ xd_a(t-1) & \textit{if } xd_a(t) < D_{\min} \\ xd_a(t-1) & f\ xd_a(t) > D_{\max} \\ xd_a(t) & \textit{else} \end{cases} \quad (4.10)$$

In short, the clamping and buffering operator, $C_{xd}[.,.]$, allows a dimensional jump only if the target dimension is within the dimensional range and there is a room for a newcomer.

Accordingly, the general pseudo-code of the MD PSO technique is given in Table 4.1.

It is easy to see that the random initialization of the swarm particles' performed in step 1 is similar to the initialization of the *bPSO* (between the same steps) given in Table 3.1. The dimensional PSO is initialized in the same way; however, there is a difference in the initialization of the positional PSO, that is, particle positions are randomized for all solution space dimensions ($\forall d \in [D_{\min}, D_{\max}]$) instead of a single dimension. After the initialization phase, step 3 first evaluates each particle a, which is residing in its current dimension, $xd_a(t)$, (1) to validate its personal best position in that dimension (step 3.1.1.1), (2) to validate (and update if improved)

Table 4.1 Pseudo-code of MD PSO algorithm

MD PSO (*termination criteria:{IterNo,* ε_C *, ...}*)

1. For $\forall a \in [1, S]$ do:

   1.1. Randomize $xd_a(1), vd_a(1)$

   1.2. Initialize $\tilde{xd}_a(0) = xd_a(1)$

   1.3. For $\forall d \in [D_{\min}, D_{\max}]$ do:

      1.3.1. Randomize $xx_a^d(1)$, $xv_a^d(1)$

      1.3.2. Initialize $xy_a^d(0) = xx_a^d(1)$

      1.3.3. Initialize $x\hat{y}^d(0) = xx_a^d(1)$

   1.4. End For.

2. End For.

3. For $\forall t \in [1, IterNo]$ do:

   3.1. For $\forall a \in [1, S]$ do:

      3.1.1. If ($f(xx_a^{xd_a(t)}(t)) < \min\left(f(xy_a^{xd_a(t)}(t-1), \min_{p \in S-\{a\}} \left(f(xx_p^{xd_a(t)}(t)) \right) \right)$) then do:

         3.1.1.1. $xy_a^{xd_a(t)}(t) = xx_a^{xd_a(t)}(t)$

         3.1.1.2. If ($f(xx_a^{xd_a(t)}(t)) < f(xy_{gbest(xd_a(t))}^{xd_a(t)}(t-1))$) then $gbest(xd_a(t))=a$

         3.1.1.3. If ($f(xx_a^{xd_a(t)}(t)) < f(xy_a^{\tilde{d}_a(t-1)}(t-1))$) then $\tilde{xd}_a(t) = xd_a(t)$

         3.1.1.4. If ($f(xx_a^{xd_a(t)}(t)) < f(x\hat{y}^{dbest}(t-1))$) then $dbest = xd_a(t)$

      3.1.2. End If.

   3.2. End For.

   3.3. If the *termination criteria* are met, then **Stop**.

   3.4. For $\forall a \in [1, S]$ do:

      3.4.1. For $\forall j \in [1, xd_a(t)]$ do:

         3.4.1.1. Compute $vx_{a,j}^{xd_a(t)}(t+1)$ and $xx_{a,j}^{xd_a(t)}(t+1)$ using (4.6)

      3.4.2. End For.

      3.4.3. Compute $vd_a(t+1)$ and $xd_a(t+1)$ using Eq. (4.8)

   3.5. End For.

4. End For.

---

the *gbest* particle in that dimension (step 3.1.1.2), (3) to validate (and update if improved) its personal best dimension, (step 3.1.1.3) and finally, (4) to validate (and update if improved) the global best dimension, *dbest* (step 3.1.1.4). Step 3.1 in fact evaluates the current position and dimension of each particle with its personal best values that will be updated if any improvement is observed. The (new) personal best position/dimension updates will then lead to the computation of the (new) global best elements such as *dbest* and *gbest(d)* ($(\forall d \in [D_{\min}, D_{\max}])$). At any time, *t*, the optimum solution will be $x\hat{y}^{dbest}$ at the optimum dimension, *dbest*, achieved by the particle *gbest(dbest)*, and finally the best (fitness) score achieved will naturally be $f(x\hat{y}^{dbest})$. This (best) fitness score so far achieved can then be used to determine whether the termination criteria is met in step 3.3. If not, step 3.4 performs first the positional PSO (step 3.4.1) and then the dimensional

PSO (step 3.4.3) to perform positional and dimensional updates for each particle, respectively. Once each particle moves to a new position and (jumps to a new) dimension, then in the next iteration its personal best updates will be performed and so on, until the termination criteria is met.

4.4 Programming Remarks and Software Packages

As this is the second chapter including software packages and programming remarks, in the next subsection we shall first complete the description of the *PSO_MDlib* test bed application to show how to apply MD PSO for multi-dimensional nonlinear function minimization, and then start with another major MD PSO application with GUI support, namely *PSOtestApp*.

4.4.1 MD PSO Operation in PSO_MDlib *Application*

The test bed application, **PSO_MDlib,** is initially designed for regular MD PSO operations for the sole purpose of multi-dimensional nonlinear function minimization. Recall that in the previous chapter, we fixed the dimension of any function as **_dMin = _dMax − 1**. Recall further that both **_dMin** and **dMax** correspond to the dimensional range $\{D_{min}, D_{max}\}$ and thus any logical range values can be assigned for them within the **PSOparam** structure (default: {2, 101}). The target dimension can be set in at the beginning of the **main()** function, i.e.,

int **tar_dim[3] = {20, 50, 80};**

...

pDim = &tar_dim[2];//The target dimension.

which points to the third entry of the **tar_dim[2]** array (**pDim = 80*). The dimensional bias stored in the pointer **pDim** will then be used as the dimensional bias in all nonlinear functions implemented in **MYFUN.cpp** which makes them biased to all dimensions except the target dimension (**i.e., *pDim = 80**). In other words, the nonlinear function has a unique global minimum (i.e., 0 for all functions) *only* in the target dimension. Then MD PSO can be tested accordingly to find out whether it can converge to the global optimum that resides in the target (true) dimension. Once the MD PSO object has been initialized with a proper dimensional range and dimensional PSO parameters (i.e., **_psoDef._dMin, _psoDef._dMax, _psoDef._vdMin, _psoDef._vdMax**) then the rest of the code is identical for both operations. At the end, recall that MD PSO is just the multi-dimensional extension of the basic PSO.

As initially explained, in the main MD PSO process loop, i.e., the step 3 of the MD PSO pseudo-code given in Table 4.1, is performed within Table 3.7 given in

Sect. 3.5. In that section we explained it in parallel with the basic PSO run, more details can now be given with respect to the parameters of the MD PSO swarm particles. First of all, note that there is an individual *gbest* particle for each dimension and whenever the particle *a* visits resides in the current dimension, **cur_dim_a**, it can be the new *gbest* particle in that dimension if its current position surpasses *gbest*'s. Note that **cur_dim_a** can now be any integer number between **m_xdMin** and **m_xdMax-1**. Another MD PSO parameter, *dbest* (**m_dbest** in the code), which can be updated only if the particle becomes the new *gbest* in **cur_dim_a**, and it can achieve a fitness score even better than the best fitness score achieved in the dimension *dbest*.

The termination criteria (by either *IterNo* or ε_C) were explained in Sect. 3.5 with the code given in Table 3.9. The best performance achieved by the MD PSO swarm can be gathered from the personal best position of the *gbest* particle in *dbest* dimension, i.e., **m_gbest[m_dbest - m_xdMin]** and the call in the **if()** statement, which compares it with the **m_eCutOff**, i.e., **m_pPA[m_gbest[m_dbest - m_xdMin]]- > GetPBScore(m_dbest)**.

If neither of the termination criteria is reached, then step 3.4 is executed to update the position and the dimension of each swarm particle as given in Table 4.2. As shown in Table 3.10 in Sect. 3.5, first in the positional update, each particle's cognitive (**cogn**) and social (**social**) components are computed and the velocity update is composed over the previous velocity weighted with the inertia factor. The velocity is clamped (**vx- > CheckLimit()**) and finally, the particle's current position is updated to a new position with the composed and clamped

Table 4.2 Implementation of Step 3.4 of the MD PSO pseudo-code given in Table 4.1

```
...
for (a=0;a<m_noP;a++)
{ // Update loop..
        // 1. Particle speed & location update in the prev. dimension..
        ... // See Section Error! Reference source not found.
        // The best swarm pos. in current dim..
        ... // See Section Error! Reference source not found.
        // 2. MD PSO DIMENSION UPDATE for the NEXT iteration..
        int xdim =  m_pPA[a]->GetDim(); // dim. of the particle..
        int vdim =  m_pPA[a]->GetDVel(); // dim. velocity of the particle..
        int dbestx = m_pPA[a]->GetPBDim(); // Personal best dim. of the particle..

        vdim = (int) (vdim+2*RAND*(dbestx-xdim)+ 2*RAND*(m_dbest-xdim)); // dim. vel. update..
        if (vdim>m_vdMax) vdim=m_vdMax;
        if (vdim<m_vdMin) vdim=m_vdMin;
        m_pPA[a]->SetDVel(vdim);

        xdim += vdim; // update dim..
        if( xdim < m_xdMax && xdim >= m_xdMin && m_pDimHist[xdim] < xdim)
                m_pPA[a]->SetDim(xdim);
        ++m_pDimHist[m_pPA[a]->GetDim()];
} // for all members..
...
```

velocity term. As the last step, this new position is also verified whether it falls within the problem boundaries specified in advance ($\pm x_{max}$ by **xx- > Check-Limit()**), all of which are performed within step 3.4.1 of the MD PSO pseudo-code. The dimensional PSO within step 3.4.3 of the MD PSO pseudo-code is then performed and this is simply a PSO implementation in 1D, over the integer numbers to find out the optimal dimension using the PSO mechanism. Within the loop, each particle's cognitive (**dbestx-xdim**) and social components (**m_dbest-xdim**) are computed and the velocity update is composed over the previous dimensional velocity component, **vdim**. The velocity is clamped and the current dimension of the particle is updated. As the last step, this new dimension is also verified whether it falls within the problem boundaries specified in advance. Recall that $C_{xd}[.,.]$ and $C_{vd}[.,.]$ are the clamping operators applied over dimensional components, $xd_a(t)$ and $vd_a(t)$, respectively. Note that in the code both clamping operators are simply implemented by four member variables of the class: **m_xdMin, m_xdMax** and **m_vdMin, m_vdMax.**

4.4.2 MD PSO Operation in **PSOTestApp** *Application*

The major MD PSO test bed application is **PSOTestApp**. It is a single dialog-based Windows application over which several MD PSO applications, mostly based on data (or feature) clustering, are implemented. In this section, its programming basics and use will only be introduced whilst keeping the details of the individual applications to the following chapters. **PSOTestApp** is a multithread application where there is a dedicated thread for its GUI (Dialog) implementation as shown in Fig. 4.4. In the figure, the snapshot on the top is the main GUI where a selection of MD PSO applications is shown and the one in the bottom shows also a pop-up menu dialog with MD PSO parameters. This pop-up dialog is activated whenever the "Run" button is pressed. So the user first selects a proper (MD PSO) application from the main combo box, then opens the proper source file(s) (to process or to read data), and finally presses the "Run" button to initiate the MD PSO application with the proper parameters that can be specified in the pop-up dialog "MD PSO Parameters".

PSOTestApp workspace consists of eight projects: *Converters, DbgRouter, Frame, FrameRenderer, Image, MTL, PSOCluster,* and *PSOTestApp.* The first six are the supplementary projects (description of which is skipped as it is beyond the scope of this book), which are used to render an image, create a thread, output debug messages during run-time, etc. *PSOCluster* is the main DLL project, which executes all MD PSO applications [1–8 in Fig. 4.4 (top)]. As shown in the figure, there are eight applications listed in the combo list, six of which are implemented: "1. 2D Clustering over binary (B\W) images…", "2. 3D Color Quantization over color images…", "4. 2D Clustering via GMM over binary (B\W) images…", "6. N-D Clustering over *.dt (data) files…", "7. RBF Network Training via N-D Clustering…", and "8. Feature Synthesis…". In this section, we shall give an

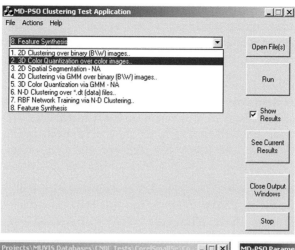

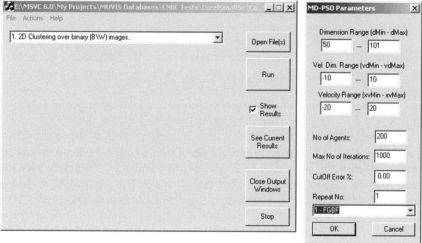

Fig. 4.4 GUI of **PSOTestApp** with several MD PSO applications (*top*) and MD PSO Parameters dialog is activated when pressed "Run" button (*bottom*)

overview and data structures for the *PSOTestApp* and *PSOCluster* projects, and then focus on the first application (*1. 2D Clustering over binary (B\W) images*). The rest of the clustering applications (except eighth application *Feature Synthesis*) will be covered in Chaps. 5 and 6. The eighth application, Evolutionary Feature Synthesis, will be covered in Chap. 10.

The entry point of a Windows Dialog workspace created by Visual Studio 6.5 is the *[nameofApp]Dlg.cpp*. For **PSOTestApp** it is therefore, **PSOtestAppDlg.cpp** where all Windows dialog-based control and callback functions are implemented. In this source file, a separate thread operates all the user's actions by the callback functions of the class **CPSOtestAppDlg** such as:

- void **CPSOtestAppDlg::OnButtonOpen()** //Open button
- void **CPSOtestAppDlg::OnDopso()** //Run button
- void **CPSOtestAppDlg::OnStop()** //Stop button
- etc.

When "Run" button is pressed, the MD PSO application (selected in the combo list) is initiated within the **OnDopso()** function and the request is conducted to the proper interface class implemented within the **PSOCluster** library (DLL). For this, in the header file **PSOtestAppDlg.h,** the class **CPSOtestAppDlg** contains six interface classes, each of which has (the one and only) member object to accomplish the task initiated. These member objects are as follows:

- **CPSOcluster m_PSOt;** //The generic 2-D PSO clustering obj..
- **CPSOclusterND m_PSOn;** //The generic N-D PSO clustering obj..
- **CPSO_RBFnet m_PSOr;** //The generic PSO RBF Net. training obj..
- **CPSOcolorQ m_PSOc;** //The generic PSO colorQ obj..
- **CPSOclusterGMM m_PSOgmmt;** //The GMM PSO clustering obj..
- **CPSOFeatureSynthesis m_PSOfs;** //The Feat. Synth. obj..

Each interface class is responsible of one of the MD PSO applications listed above and each object executes its task in a dedicated thread created at the beginning of the operation. Assume for instance that a user selects "2D Clustering" application. Then in the **OnDopso()** function the following code in Table 4.3 activates the API function, **ApplyPSO()**, of the **CPSOcluster** class.

Recall that **CPSOcluster** class contains the API as well as the implementer functions for the 2D clustering operations using the MD PSO. It has the following member functions as presented in Table 4.4. The **Init()** function initializes the object with the window handler of the dialog, which is used to send Windows messages to the GUI thread (handled within the **CPSOtestAppDlg** class). The

Table 4.3 The callback function **OnDopso()** activated when pressed "Run" button on **PSOtestApp** GUI

```
...
void CPSOtestAppDlg::OnDopso()
{
...
        switch (op_index) {
        case 0: // 2D Clustering..
                m_PSOt.ApplyPSO(m_pXImList, m_psoParam, m_saParam);
                SetWindowText("2D Clustering initiated.. Please Wait..");
                break;
        case 1: // Color Quantization..
        default:
                ...
        }
}
...
```

Table 4.4 Member functions of **CPSOcluster** class

```
// This class is exported from the PSOtester.dll
class PSOTESTER_API CPSOcluster
{
public:
        CPSOcluster(void);
        ~CPSOcluster(void) {;}
        // API Functions..
        void    Init (HWND hWnd) {m_hWnd = hWnd;}
        void    ApplyPSO(CQueueList<CFileName>  *pFL,  PSOparam  psoParam,  SPSAparam
saParam);
        void    Stop();
        void    ShowResults();

protected: // LOCALS..
        void    PSOThread();
        void    GetResults(FILE *fp);

// STATICS
        static double       ValidityIndex(CSolSpace<CPixel2D> *pPos, int);
        static double       ValidityIndex2(CSolSpace<CPixel2D> *pPos, int);
        static double       GetDist(CPixel2D *pPos1, CPixel2D *pPos2);
        static  void        FindGBDim(CParticle<CSolSpace<CPixel2D>,CPixel2D>** pPA, int noP);

        static CQueueList<CPixel2D>*     s_pPixelQ ;
...
```

Stop() function can stop an ongoing MD PSO application anytime and abruptly. The **ShowResults()** function will show the 2D clustering results on a separate dialog, which is created and controlled within the **CPSOtestAppDlg** class. Finally, as presented in Table 4.5 the API function, **ApplyPSO()**, is called within the callback function **CPSOtestAppDlg::OnDopso()** and creates a dedicated thread in which the MD PSO application (2D clustering) is executed. This thread function is called **PSOThread()** and created as:

    CMThread < CPSOcluster > (PSOThread,   this,   THREAD_PRIOR-
ITY_NORMAL).Begin ();

There are *three* set of parameters of this API function, the list of input filenames (**pFL**), MD PSO parameters (**psoParam**), and SPSA parameters (**saParam**), which will be explained in the next chapter. The input files for 2D clustering application are black and white images similar to the ones shown in Figs. 3.7 and 4.1, and 2D data points are represented by white pixels. As the MD PSO process terminates, the resultant output images consist of the clusters extracted from the image, each of which rendered with three-color representation, as typical examples are shown in Figs. 3.8 and 3.9. The structure, **psoParam** contains all MD PSO parameters, which can be edited by the user via MD PSO parameters dialog. Both **psoParam** and **saParam** are copied into the member structures (**m_psoParam** and **m_saParam**) and as a result, the MD PSO process can now be executed by a separate thread within the function **PSOThread()**.

Table 4.5 The **ApplyPSO**() API function of the **CPSOcluster** class

```
...
void  CPSOcluster::ApplyPSO(CQueueList<CFileName>  *pFL, PSOparam  psoParam,  SPSAparam
saParam)
{
        if(!(m_pFL = pFL)) return;
        memcpy(&m_psoParam, &psoParam, sizeof(PSOparam));
        memcpy(&m_saParam, &saParam, sizeof(SPSAparam));

        // Create the MD PSO thread..
        CMThread<CPSOcluster> (PSOThread, this, THREAD_PRIORITY_NORMAL).Begin ();
}
...
```

As this is a 2D clustering application of the MD PSO, the data space (**class X**) is implemented by the **CPixel2D** class declared in the **Pixel2D.h** file. This class basically contains all the arithmetic operations that can be performed over 2D pixels with coordinates (**m_x** and **m_y**). Recall that **COneClass** was used as the data space class (**class X**) in Sect. 3.5 for nonlinear function minimization and now as the application is changed to 2D clustering, a proper <**class X**> implementation is, first of all, required. The second required input to MD PSO for 2D clustering application is a proper fitness function. Recall that for nonlinear function minimization, one of the benchmark functions was given as the fitness function by calling, **pPSO- > SetFitnessFn(Sphere),** and now the application in hand is 2D clustering, as discussed in Sect. 3.4.2, the so-called "Clustering Validity Index" (CVI) function should be implemented in accordance with the MD PSO particles and given to the MD PSO object in the same way. In the **CPSOcluster** object, there are two sample CVI functions implemented: **ValidityIndex**() and **ValidityIndex2**(). The theory and design considerations of the MD PSO data clustering operations will be covered in Chap. 6 in detail, and thus are skipped in this section. Since both CVI functions are static (to be able to use their function pointers), the entire 2D data space (the white pixels) is stored into the following link list structure declared in **CPSOcluster** class: **static CQueueList < CPixel2D > * s_pPixelQ;** so that both CVI functions can use it to access real data points. Therefore, for each 2D black&white input image containing the white pixels for 2D clustering, the image is first decoded (Step 1), **s_pPixelQ** is created and 2D white pixels are stored (Step 2) and finally, MD PSO object is created, initialized properly, and executed to perform the clustering (Step 3). Each MD PSO clustering operation can be repeated several times, the number of which is set by the user and stored in the **m_psoParam._repNo** parameter. All three steps can easily be recognized in the **CPSOcluster::PSOThread**() function, as given in Table 4.6. Note further that the clustering operation is performed for each image in the list of input filenames (**m_pFL**).

Table 4.6 The function **CPSOcluster::PSOThread()**

```
m_pFL->Reset();
for(pFL = NULL; (pFL = m_pFL->Next()); )
{ // For all the image files, apply PSO clustering..

        // Step 1: Decode & Get the YUV frame buffer ready for PSO indexing..
        sprintf(image_name, "%s\\%s.%s", pFL->m_dir, pFL->m_title, pFL->m_ext);
        DecodeImage(theImage, image_name);
        ...
        ...
        // Step 2: Create a pixel queue and append points from image into it..
        s_pPixelQ = new CQueueList<CPixel2D>; // Create the pixel queue..
        for(int y = 0, i=0; y<Ys; y++)
                for(int x = 0; x<Xs; x++)
                        if(pYUVimg[y*Xs+x])
                        { // A Point to be indexed..
                                CPixel2D *pPix = new CPixel2D(x, y, i++); // Create a new pixel..
                                s_pPixelQ->Insert((CPixel2D *) pPix); // Insert this point into queue..
                        } // if..
        ...
        ...
        for(m_rep = 0; m_rep < m_psoParam._repNo; ++m_rep)
        {
                // Step 3: Create & Init. PSO object..
                m_pPSO = new CPSO_MD<CSolSpace<CPixel2D>,CPixel2D> (m_psoParam._noAgent,
m_psoParam._maxNoIter, m_psoParam._eCutOff, m_psoParam._mode);

                // SET 3 Fn. ptrs for Fitness, Dist (for MST) and GBDim..
                m_pPSO->SetFitnessFn(ValidityIndex2);
                m_pPSO->SetDistFn(GetDist);
                m_pPSO->SetFindGBDimFn(FindGBDim);
                m_pPSO->SetSAparam(m_saParam); // Set SA params..
                ...
                ...
```

References

1. A. Abraham, S. Das and S. Roy, "Swarm Intelligence Algorithms for Data Clustering", in *Soft Computing for Knowledge Discovery and Data Mining book,* Part IV, pp. 279-313, Oct. 25, 2007
2. M.G. Omran, A. Salman, A.P. Engelbrecht, Dynamic Clustering using Particle Swarm Optimization with Application in Image Segmentation. In Pattern Analysis and Applications **8**, 332–344 (2006)
3. G-J Qi, X-S Hua, Y. Rui, J. Tang, H.-J. Zhang, "Image Classification With Kernelized Spatial-Context," *IEEE Trans. on Multimedia,* vol.12, no.4, pp.278-287, June 2010. doi: 10.1109/TMM.2010.2046270
4. M. G. Omran, A. Salman, and A.P. Engelbrecht, Particle Swarm Optimization for Pattern Recognition and Image Processing, Springer Berlin, 2006

Chapter 5
Improving Global Convergence

Like they say, you can learn more from a guide in one day than
you can in three months fishing alone.

Mario Lopez

As a natural extension of PSO, MD PSO may also have a serious drawback of
premature convergence to a local optimum, due to the direct link of the information flow between particles and *gbest*, which "guides" the rest of the swarm
resulting in possible loss of diversity. Hence, this phenomenon increases the
probability of being trapped in local optima [1] and it is the main cause of the
premature convergence problem especially when the search space is of high
dimensions [2] and the problem to be optimized is multimodal [1]. Another reason
for the premature convergence is that particles are flown through a single point
which is (randomly) determined by *gbest* and *pbest* positions and this point is not
even guaranteed to be a local optimum [3]. Various modifications and PSO
variants have been proposed in order to address this problem such as [1, 3–28]. As
briefly discussed in Sect. 3.3, such methods usually try to improve the diversity
among the particles and the search mechanism either by changing the update
equations toward a more diversified version, by adding more randomization to the
system (to particle velocities, positions, etc.), or simply resetting some or all
particles randomly when some conditions are met. On the one hand, most of these
variants require additional parameters to accomplish the task and thus making the
algorithms even more parameter dependent. On the other hand, the main problem
is in fact the inability of the algorithm to use available diversity in one or more
positional components of a particle. Note that one or more components of any
particle may already be in a close vicinity of the global optimum. This potential is
then wasted with the (velocity) update in the next iteration, which changes all the
components at once. In this chapter, we shall address this drawback of global
convergence by developing two efficient techniques. The first one, the so-called
Fractional Global Best Formation (FGBF), collects all such promising (or simply
the best) components from each particle and fractionally creates an artificial global
best (GB) candidate, the *aGB*, which will be the swarm's global best (GB) particle
if it is better than the previous GB and the just-computed *gbest*. Note that
whenever a better *gbest* particle or *aGB* particle emerges, it will replace the current
GB particle. Without any additional change, we shall show that FGBF can avoid
local optima and thus yield the optimum (or near optimum) solution efficiently
even in high dimensional search spaces. Unfortunately FGBF is not an entirely

S. Kiranyaz et al., *Multidimensional Particle Swarm Optimization for Machine Learning
and Pattern Recognition*, Adaptation, Learning, and Optimization 15,
DOI: 10.1007/978-3-642-37846-1_5, © Springer-Verlag Berlin Heidelberg 2014

generic technique, which should be specifically adapted to the problem at hand (we shall return to this issue later). In order to address this drawback efficiently, we shall further present two generic approaches, one of which moves *gbest* efficiently or simply put, "guides" it with respect to the function (or error surface) it resides on. The idea behind this is quite simple: since the velocity update equation of *gbest* is quite poor, we shall replace it with a simple yet powerful stochastic search technique to *guide* it instead. We shall henceforth show that due to the stochastic nature of the search technique, the likelihood of getting trapped into a local optimum can significantly be decreased.

5.1 Fractional Global Best Formation

5.1.1 The Motivation

FGBF is designed to avoid premature convergence by providing a significant diversity obtained from a proper *fusion* of the swarm's best components (the individual dimension(s) of the current position of each particle in the swarm). At each iteration in a *bPSO* process, an *aGB* is *fractionally* formed by selecting the most promising or simply the best particle positional vector components from the entire swarm.

Therefore, especially during the initial steps, the FGBF can most of the time be a better alternative than the native *gbest* particle, since it has the advantage of assessing each component of every particle in the swarm individually, and forming the *aGB* particle fractionally by using the most promising components among them. This process naturally uses the available diversity among individual components and thus it can prevent the swarm from trapping in local optima. Take for instance, the function minimization problem as illustrated in Fig. 5.1 where 2D space is used for illustration purposes. In the figure, three particles in a swarm are ranked as the 1st (or the *gbest*), the 3rd, and the 8th with respect to their proximity to the target position (or the global solution) of some function. Although *gbest* particle (i.e., 1st ranked particle) is the closest in the overall sense, the particles ranked 3rd and 8th provide the best x and y dimensions (closest to the target's respective dimensions) in the entire swarm, and hence the *aGB* particle via FGBF yields a closer particle to the target than the swarm's *gbest*.

5.1.2 PSO with FGBF

Suppose for a swarm ξ, FGBF is performed in a PSO process at a (fixed) dimension N. Recall from the earlier discussion that at iteration, t, each PSO particle, a, has the following components: position, $x_{a,j}(t)$, velocity, $v_{a,j}(t)$ and the personal best

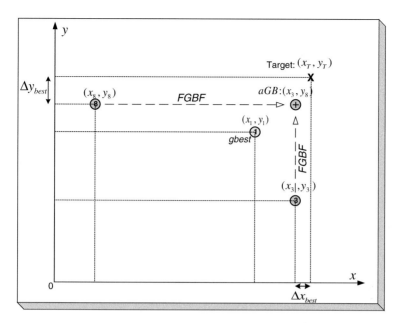

Fig. 5.1 A sample FGBF in 2D space

position, $y_{a,j}(t), j \in [1, N]$. The aGB particle, obtained through the FGBF process, is fractionally formed from the components of some swarm particles, and therefore it does not use any velocity term. Consequently, $y_{aGB}(t)$ is set to the best of $x_{aGB}(t)$ and $y_{aGB}(t-1)$. As a result, the FGBF process creates one aGB particle providing a potential GB solution, $y_{aGB}(t)$. Let $f(a, j)$ be the dimensional fitness score of the jth dimensional component of the position vector of particle a. Suppose that all dimensional fitness scores $(f(a, j), \forall a \in [1, S], \forall j \in [1, N])$ can be computed in step 3.1 and FGBF pseudo-code as given in Table 5.1 can then be plugged in between steps 3.3 and 3.4 of $bPSO$'s pseudo-code.

Step 2 in the FGBF pseudo-code along with the computation of $f(a, j)$ depends entirely on the optimization problem. It keeps track of partial fitness contributions from each individual component from each particle's position. For those problems

Table 5.1 Pseudo-code of FGBF in $bPSO$

FGBF in $bPSO$ $(\xi, f(a, j))$
1. Create a new aGB particle, $\{x_{aGB,j}(t), y_{aGB,j}(t)\}$ for $\forall j \in [1, N]$
2. Let $a[j] = \arg \min\limits_{a \in \xi j \in [1, N]} (f(a, j))$ be the index array of particles yielding the minimum $f(a, j)$ for the jth dimensional component
3. $x_{aGB,j}(t) = x_{a[j],j}(t)$ for $\forall j \in [1, N]$
4. If $(f(x_{aGB}(t)) < f(y_{aGB}(t-1)))$ then $y_{aGB}(t) = x_{aGB}(t)$
5. Else $y_{aGB}(t) = y_{aGB}(t-1)$
6. If $(f(y_{aGB}(t)) < f(\tilde{y}(t)))$ then $\tilde{y}(t) = y_{aGB}(t)$

without any constraints (e.g., nonlinear function minimization), the best dimensional components can simply be selected, whereas in others (e.g., clustering), some promising components, which satisfy the problem constraints or certain criteria required, are first selected, grouped, and the most suitable one in each group is then used for FGBF. Here, the internal nature of the problem will determine the "suitability" of the selection.

5.1.3 MD PSO with FGBF

The previous section introduced the principles of FGBF when applied in a *bPSO* process on a single dimension. In this section, we present its generalized form with the proposed MD PSO where there is one *gbest* particle per (potential) dimension of the solution space. For this purpose, recall that at a particular iteration, t, each MD PSO particle, a, has the following components: position ($xx_{a,j}^{xd_a(t)}(t)$), velocity ($vx_{a,j}^{xd_a(t)}(t)$) and the personal best position ($xy_{a,j}^{xd_a(t)}(t)$) for each potential dimensions in solution space (i.e., $xd_a(t) \in [D_{min}, D_{max}]$ and $j \in [1, xd_a(t)]$) and their respective counterparts in the dimensional PSO process (i.e., $xd_a(t)$, $vd_a(t)$, and $\tilde{xd}_a(t)$). The *aGB* particle does not need dimensional components where a single positional component with the maximum dimension D_{max} is created to cover all dimensions in the range, $\forall d \in [D_{min}, D_{max}]$, and as explained earlier, there is no need for the velocity term either, since *aGB* particle is fractionally (re-) formed from the dimensions of some swarm particles at each iteration.

Furthermore, the aforementioned competitive selection ensures that $xy_{aGB}^d(t)$, $\forall d \in [D_{min}, D_{max}]$ is set to the best of the $xx_{aGB}^d(t)$ and $xy_{aGB}^d(t-1)$. As a result, the FGBF process creates one *aGB* particle providing (potential) GB solutions ($xy_{aGB}^d(t)$) for all dimensions in the given range (i.e., $\forall d \in [D_{min}, D_{max}]$). Let $f(a, j)$ be the dimensional fitness score of the jth component of particle a, which has the current dimension, $xd_a(t)$ and $j \in [1, xd_a(t)]$. At a particular time t, all dimensional fitness scores ($f(a, j)$, $\forall a \in [1, S]$) can be computed in step 3.1 and FGBF pseudo-code for MD PSO as given in Table 5.2 can then be plugged in between steps 3.2 and 3.3 of the MD PSO's pseudo-code. Next, we will present the application of MD PSO with FGBF to nonlinear function minimization and other applications will be presented in detail in the following chapters.

5.1.4 Nonlinear Function Minimization

A preliminary discussion and experimental results of the application of the basic PSO over four nonlinear functions were given in Sect. 3.4.1. We now selected seven benchmark functions and biased them with a dimensional term in order to

Table 5.2 Pseudo-code for FGBF in MD PSO

FGBF in MD PSO $(f(a, j))$

1. Create a new *aGB* particle, $\{xx^d_{aGB, j}(t), xy^d_{aGB, j}(t)\}$ for $\forall d \in [D_{min}, D_{max}]$, $\forall j \in [1, d]$
2. Let $a[j] = \arg \min\limits_{a\in[1,S]j\in[1,D_{max}]} (f(a, j))$ be the index array of particles yielding the minimum
   $f(a, j)$ for the *j*th dimension
3. For $\forall d \in [D_{min}, D_{max}]$ do:
   3.1. $xx^d_{aGB, j}(t) = xx^d_{a[j],j}(t)$ for $\forall j \in [1, d]$
   3.2. If $(f(xx^d_{aGB}(t)) < f(xy^d_{aGB}(t-1)))$ then $xy^d_{aGB}(t) = xx^d_{aGB}(t)V$
   3.3. Else $xy^d_{aGB}(t) = xy^d_{aGB}(t-1)$
   3.4. If $(f(xy^d_{aGB}(t)) < f(xy^d_{gbest(d)}(t)))$ then $xy^d_{gbest(d)}(t) = xy^d_{aGB}(t)$
4. End for
5. Re-assign *dbest*: dbest = $\arg \min\limits_{d\in[D_{min}, D_{max}]} (f(xy^d_{gbest(d)}(t)))$

test the performance of MD PSO. The functions given in Table 5.3 provide a good mixture of complexity and modality and have been widely studied by several researchers, e.g., see [10, 16, 22, 29–31]. The dimensional bias term, $\Psi(d)$, has the form of $\Psi(d) = K|d - d_0|^\alpha$ where the constants K and α are properly set with respect to the dynamic range of the function to be minimized. Note that the variable d_0, $D_{min} \le d_0 \le D_{max}$, is the target dimension in which the global minimum resides and hence all functions have the global minimum $F_n(x, d_0) = 0$, when $d = d_0$. *Sphere*, *De Jong*, and *Rosenbrock* are the unimodal functions and the rest are multimodal, meaning that they have many deceiving local minima. On the macroscopic level, *Griewank* demonstrates certain similarities with unimodal functions especially when the dimensionality is above 20; however, in low dimensions it bears a significant noise, which creates many local minima due to the second multiplication term with *cosine* components. Yet with the addition of dimensional bias term $\Psi(d)$, even unimodal functions eventually become multimodal, since they now have a local minimum at every dimension (which is their global minimum at that dimension without $\Psi(d)$) but only one global minimum at dimension, d_0.

Recall from the earlier remarks that a MD PSO particle *a* represents a potential solution at a certain dimension, and therefore the *j*th component of a *d*-dimensional point $(x_j, j \in [1, d])$ is stored in its positional component, $xx^d_{a, j}(t)$ at time *t*. Step 3.1 in MD PSO's pseudo-code computes the (dimensional) fitness score $(f(a,j))$ of the *j*th component (x_j) and at step 2 in the FGBF process, the index of the particle with those x_j's yielding minimum $f(a, j)$ is then stored in the array $a[j]$. Except the nonseparable functions, *Rosenbrock* and *Griewank*, the assignment of $f(a,j)$ for particle *a* is straightforward (e.g., $f(a,j) = x_j^2$ for *Sphere*, $f(a, j) = x_j\sin\left(\sqrt{|x_j|}\right)$ for *Schwefel*, etc., simply using the term with the *j*th component of the summation). For *Rosenbrock*, we can set $f(a, j) = (x_{j+1} - x_j^2)^2 + (x_j - 1)^2$ since the *aGB* particle, which is fractionally formed by those x_j's minimizing the *j*th summation

Table 5.3 Benchmark functions with dimensional bias

Function	Formula	Initial range $\pm x_{max}$	Dimension range $[D_{min}, D_{max}]$
Sphere	$F_1(x, d_0) = \left(\sum_{i=1}^{d} x_i^2\right) + (d - d_0)^4$	± 150	[2, 100]
De Jong	$F_2(x, d_0) = \left(\sum_{i=1}^{d} i x_i^4\right) + (d - d_0)^4$	± 50	[2, 100]
Rosenbrock	$F_3(x, d) = \left(\sum_{i=1}^{d} 100\left(x_{i+1} - x_i^2\right)^2 + (x_i - 1)^2\right) + (d - d_0)^4$	± 50	[2, 100]
Rastrigin	$F_4(x, d_0) = \left(\sum_{i=1}^{d} 10 + x_i^2 - 10\cos(2\pi x_i)\right) + (d - d_0)^4$	± 50	[2, 100]
Griewank	$F_5(x, d_0) = \left(\frac{1}{4,000}\sum_{i=1}^{d} x_i^2 - \prod_{i=1}^{d}\cos\left(\frac{x_i}{\sqrt{i+1}}\right)\right) + 0.2\,(d - d_0)^2$	± 500	[2, 100]
Schwefel	$F_6(x, d_0) = \left(418.9829d + \sum_{i=1}^{d} x_i \sin\left(\sqrt{|x_i|}\right)\right) + 40\,(d - d_0)^2$	± 500	[2, 100]
Giunta	$F_7(x, d_0) = \left(\sum_{i=1}^{d}\sin\left(\frac{16}{15}x_i - 1\right) + \sin^2\left(\frac{16}{15}x_i - 1\right) + \frac{1}{50}\sin\left(4\left(\frac{16}{15}x_i - 1\right)\right) + \frac{268}{1,000}\right) + \sqrt{|d - d_0|}$	± 500	[2, 100]

term, eventually minimizes the function. Finally for *Griewank* one can approximate $f(a, j) \approx x_j^2$ for particle a and the FGBF operation then finds and uses such x_j that can come to a close vicinity of the global minimum at dimension j on a macroscopic scale, so that the native PSO process can then have a higher chance of avoiding those noise-like local optima, and thus eventually converge to the global optimum.

We use the termination criteria as the combination of the maximum number of iterations allowed (iterNo = 5,000) and the cut-off error ($\varepsilon_C = 10^{-4}$). Table 5.3 also presents both positional, $\pm x_{max}$, and dimensional $[D_{min}, D_{max}]$ range values, whereas the other parameters are empirically set as $V_{max} = x_{max}/2$ and $VD_{max} = 18$. Unless stated otherwise, these range values are used in all experiments presented in this section. The first set of experiments was performed for comparative evaluation of the standalone MD PSO versus *bPSO* over both uni- and multimodal functions. Figure 5.2 presents typical plots where both techniques are applied over the unimodal function, *De Jong* using the swarm size, $S = 160$. The red curves of both plots in Fig. 5.2 and all the rest of the figures in this section represent the behavior of the GB particle (whether it is a new *gbest* or the *aGB* particle created by FGBF) and the corresponding blue curves represent the behavior of the *gbest* particle when the termination criteria are met (e.g., gbest = 74 for *bPSO* and $f[y_{74}(158)] = 9.21 \times 10^{-5} < \varepsilon_C$). Naturally, the true dimension ($d_0 = 20$) is set in advance for the *bPSO* process and it converges to the global optima within 158 iterations as shown in the right plot, whereas MD PSO takes 700 iterations to finally place the GB particle at the target dimension ($d_0 = 20$) and then only 80 iterations more to satisfy the termination criteria. Recall that its objective is to find the global minimum of the function at the true dimension. Overall, the standalone MD PSO is slower compared to *bPSO*, but over an extensive set of experiments, its convergence behavior to the global optimum is found similar to that of the *bPSO*. For instance, their performance is degraded in

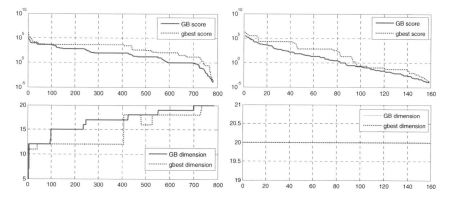

Fig. 5.2 Fitness score (*top* in log-scale) and dimension (*bottom*) plots vs. iteration number for MD PSO (*left*) and *bPSO* (*right*) operations both of which run over *De Jong* function

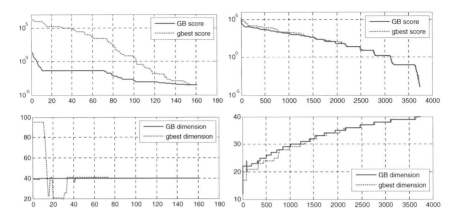

Fig. 5.3 Fitness score (*top* in log-scale) and dimension (*bottom*) plots vs. iteration number for a MD PSO run over *Sphere* function with (*left*) and without (*right*) FGBF

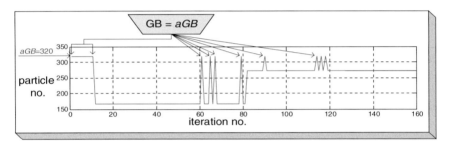

Fig. 5.4 Particle index plot for the MD PSO with FGBF operation shown in Fig. 5.3

higher dimensions, e.g., for the same function but at $d_0 = 50$, both require—on the average—five times more iterations to find the global minimum.

A significant speed improvement can be achieved when MD PSO is performed with FGBF. A typical MD PSO run using the swarm size, $S = 320$, over another unimodal function, *Sphere*, but at a higher (target) dimension, is shown in Fig. 5.3. Note that the one with FGBF (left) took only 160 iterations, whereas the stand-alone MD PSO (right) is completed within 3,740 iterations. Note also that within a few iterations, the process with FGBF already found the true dimension, $d_0 = 40$, and after only 10 iterations, the aGB particle already came in a close vicinity of the global minimum (i.e., $f(xy_{aGB}^{40}(10)) \cong 4 \times 10^{-2}$). As shown in Fig. 5.4, the particle index plot for this operation clearly shows the time instances where aGB (with index number 320) becomes the GB particle, e.g., the first 14 iterations and then occasionally in the rest of the process.

Besides the significant speed improvement for unimodal functions, the primary contribution of FGBF technique becomes most visible when applied over multi-modal functions where the *bPSO* (and the standalone MD PSO) are generally not

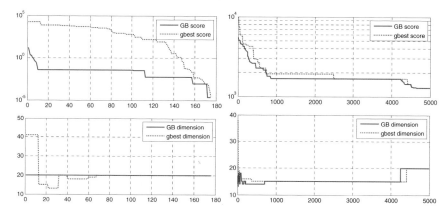

Fig. 5.5 Fitness score (*top* in log-scale) and dimension (*bottom*) plots vs. iteration number for a MD PSO run over *Schwefel* function with (*left*) and without (*right*) FGBF

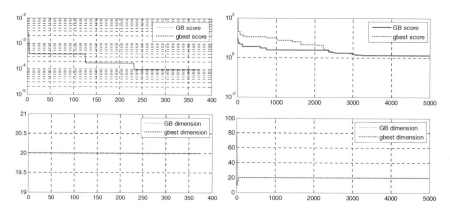

Fig. 5.6 Fitness score (*top* in log scale) and dimension (*bottom*) plots versus iteration number for a MD PSO run over *Giunta* function with (*left*) and without (*right*) FGBF

able to converge to the global optimum even at the low dimensions. Figures 5.5 and 5.6 present two (standalone MD PSO vs. MD PSO with FGBF) applications (using a swarm size 320) over *Schwefel* and *Giunta* functions at $d_0 = 20$. Note that when FGBF is used, MD PSO can directly have the *aGB* particle in the target dimension ($d_0 = 20$) at the beginning of the operation. Furthermore, the PSO process benefits from having an aGB particle that is indeed in a close vicinity of the global minimum. This eventually helps the swarm to move toward the *right* direction thereafter. Without this mechanism, both standalone PSO applications are eventually trapped into local minima due to the highly multimodal nature of these functions. This is quite evident in the right-hand plots of both figures, and except for few minority cases, it is also true for other multimodal functions. In higher dimensions, standalone MD PSO applications over multimodal functions

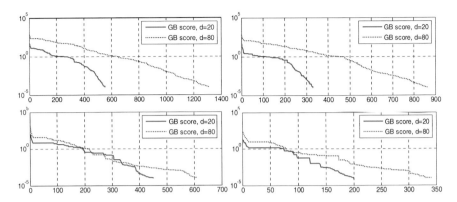

Fig. 5.7 MD PSO with FGBF operation over *Griewank* (*top*) and *Rastrigin* (*bottom*) functions with $d_0 = 20$ (*red*) and $d_0 = 80$ (*blue*) using the swarm size, $S = 80$ (*left*) and $S = 320$ (*right*)

yield even worse results such as earlier traps in local minima and possibly at the wrong dimension. For example, in standalone MD PSO operations over *Schwefel* and *Giunta* with $d_0 = 80$, the GB scores at $t = 4,999$ ($f(x\hat{y}^{80})$) are 8,955.39 and 1.83, respectively.

An observation worth mentioning here is that MD PSO with FGBF is usually affected by the higher dimensions, but its performance degradation usually entails a slower convergence as opposed to an entrapment at a local minimum. For instance, when applied over *Schwefel* and *Giunta* at $d_0 = 80$, the convergence to the global optima is still achieved after only a slight increase in the number of iterations, i.e., 119 and 484 iterations, respectively. Moreover, Fig. 5.7 presents the fitness plots for MD PSO with FGBF using two different swarm sizes over two more multimodal functions, *Griewank* and *Rastrigin*. Similar to earlier results, the global minimum at the true dimension is reached for both functions; however, for $d_0 = 20$ (red curves) the process takes a few hundreds iterations less than the one at $d_0 = 80$ (blue curves).

The swarm size has a direct effect on the performance of MD PSO with FGBF, that is, a larger swarm size increases the speed of convergence, which is quite evident in Fig. 5.7 between the corresponding plots on the left- and the right side. This is due to the fact that with the larger swarm size, the probability of having better swarm components (which are closer to the global optimum) of the *aGB* particle increases, thus yielding a better *aGB* particle formation in general. Note that this is also clear in the plots at both sides, i.e., at the beginning (within the first 10–15 iterations when *aGB* is usually the GB particle) the drop in the fitness score is much steeper on the right-hand plots with respect to the ones on the left.

For an overall performance evaluation, both proposed methods are tested over seven benchmark functions using three different swarm sizes (160, 320, and 640) and target dimensions (20, 50, and 80). For each setting, 100 runs are performed and the first- and second-order statistics (mean, μ and standard deviation, σ) of the operation time (total number of iterations) and the two components of the solution, the fitness score achieved in the resulting dimension (*dbest*), are presented in Table 5.4. During each run, the operation terminates when the fitness score drops below the cut-off error ($\varepsilon_C = 10^{-4}$) and it is assumed that the global minimum of the function in the target dimension is reached, henceforth, the score is set to 0 and obviously, dbest $= d_0$. Therefore, for a particular function, the target dimension, d_0 and swarm size, S, obtaining $\mu = 0$ as the average score means that the method converges to the global minimum in the target dimension at every run. On the other hand, having the average iteration number as 5,000 indicates that the method cannot converge to the global minimum at all, instead it gets trapped in a local minimum. The statistical results enlisted in Table 5.4 approve earlier observations and remarks about the effects of modality, swarm size, and dimension over the performance (both speed and accuracy). Particularly for the standalone MD PSO application, increasing the swarm size improves the speed of convergence wherever the global minimum is reached for unimodal functions, *Sphere* and *De Jong*, whilst reducing the score significantly for the others.

The score reduction is particularly visible on higher dimensions, e.g., for $d_0 = 80$, compare the highlighted average scores of the top five functions. Note that especially for *De Jong* at $d_0 = 50$, none of the standalone MD PSO runs with $S = 160$ converges to the global minimum whilst they all converge with a higher swarm population (i.e., 320 or 640).

Both dimension and modality have a direct effect on the performance of the standalone MD PSO. For unimodal functions, its convergence speed decreases with increasing dimension, e.g., see the highlighted average values of the iteration numbers for *Sphere* at $d_0 = 20$ versus $d_0 = 50$. For multimodal functions, regardless of the dimension and the swarm size, all standalone MD PSO runs get trap in local minima (except perhaps few runs on *Rosenbrock* at $d_0 = 20$); however, the fitness performance still depends on the dimension, that is, the final score tends to increase in higher dimensions indicating an earlier entrapment at a local minimum. Regardless of the swarm size, this can easily be seen in all multimodal functions except *Griewank* and *Giunta* both of which show higher modalities in lower dimensions. Especially, *Griewank* becomes a plain *Sphere* function when the dimensionality exceeds 20. This is the reason of the performance improvement (or score reduction) from $d_0 = 20$ to $d_0 = 50$ but note that the worst performance (highest average score) is still encountered at $d_0 = 80$.

As the entire statistics in the right side of Table 5.4 indicate, MD PSO with FGBF finds the global minimum at the target dimension for all runs over all functions regardless of the dimension, swarm size and modality, and without any exception. Moreover, the mutual application of MD PSO and FGBF significantly improves the convergence speed, e.g., compares the highlighted average iteration numbers with the results of the standalone MD PSO. Dimension, modality, and swarm size might still be important factors over the speed and have the same effects as mentioned earlier, i.e., the speed degrades with modality and dimensionality, whereas it improves with increasing swarm size. Their effects, however, vary significantly among the functions, e.g., as highlighted in Table 5.4, the swarm

Table 5.4 Statistical results from 100 runs over seven benchmark functions

Functions			Standalone MD PSO						MD PSO with FGBF					
			Score		Iter. No		dbest		Score		Iter. No		dbest	
	S	d_0	μ	σ	μ	σ	μ	σ	μ	σ	μ	σ	μ	σ
Sphere	160	20	**0**	0	**1,475**	117	20	0	0	0	**166**	27	20	0
		50	0	0	**4,605**	280	50	0	0	0	**172**	37	50	0
		80	**45.066**	61.23	4,659	1,046	78.5	1.234	0	0	169	34	80	0
	320	20	0	0	**1,024**	58	20	0	0	0	**133**	16	20	0
		50	0	0	**3,166**	631	50	0	0	0	**131**	25	50	0
		80	**5.949**	7.584	4,712	626	79	0.858	0	0	126	24	80	0
	640	20	0	0	**932**	161	20	0	0	0	**93**	15	20	0
		50	0	0	**2,612**	701	50	0	0	0	**101**	15	50	0
		80	**0.317**	0.462	4,831	343	79.7	0.47	0	0	95	15	80	0
De Jong	160	20	0	0	**1,047**	74	20	0	0	0	**6**	1	20	0
		50	0.705	0.462	5,000	0	49.3	0.47	0	0	18	28	50	0
		80	**5,184.4**	1,745.7	5,000	0	71.65	0.745	0	0	102	24	80	0
	320	20	0	0	**833**	66	20	0	0	0	**4**	1	20	0
		50	0	0	**4,004**	166	50	0	0	0	**6**	1	50	0
		80	**811.02**	323.37	5,000	0	74.85	0.587	0	0	19	20	80	0
	640	20	0	0	**643**	69	20	0	0	0	**2**	1	20	0
		50	0	0	**3,184**	145	50	0	0	0	**4**	1	50	0
		80	**67.452**	24.75	5,000	0	77.3	0.47	0	0	7	1	80	0

(continued)

Table 5.4 (continued)

Functions			Standalone MD PSO						MD PSO with FGBF					
			Score		Iter. No		dbest		Score		Iter. No		dbest	
Rosenbrock	160	20	0.0008	0.0004	4,797	903	20	0	0	0	1,619	1,785	20	0
		50	0.009	0.003	5,000	0	50	0	0	0	398	149	50	0
		80	**7.617**	7.834	500	0	78.75	0.786	0	0	367	98	80	0
	320	20	0.001	0.002	4,317	1,221	20	0	0	0	325	239	20	0
		50	0.009	0.004	5,000	0	50	0	0	0	257	33	50	0
		80	**0.49**	0.495	5,000	0	79.65	0.587	0	0	258	47	80	0
	640	20	0.0009	0.002	4,560	982	20	0	0	0	160	37	20	0
		50	0.006	0.003	5,000	0	50	0	0	0	193	58	50	0
		80	**0.018**	0.005	5,000	0	80	0	0	0	189	23	80	0
Rastrigin	160	20	2.137	0.565	5,000	0	19.35	0.49	0	0	**304**	40	20	0
		50	24.051	5.583	5,000	0	48.2	0.41	0	0	**392**	49	50	0
		80	**212.375**	165.196	5,000	0	77.15	1.871	0	0	**375**	55	80	0
	320	20	1.506	0.531	5,000	0	19.35	0.489	0	0	228	21	20	0
		50	9.788	4.181	5,000	0	48.95	0.224	0	0	277	47	50	0
		80	**77.317**	31.909	5,000	0	77.45	0.605	0	0	272	65	80	0
	640	20	0.96	0.527	5,000	0	19.501	0.513	0	0	**164**	15	20	0
		50	7.308	2.581	5,000	0	49.3	0.657	0	0	**206**	30	50	0
		80	**28.709**	8.399	5,000	0	78.3	0.933	0	0	**195**	47	80	0
Griewank	160	20	3.262	14.485	4,163	1,486	19.101	4.025	0	0	438	78	20	0
		50	0.232	0.271	4,779	673	49.2	0.696	0	0	725	36	50	0
		80	**9.499**	5.731	5,000	0	73.9	2.435	0	0	1,029	57	80	0
	320	20	0.019	0.016	4,219	1,601	20	0	0	0	395	76	20	0
		50	0.007	0.009	4,537	454	50	0	0	0	618	34	50	0
		80	**3.059**	1.857	5,000	0	76.45	1.099	0	0	866	43	80	0
	640	20	0.017	0.017	4,205	1,638	20	0	0	0	325	74	20	0
		50	0.016	0.02	4,306	877	50	0	0	0	531	37	50	0
		80	**0.675**	0.385	5,000	0	78.25	0.55	0	0	710	44	80	0

(continued)

Table 5.4 (continued)

Functions			Standalone MD PSO					MD PSO with FGBF				
			Score		Iter. No	dbest		Score	Iter. No		dbest	
Schwefel	160	20	1,432.3	336.76	5,000	16.30	1.418	0	215	33	20	0
		50	6,036.2	597.33	5,000	45.45	1.234	0	199	29	50	0
		80	11,304	1,489.8	5,000	74.7	2.226	0	161	35	80	0
	320	20	1,261.7	320.709	5,000	16.20	1.508	0	168	36	20	0
		50	5,288.8	576.132	5,000	45.55	1.394	0	146	24	50	0
		80	8,882.8	1,434.3	5,000	75.8	1.814	0	120	22	80	0
	640	20	946.612	264.635	5,000	16.85	1.268	0	121	21	20	0
		50	4,412.2	921.062	5,000	45.2	2.067	0	103	15	50	0
		80	8,032.1	1,180.9	5,000	75.2	2.447	0	85	12	80	0
Giunta	160	20	1.488	0.69	5,000	20	0	0	**793**	626	20	0
		50	0.776	0.207	5,000	50	0	0	**699**	281	50	0
		80	1.131	0.14	5,000	80	0	0	**863**	293	80	0
	320	20	1.003	0.582	5,000	20	0	0	128	92	20	0
		50	0.543	0.127	5,000	50	0	0	283	113	50	0
		80	1.140	0.756	5,000	80	0	0	456	115	80	0
	640	20	0.675	0.517	5,000	20	0	0	**1**	1	20	0
		50	0.418	0.140	5,000	50	0	0	**4**	4	50	0
		80	0.789	0.145	5,000	80	0	0	**20**	6	80	0

size can enhance the speed radically for *Giunta* but only merely for *Griewank*. The same statement can be made concerning the dimension of *De Jong* and *Sphere*.

Based on the results in Table 5.4, we can perform comparative evaluations with some of the promising PSO variants such as [1, 30–32] where similar experiments were conducted over some or all of these benchmark functions. They have, however, the advantage of fixed dimension, whereas MD PSO with FGBF finds the true dimension as part of the optimization process. Furthermore, it is rather difficult to make speed comparisons since none of them really find the global minimum for most functions; instead, they have demonstrated some incremental performance improvements in terms of score reduction with respect to some other competing technique(s). For example in Angeline [32], a tournament selection mechanism is formed among particles and the method is applied over four functions (*Sphere, Rosenbrock, Rastrigin,* and *Griewank*). Although the method is performed over a reduced positional range, ±15, and at low dimensions (10, 20, and 30), they got varying average scores between the range {0.3, 1,194}. As a result, both better and worse performances than the *bPSO* were reported depending on the function. In Esquivel and Coello Coello [30], *bPSO* and two PSO variants, *GCPSO* and mutation-extended PSO over three neighborhood topologies were applied to some common multimodal functions, *Rastrigin, Schwefel,* and *Griewank*. Although the dimension is rather low (30), none of the topologies over any PSO variant converged to the global minimum and average scores varying in the range of {0.0014, 4,762} were reported. In Riget and Vesterstrom [1], a diversity guided PSO variant, *ARPSO*, along with two competing methods, *bPSO* and *GA* were applied over the multimodal functions *Rastrigin, Rosenbrock,* and *Griewank*, at three different dimensions (20, 50, and 100). The range was kept quite low for *Rosenbrock* and *Rastrigin*, ±100 and ±5.12, respectively, and for each run; the number of evaluations (product of iterations and the swarm size) was kept in the range of 400,000–2,000,000, depending on the dimension. The experimental results have shown that none of the three methods converged to the global minimum except ARPSO over (only) *Rastrigin* at dimension 20. Only when ARPSO runs until stagnation is reached after 200,000 evaluations, it can find the global minimum over *Rastrigin* at higher dimensions (50 and 100). However, in practical sense, this indicates that the total number of iterations might be in the magnitude of 10^5 or even higher. Recall that the number of iterations required for MD PSO with FGBF to convergence to the global minimum is less than 400 for any dimension. *ARPSO* performed better than *bPSO* and *GA* over *Rastrigin* and *Rosenbrock* but worse over *Griewank*. The CPSO proposed in Bergh and Engelbrecht [24] was applied over five functions of which four are common (*Sphere, Rastrigin, Rosenbrock,* and *Griewank*). The dimension of all functions was fixed to 30 and in this dimension, CPSO performed better than *bPSO* in 80 % of the experiments. Finally in Richards and Ventura [31], dynamic sociometries via *ring* and *star* were introduced among the swarm particles and the performance of various combinations of swarm size and sociometry over six functions (the ones used in this section except *Schwefel*) was reported. Although the tests were performed over comparatively reduced positional ranges and at a low dimension (30), the experimental results indicate that none of the sociometry and swarm size combination converged to the global minimum for multimodal functions except only for some dimensions of the *Griewank* function.

5.2 Optimization in Dynamic Environments

Particle swarm optimization was proposed as an optimization technique for static environments; however, many real problems are dynamic, meaning that the environment and the characteristics of the global optimum can change in time. In this section, we shall first adapt PSO with FGBF over multimodal, nonstationary environments. To establish the follow up of local optima, we will also introduce a multiswarm algorithm, which enables each swarm to converge to a different optimum and use FGBF technique, distinctively. Finally for the multidimensional dynamic environments where the optimum dimension itself changes in time, we utilize MD PSO, which eventually pushes the frontier of the optimization problems in dynamic environments toward a global search in a multidimensional space, where there exists a multimodal problem possibly in each dimension. We shall present both the standalone MD PSO and the mutual application of MD PSO and FGBF over the moving peaks benchmark (MPB), which originally simulates a dynamic environment in a unique (fixed) dimension. MPB is appropriately extended to accomplish the simulation of a multidimensional dynamic system, which contains dynamic environments active in multiple dimensions.

5.2.1 Dynamic Environments: The Test Bed

Conceptually speaking, MPB developed by Branke in [33] is a simulation of a configurable dynamic environment changing over time. The environment consists of a certain number of peaks with varying locations, heights, and widths. The dimension of the fitness function is fixed in advance and thus is an input parameter of the benchmark. An N-dimensional fitness function with m peaks is expressed as:

$$F(\vec{x},\ t) = \max\left(B(\vec{x}),\ \max_{p=1...m}\ P(\vec{x},\ h_p(t),\ w_p(t),\ \vec{c}_p(t))\right) \qquad (5.1)$$

where $B(\vec{x})$ is time invariant basis landscape, whose utilization is optional, and P is the function defining the height of the pth peak at location $\vec{x}$, where each of the m peaks can have its own dynamic parameters such as height, $h_p(t)$, width, $w_p(t)$ and location vector of the peak center, $\vec{c}_p(t)$. Each peak parameter can be initialized randomly or set to a certain value and after a time period (number of evaluations), T_e, at a time (evaluation) t, a change over a single peak, p, can be defined as follows:

$$
\begin{aligned}
h_p(t) &= h_p(t - T_e) + r_1 \Delta h \\
w_p(t) &= w_p(t - T_e) + r_2 \Delta w \\
\vec{c}_p(t) &= \vec{c}_p(t - T_e) + \vec{v}_p(t)
\end{aligned}
\tag{5.2}
$$

where r_1, $r_2 \in N(0, 1)$, Δh and Δw are the height and width severities and $\vec{v}_p(t)$ is the normalized shift vector, which is a linear combination of a random vector and the previous shift vector, $\vec{v}_p(t - T_e)$. The type and number of peaks along with their initial heights and widths, environment (search space) dimension and size, change severity, level of change randomness, and change frequency can be defined [33]. To allow comparative evaluations among different algorithms, three standard settings of such MPB parameters, the so-called *Scenarios*, have been defined. *Scenario 2* is the most widely used. Each scenario allows a range of values, among them the following are commonly used: number of peaks $= 10$, change severity vlength $= 1.0$, correlation lambda $= 0.0$, and peak change frequency $= 5,000$. In *Scenario 2*, no basis landscape is used and peak type is a simple *cone* with the following expression

$$
P(\vec{x}, h_p(t), w_p(t), \vec{c}_p(t)) = h_p(t) - s_p(t) \left\| \vec{x} - \vec{c}_p(t) \right\| \quad \text{where}
$$

$$
s_p(t) = \frac{h_p(t)}{w_p(t)} \quad \text{and} \quad \left\| \vec{x} - \vec{c}_p(t) \right\| = \sqrt{\sum_{i=1}^{N} (x_i - c_{pi})^2} \quad \forall x_i \in \vec{x}, \forall c_{pi} \in \vec{c}_p(t)
$$

$$
\tag{5.3}
$$

where $s_p(t)$ is the slope and $\|\,.\,\|$ is the *Euclidean* distance. More detailed information on MPB and the rest of the parameters used in this benchmark can be obtained from Branke [33].

5.2.2 Multiswarm PSO

The main problem of using the basic PSO algorithm in a dynamic environment is that eventually the swarm will converge to a single peak—whether global or local. When another peak becomes the global maximum as a result of an environmental change, it is likely that the particles keep circulating close to the peak to which the swarm has converged and thus they cannot find the new global maximum. Blackwell and Branke have addressed this problem in [34] and [35] by introducing multiswarms that are actually separate PSO processes. Recall from Sect. 3.3.2 that each particle is now a member of one of the swarms only and it is unaware of other swarms. Hence in this problem domain, the main idea is that each swarm can converge to a separate peak. Swarms interact only by mutual repulsion that keeps them from converging to the same peak. For a single swarm it is essential to maintain enough diversity, so that the swarm can track small location changes of the peak to which it is converging. For this purpose Blackwell and Branke introduced charged and quantum swarms, which are analogs to an atom having a nucleus and charged particles randomly orbiting it. The particles in the nucleus take care of the fine tuning of the result while the charged particles are responsible of detecting the position changes. However, it is clear that, instead of charged or quantum swarms, some other method can also be used to ensure sufficient diversity among particles of a single swarm, so that the peak can be tracked despite of small location changes.

As one might expect, the best results are achieved when the number of swarms is set equal to the number of peaks. However, it is then required that the number of peaks is known beforehand. In [5], Blackwell presents self-adapting multiswarms, which can be created or removed during the PSO process, and therefore it is not necessary to fix the number swarms beforehand.

The repulsion between swarms is realized by simply reinitializing the worse of two swarms if they move within a certain range from each other. Using physical repulsion could lead to equilibrium, where swarm repulsion prevents both swarms from getting close to a peak. A proper proximity threshold, r_{rep} can be obtained by using the average radius of the peak basin, r_{bas}. If p peaks are evenly distributed in N-dimensional cube, X^N, then $r_{\text{rep}} = r_{\text{bas}} = X/p^{1/N}$.

5.2.3 FGBF for the Moving Peak Benchmark for MPB

The previous section introduced how FGBF process works within a $bPSO$ at a fixed dimension and referred to some applications in other domains representing static environments. However, in dynamic environments, this approach eventually leads the swarm to converge to a single peak (whether global or local), and therefore it may lose its ability to track other peaks. As any of the peaks can become the optimum peak as a result of environmental changes, $bPSO$ equipped with FGBF is likely to lead to suboptimal solutions. This is the basic motivation of using multiswarms along with the FGBF operation. As described earlier, mutual repulsion between swarms is applied, where the distance between the swarms' global best locations is used to measure the distance between two swarms. Instead of using charged or quantum swarms, FGBF is sufficient to collect the necessary diversity and thus enable peak tracking if the peaks' locations are changed. Particle velocities are also reinitialize after each environment change to enhance diversity.

Each particle with index a in a swarm ξ, represents a potential solution and therefore, the jth component of an N-dimensional point $(x_j, j \in \{1, N\})$ is stored in its positional component, $x_{a,j}(t)$, at a time t. The aim of the PSO process is to search for the global optimum point, which maximizes $P(\vec{x}, h_p(t), w_p(t), \vec{c}_p(t))$, in other words, finding the global (highest) peak in MPB environment. Recall that in *Scenario 2* of MPB the peaks used are all in *cone* shape, as given in Eq. (5.3).

Since in Eq. (5.3), $h_p(t)$ and $s_p(t)$ are both set by MPB, finding the highest peak is equivalent to minimizing the $\left\| \vec{x} - \vec{c}_p(t) \right\|$ term, yielding $f(a, j) = -\left(x_j - c_{pj} \right)^2$. Step 3.1 in *bPSO*'s pseudo-code computes the (dimensional) fitness scores $(f(a, j), f(\text{gbest}, j))$ of the *j*th components $(x_{a, j}, y_{\text{gbest}, j})$ and in step 1 of the FGBF process, the dimensional component yielding maximum $f(a, j)$ is then placed in *aGB*. In step 3, these dimensional components are replaced by dimensional components of the personal best position of the *gbest* particle, if they yield higher dimensional fitness scores. We do not expect that dimensional fitness scores can be evaluated with respect to the optimum peak, since this requires the a priori knowledge of the global optimum, instead we use either the *current* peak where the particle resides on or the peak to which the swarm is converging (swarm peak). We shall thus consider and evaluate both *modes* separately.

5.2.4 Optimization over Multidimensional MPB

For testing the multidimensional optimization technique Branke's MPB is extended to a multidimensional version, in which there exists multiple search space dimensions within a dimensional range $\{D_{\min}, D_{\max}\}$, and the optimal dimension changes over time in addition to the dynamic nature of the conventional MPB. Peak locations in different dimensions share the common coordinates in the fitness function, and thus such an extension further allows exploitation of the information gathered in other search space dimensions.

The multidimensional extension of the MPB is straightforward. The initialization and changes of peak locations must now be done in the highest possible search space dimension, $D_{\max}$. Locations in other dimensions can be obtained simply by leaving out the unused coordinates (nonexisting dimensions). The optimal dimension is chosen randomly every time the environment is changed. Therefore, the fitness function with *m* peaks in a multidimensional environment can be expressed as:

$$F(\vec{x}^d, t) = \max\left(B(\vec{x}^d),\ \max_{p=1...m}\ P\left(\vec{x}^d, d, h_p(t), w_p(t), \vec{c}_p^d(t) \right) \right) \qquad (5.4)$$

where $d \in \{D_{\min}, D_{\max}\}$ is the dimension of position $\vec{x}^d$ and $\vec{c}_p^d(t)$ refers to first *d* coordinates (dimensions) of the peak center location. A *cone* peak is now expressed as follows:

$$P\left(\vec{x}^d, h_p(t), w_p(t), \vec{c}_{pd}(t) \right) = h_p(t) - s_p(t) * \left\| \vec{x}^d - \vec{c}_p^d(t) \right\| / d - (D_{\text{opt}} - d)^2 \quad \text{where}$$

$$s_p(t) = \frac{h_p(t)}{w_p(t)} \quad \text{and} \quad \left\| \vec{x}^d - \vec{c}_p^d(t) \right\| = \sqrt{\sum_{i=1}^d \left(x_i^d - c_{pi}^d \right)^2} \quad \forall x_i^d \in \vec{x}^d, \forall c_{pi}^d \in \vec{c}_p^d(t)$$

$$(5.5)$$

where D_{opt} is the current optimal dimension. Compared with Eq. (5.3), now for all nonoptimal dimensions a penalty term $(D_{opt} - d)^2$ is subtracted from the whole environment. In addition, the peak slopes are scaled by the term $1/d$. The purpose of this scaling is to prevent the benchmark from favoring lower dimensions. Otherwise, a solution, whose coordinates each differs from the optimum by 1.0 would be a better solution in a lower dimension as the *Euclidian* distance is used.

Similar to the unidimensional (PSO) case, each positional component $xx_a^d(t)$ of a MD PSO particle represents a potential solution in dimension d. The only difference is that now the dimension of the optimal solution is not known beforehand, but it can vary within the defined range. Even a single particle can provide potential solutions in different dimensions as it makes interdimensional passes as a result of MD PSO process. This dynamic multidimensional optimization algorithm combines multiswarms and FGBF with MD PSO. As in the different dimensions, the common coordinates of the peak locations are the same, it does not seem purposeful for two swarms to converge to a same peak in different dimensions. Therefore, the mutual repulsion between swarms is extended to affect swarms that are in different dimensions. Obviously, only the common coordinates are considered when the swarm distance is computed.

FGBF naturally exploits information gathered in other dimensions. When the *aGB* particle is created, FGBF algorithm is not limited to use dimensional components from only those particles which are in a certain dimension, but it can combine dimensional coordinates of particles in different dimensions. Note that as we still use the dimensional fitness score, $f(a, j) = -(x_j - c_{pj})^2$, the common coordinates of the positional components of the *aGB* particle created in different search space dimensions, $d \in \{1, D_{max}\}$, shall be the same. In other words, it is not necessary to create the positional components of the *aGB* particle from scratch in every search space dimension, $d \in \{1, D_{max}\}$, instead one (new) coordinate (dimension) to the *aGB* particle is created and added. Note also that it is still possible that in some search space dimensions *aGB* beats the native *gbest* particle, while in other dimensions it does not. In multidimensional version also the dimension and dimensional velocity of each particle are reinitialized after an environmental change in addition to the particle velocities in each dimension.

5.2.5 Performance Evaluation on Conventional MPB

We conducted an exhaustive set of experiments over the MPB *Scenario 2* using the settings given earlier. In order to investigate the effect of multiswarm settings, different numbers of swarms and particles in a swarm are used. Both FGBF modes are applied using the current and swarm peaks. In order to investigate how FGBF and multiswarms individually contribute to the results, experiments with each of them performed separately will also be presented.

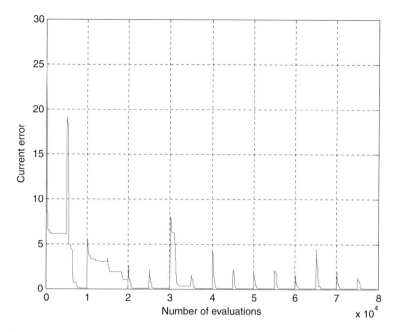

Fig. 5.8 Current error at the beginning of a run

Figure 5.8 presents the current error plot, which shows the difference between the global maximum and the current best result during the first 80,000 function evaluations, when 10 swarms, each with four particles, are used and the swarm peak mode is applied for the FGBF operation. It can be seen from the figure that as the environment changes after every 5,000 evaluations, it causes the results to temporarily deteriorate. However, it is clear that after these environment changes, the results improve (i.e., the error decreases quite rapidly), which shows the benefit of tracking the peaks instead of randomizing the swarm when a change occurs. The figure also reveals other typical behavior of the algorithm. First of all, after the first few environmental changes, the algorithm does not behave as well as at later stages. This is because the swarms have not yet converged to a peak. Generally, it is more difficult to initially converge to a narrow or low peak than to keep tracking a peak that becomes narrow and/or low. It can also be seen that typically the algorithm gets close to the optimal solution before the environment is changed again. In few cases, where the optimal solution is not found, the algorithm has been unable to keep a swarm tracking that peak, which is too narrow.

In Figs. 5.9 and 5.10, the contributions of multiswarms with FGBF are demonstrated. The algorithm is run on MPB applying the same environment changes, first with both using multiswarms and FGBF, then without multiswarms and finally without FGBF. Same settings are used as before. Without multiswarms, the number of particles is set to 40 to keep the total number of particles unchanged.

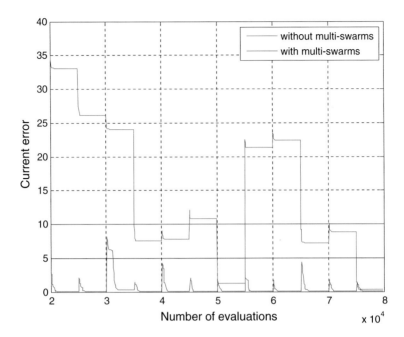

Fig. 5.9 Effect of multi-swarms on results

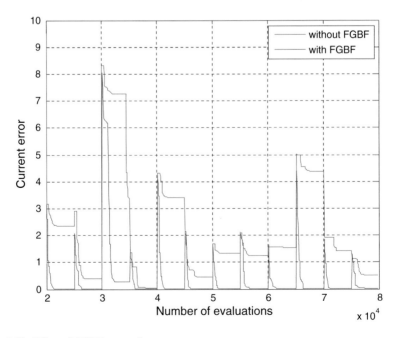

Fig. 5.10 Effect of FGBF on results

Table 5.5 Best known results on the MPB

Source	Base algorithm	Offline error
Blackwell and Branke [34]	PSO	2.16 ± 0.06
Li et al. [36]	PSO	1.93 ± 0.06
Mendes and Mohais [37]	Differential evolution	1.75 ± 0.03
Blackwell and Branke [35]	PSO	1.75 ± 0.06
Moser and Hendtlass [38]	Extremal optimization	0.66 ± 0.02

As expected, the results without multiswarms are significantly deteriorated due to the aforementioned reasoning. When the environment is changed, the highest point of the peak to which the swarm is converging can be found quickly, but that can provide good results only when that peak happens to be the global optimum. When multiswarms are used, but without using the FGBF, it is clear that the algorithm can still establish some kind of follow-up of peaks as the results immediately after environment changes are only slightly worse than with FGBF. However, if FGBF is not used, the algorithm can seldom find the global optimum. Either there is no swarm converging to the highest peak or the peak center just cannot be found fast enough.

The reported performance of five methods from the literature using the same, MPB with the same settings is listed in Table 5.5.

The best results were achieved by the extremal optimization algorithm [38]; however, this algorithm is specifically designed *only* for MPB and its applicability to other practical dynamic problems is not clear. The best results by a PSO-based algorithm were achieved by Blackwell and Branke's multiswarm algorithm described earlier. For evaluation and comparison purposes, the performance results of multiple swarms with FGBF in terms of the offline error are listed in Table 5.6. Each result given is the average of 50 runs, where each run consists of 500,000 function evaluations. As expected the best results are achieved when 10 swarms are used. Four particles in a swarm turned out to be the best setting. Between the

Table 5.6 Offline error using *Scenario* 2

No. of swarms	No. of particles	Swarm peak	Current peak
10	2	1.81 ± 0.50	2.58 ± 0.55
10	3	1.22 ± 0.43	1.64 ± 0.53
10	4	**1.03 ± 0.35**	**1.37 ± 0.50**
10	5	1.19 ± 0.32	1.52 ± 0.44
10	6	1.27 ± 0.41	1.59 ± 0.57
10	8	1.31 ± 0.43	1.61 ± 0.45
10	10	1.40 ± 0.39	1.70 ± 0.55
8	4	1.50 ± 0.41	1.78 ± 0.57
9	4	1.31 ± 0.54	1.66 ± 0.54
11	4	1.09 ± 0.35	1.41 ± 0.42
12	4	1.11 ± 0.30	1.46 ± 0.43

two FGBF modes, better results are obtained when the swarm peak is used as the peak $\vec{c}_p(t)$.

5.2.6 Performance Evaluation on Multidimensional MPB

The extended multidimensional MPB uses similar settings as in the case of conventional MPB (at fixed dimension) except that the change frequency is set to 15,000. The search space dimension range used is $d \in [5, 15]$. Figure 5.11 shows how the global optimal dimension changes over time and how MD PSO is tracking these changes. The current best dimension represents the dimension, where the best solution is achieved among all swarms' *dbest* dimensions. Ten multiswarms are used with seven particles in each. FGBF is used with the swarm peak mode. It can be seen that the algorithm always finds the optimal dimension, even though the difference in peaks heights between the optimal dimension and its neighbor dimensions is quite insignificant (=1) compared to the peak heights (30–70). Figure 5.12 shows how the current error behaves during the first 250,000 evaluations, when the same settings are used. It can be seen that the algorithm behavior is similar to the unidimensional case, but now the initial phase, when the algorithm has not been yet behaving at its best is longer. Similarly, it takes a longer time to

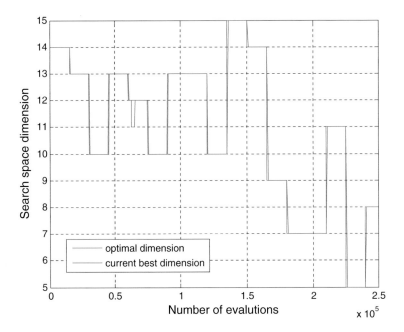

Fig. 5.11 Optimum dimension tracking in a MD PSO run

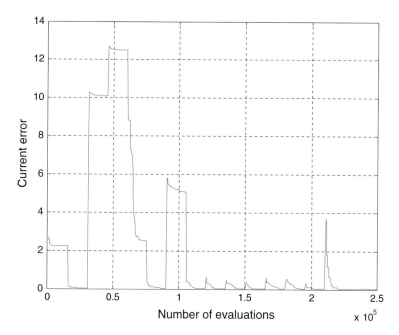

Fig. 5.12 Current error at the beginning of a MD PSO run

regain the optimal behavior if the follow up of some peaks is lost for some reason (it is, for example, possible that higher peaks hide other lower peaks under them).

Figures 5.13 and 5.14 illustrate the effect of using multiswarms on the performance of the algorithm. Without multiswarms the number of particles is set to 70. Figure 5.13 shows that a single swarm can also find the optimal dimension easily; however, as in the unidimensional case, without use of multiswarms, the optimal peak can be found only if it happens to be the peak to which the swarm is converging. This can be seen in Fig. 5.14. During the initial phase of the multiswarm algorithm results with and without multiswarms are similar. This indicates that both algorithms initially converge to the same peak (highest) and as a result of the first few environmental changes some peaks that are not yet discovered by multiswarms become the highest.

Figures 5.15 and 5.16 illustrate the effect of FGBF on the performance of the algorithm. In Fig. 5.15, it can be seen that without FGBF the algorithm has severe problems in tracking the optimal dimension. In this case, it loses the benefit of exploiting the natural diversity among the dimensional components and it is not able to exploit information gathered from other dimensions. Therefore, even if some particles visit the optimal dimension, they cannot track the global peak *fast* enough that would hence surpasses the best results in other dimensions. Therefore, the algorithm gets trapped in some suboptimum dimension where it happens to find the best solution in an early phase. Such reasons also cause the current error to be generally higher without FGBF, as can be seen in Fig. 5.16.

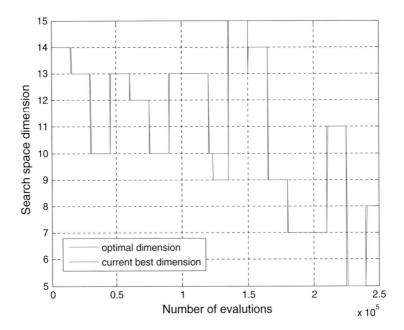

Fig. 5.13 Optimum dimension tracking without multi-swarms in a MD PSO run

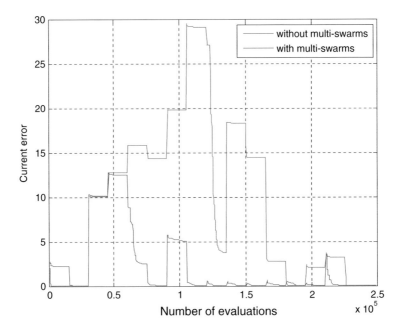

Fig. 5.14 Effect of multi-swarms on the performance

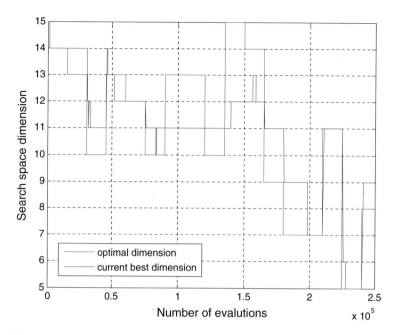

Fig. 5.15 Optimum dimension tracking *without* FGBF in a MD PSO run

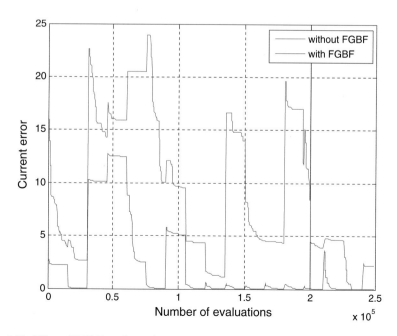

Fig. 5.16 Effect of FGBF on the performance

Table 5.7 Offline error on extended MPB	No. of swarms	No. of particles	Swarm peak	Current peak
	10	4	2.01 ± 0.98	**3.29 ± 1.44**
	10	5	1.77 ± 0.83	3.41 ± 1.69
	10	6	1.79 ± 0.98	3.64 ± 1.60
	10	7	**1.69 ± 0.75**	3.71 ± 1.74
	10	8	1.84 ± 0.97	4.21 ± 1.83
	10	10	1.96 ± 0.94	4.20 ± 2.03
	8	7	1.79 ± 0.91	3.72 ± 1.86
	9	7	1.83 ± 0.84	4.30 ± 2.15
	11	7	1.75 ± 0.91	3.52 ± 1.40
	12	7	2.03 ± 0.97	4.01 ± 1.97

The numerical results in terms of offline errors are given in Table 5.7. Each result given is the average of 50 runs, where each run consists of 500,000 function evaluations. As in the unimodal case, the best results are achieved when the number of swarms is equal to the number of peaks, which is 10. Interestingly, when the swarm peak *mode* is used the optimal number of particles becomes seven while with current peak mode it is still four. Note that these results cannot be directly compared with the results on conventional MPB since the objective function of multidimensional MPB is somewhat different.

Overall, MD PSO operation with FGBF and with multiswarms fundamentally upgrades the particle structure and the swarm guidance, both of which accomplish substantial improvements in terms of speed and accuracy for dynamic, multidimensional and multimodal environments.

5.3 Who Will Guide the Guide?

Let us recall the definition of the Merriam Webster dictionary for optimization, *the mathematical procedures (as finding the maximum of a function) involved in this.* More specifically, consider now the problem of finding a root θ^* (either minimum or maximum point) of the gradient equation: $g(\theta) \equiv \frac{\partial L(\theta)}{\partial \theta} = 0$ for some differentiable function $L : R^p \to R^1$. As discussed in Chap. 2, when g is defined and L is a unimodal function, there are powerful deterministic methods for finding the global θ^* such as traditional steepest descent and Newton–Raphson methods. However, in many real problems g cannot be observed directly and/or L is multimodal, in which case the aforementioned approaches may be trapped into some deceiving local optima. This brought the *era* of stochastic optimization algorithms, which can estimate the gradient and may avoid being trapped into a local optimum due to their stochastic nature. One of the most popular stochastic optimization techniques is stochastic approximation (SA), in particular the form that is called "gradient free" SA. Among many SA variants proposed by several researchers such as

Styblinski and Tang [39], Kushner [40], Gelfand and Mitter [41], and Chin [42], the one and somewhat different SA application is called simultaneous perturbation SA (SPSA) proposed by Spall in 1992 [43]. The main advantage of SPSA is that it often achieves a much more economical operation in terms of loss function evaluations, which are usually the most computationally intensive part of an optimization process.

As discussed earlier, PSO has a severe drawback in the update of its global best (*gbest*) particle, which has a crucial role of guiding the rest of the swarm. At any iteration of a PSO process, *gbest* is the most important particle; however, it has the poorest update equation, i.e., when a particle becomes *gbest*, it resides on its personal best position (*pbest*) and thus both *social* and *cognitive* components are nullified in the velocity update equation. Although it guides the swarm during the following iterations, ironically it lacks the necessary guidance to do so effectively. In that, if *gbest* is (likely to get) trapped in a local optimum, so is the rest of the swarm due to the aforementioned direct link of information flow. We have shown that an enhanced *guidance* achieved by FGBF alone is indeed sufficient in most cases to achieve global convergence performance on multimodal functions and even in high dimensions. However, the underlying mechanism for creating the *aGB* particle, the so-called FGBF, is not generic in the sense that it is rather problem dependent, which requires (the estimate of) individual dimensional fitness scores. This may be quite hard or even not possible for certain problems.

In order to address this drawback efficiently, in this section we shall present two approaches, one of which moves *gbest* efficiently or simply put, "guides" it with respect to the function (or error surface) it resides on. The idea behind this is quite simple; since the velocity update equation of *gbest* is quite poor, SPSA as a simple yet powerful search technique is used to *drive* it instead. Due to its stochastic nature, the likelihood of getting trapped into a local optimum further decreased and with the SA, *gbest* is driven according to (an approximation of) the gradient of the function. The second approach has a similar idea with the FGBF, i.e., an *aGB* particle is created by SPSA this time, which is applied over the personal best (*pbest*) position of the *gbest* particle. The *aGB* particle will then guide the swarm instead of *gbest* if it achieves a better fitness score than the (personal best position of) *gbest*. Note that both approaches only deal with the *gbest* particle and hence the internal PSO process remains as is. That is, neither of them is a PSO variant by itself; rather a solution for the problem of the original PSO caused by the poor *gbest* update. Furthermore, we shall demonstrate that both approaches have a negligible computational cost overhead, e.g., only few percent increase of the computational complexity, which can be easily compensated with a slight reduction either in the swarm size or in the maximum iteration number allowed. Both approaches of SA-driven PSO (SAD PSO) will be tested and evaluated against the basic PSO (*bPSO*) over several benchmark uni- and multimodal functions in high dimensions. Moreover, they are also applied to the multidimensional extension of PSO, the MD PSO technique.

5.3.1 SPSA Overview

Recall that there are two common SA methods: finite difference SA (FDSA) and simultaneous perturbation SA (SPSA). As covered in Sect. 2.3.2, FDSA adopts the traditional Kiefer–Wolfowitz approach to approximate gradient vectors as a vector of p partial derivatives where p is the dimension of the loss function. On the other hand, SPSA has all elements of $\widehat{\theta}_k$ perturbed simultaneously using only two measurements of the loss (fitness) function as

$$\widehat{g}_k(\widehat{\theta}_k) = \frac{L(\widehat{\theta}_k + c_k \Delta_k) - L(\widehat{\theta}_k - c_k \Delta_k)}{2c_k} \begin{bmatrix} \Delta_{k1}^{-1} \\ \Delta_{k2}^{-1} \\ . \\ . \\ . \\ \Delta_{kp}^{-1} \end{bmatrix} \tag{5.6}$$

where the p-dimensional random variable Δ_k is usually chosen as a *Bernoulli* ± 1 distribution and c_k is a scalar gain sequence satisfying certain conditions [43].

Spall in [43] presented conditions for convergence of SPSA (i.e., $\widehat{\theta}_k \to \theta^*$) and show that under certain conditions both SPSA and FDSA have the same convergence ability –yet SPSA needs only two measurements, whereas FDSA requires $2p$ measurements. This makes SPSA the natural choice for driving *gbest* in both approaches. Table 5.8 presents the general pseudo-code of the SPSA technique.

SPSA has five parameters as given in Table 5.8. Spall in [44] recommended to use values for A (the stability constant), α, and γ as 60, 0.602, and 0.101, respectively. However, he also concluded that "the choice of both gain sequences is critical to the performance of the SPSA as with all stochastic optimization algorithms and the choice of their respective algorithm coefficients". This especially makes the choice of the gain parameters a and c critical for a particular problem. For instance, Maryak and Chin in [45] proposed to set these parameters

Table 5.8 Pseudo-code for *SPSA* technique

SPSA (IterNo, a, c, A, α, γ)

1. Initialize $\widehat{\theta}_1$
2. For $\forall k \in [1, \text{IterNo}]$do:
   2.1. Generate zero mean, p-dimensional perturbation vector: Δ_k
   2.2. Let $a_k = a/(A + k)^\alpha$ and $c_k = c/k^\gamma$
   2.3. Compute $L(\widehat{\theta}_k + c_k \Delta_k)$ and $L(\widehat{\theta}_k - c_k \Delta_k)$
   2.4. Compute $\widehat{g}_k(\widehat{\theta}_k)$ using Eq. (5.6)
   2.5. Compute $\widehat{\theta}_{k+1}$ using Eq. (2.9)
3. End for

---

with respect to the problem whilst keeping the other three (A, α, and γ) as recommended in Spall [44].

Maeda and Kuratani in [46] used SPSA with the *bPSO* in a hybrid algorithm called Simultaneous Perturbation PSO (SP-PSO) over a limited set of problems and reported some slight improvements over the *bPSO*. Both proposed SP-PSO variants involved the insertion of $\hat{g}_k\left(\hat{\theta}_k\right)$ directly over the velocity equations of all swarm particles with the intention of improving their local search capability. This may, however, present some drawbacks. First of all, performing SPSA at each iteration and for all particles will double the computational cost of the PSO, since SPSA will require an additional function evaluation at each iteration.[1] Secondly, such an abrupt insertion of SPSA's $\hat{g}_k\left(\hat{\theta}_k\right)$ term directly into the *bPSO* may degrade the original PSO workout, i.e., the collective swarm updates and interactions, and require an accurate scaling between the parameters of the two methods, PSO's and SPSA. Otherwise, it is possible for one technique to dominate the other, and hence their combination would not necessarily gain from the advantage of both. This is perhaps the reason of the limited success, if any, achieved by SP-PSO. As we discuss next and demonstrate its elegant performance experimentally, SPSA should not be combined with SP-PSO as such. A better alternative would be to use SPSA to guide only PSO's native guide, *gbest*.

5.3.2 SA-Driven PSO and MD PSO Applications

In this section, two distinct SAD PSO approaches are presented and applied only to *gbest* whilst keeping the internal PSO and MD PSO processes unchanged. Since SPSA and PSO are iterative processes, in both approaches to be introduced next, SPSA can easily be integrated into PSO and MD PSO by using the same iteration count (i.e., $t \equiv k$). In other words, at a particular iteration t in the PSO process, only the SPSA steps 2.1–2.5 in Table 5.8 are inserted accordingly into the PSO and MD PSO processes. The following subsections will detail each approach.

5.3.2.1 First SA-Driven PSO Approach: gbest Update by SPSA

In this approach, at each iteration, *gbest* particle is updated using SPSA. This requires the adaptation of the SPSA elements (parameters and variables) and integration of the internal SPSA part (within the loop) appropriately into the PSO

---

[1] In Maeda and Kuratani [46], the function evaluations are given with respect to the iteration number; however, it should have been noted that SP-PSO performs twice more evaluations than *bPSO* per iteration. Considering this fact, the plots therein show little or no performance improvement at all.

Table 5.9 Pseudo-code for the first *SA-driven PSO* approach

A1) *SA-driven PSO*Plugin(*termination criteria:{IterNo, ε_C,...}, V_{max}, a, c, A, α, γ*)

1. See Line 1 inTable 3.1
2. See Line 2 inTable 3.1
3. For $\forall t \in [1, IterNo]$ do:
  3.4. For $\forall a \in [1, S]$ do:
      3.4.1. if ($a = gbest$) then do:
          3.4.1.1.   *Let $k=t$, $\hat{\theta}_k = x_a(t)$ and L=f*
          3.4.1.2.   *Let $a_k = a/(A+k)^\alpha$ and $c_k = c/k^\gamma$*
          3.4.1.3.   *Compute $L(\hat{\theta}_k + c_k\Delta_k)$ and $L(\hat{\theta}_k - c_k\Delta_k)$*
          3.4.1.4.   *Compute $\hat{g}_k(\hat{\theta}_k)$ using Eq. (46)*
          3.4.1.5.   *Compute $x_a(t+1) = \hat{\theta}_{k+1}$ using Eq. (2.9)*
      3.4.2. Else do:
          3.4.2.1.   For $\forall j \in [1, N]$ do:
              3.4.2.1.1.   Compute $v_{a,j}(t+1)$ using Eq. (3.2)
              3.4.2.1.2.   If($|v_{a,j}(t+1)| > V_{max}$) then clamp it to $|v_{a,j}(t+1)| = V_{max}$
              3.4.2.1.3.   Compute $x_{a,j}(t+1)$ using Eq. (3.2)
          3.4.2.2.   End For.
  3.5. End For.
4. End For.

---

pseudo-code, as shown in Table 5.9. Note that such a "plug-in" approach will not change the internal PSO structure and only affects the *gbest* particle's updates. It only costs two extra function evaluations and hence at each iteration the total number of evaluations is increased from S to $S + 2$, where S is the swarm size.

Since the fitness of each particle's current position is computed within the PSO process, it is possible to further decrease this cost to only *one* extra fitness evaluation per iteration. Let $\hat{\theta}_k + c_k\Delta_k = x_a(t)$ in step 3.4.1.1. And thus $L(\hat{\theta}_k + c_k\Delta_k)$ is known a priori. Then naturally, $\hat{\theta}_k - c_k\Delta_k = x_a(t) - 2c_k\Delta_k$ which is the only (new) location where the (extra) fitness evaluation ($L(\hat{\theta}_k - c_k\Delta_k)$) has to be computed. Once the gradient $\left(\hat{g}_k, \left(\hat{\theta}_k\right)\right)$ is estimated in step 3.4.1.4, then the next (updated) location of the *gbest* will be: $x_a(t+1) = \hat{\theta}_{k+1}$. Note that the difference of this "low-cost" SPSA update is that $x_a(t+1)$ is updated (estimated) *not* from $x_a(t)$, but instead from $x_a(t) - c_k\Delta_k$.

This approach can easily be extended for MD PSO, which is a natural extension of PSO for multidimensional search within a given dimensional range, $d \in [D_{min}, D_{max}]$. The main difference is that in each dimension, there is a distinct *gbest* particle, *gbest*(d). So SPSA is applied individually over the position of each *gbest*(d) if it (re-) visits the dimension d, (i.e., $d = xd_{gbest}(t)$). Therefore, there can be a maximum of $2(D_{max} - D_{min} + 1)$ number of function evaluations, indicating a significant cost especially if a large dimensional range is used. However, this is a theoretical limit, which can *only* happen if gbest(i) $\neq$ gbest(j) for $\forall i, j \in$ $[D_{\min}, D_{\max}]$, $i \neq j$ and all particles visit the particular dimensions in which they are *gbest* (i.e., $xd_{\text{gbest}(d)}(t) = d$, $\forall t \in [1, \text{iterNo}]$). Especially in a wide dimensional range, note that this is highly unlikely due to the dimensional velocity, which makes particles move (jump) from one dimension to another at each iteration. It is straightforward to see that under the assumption of a uniform distribution for particles' movements over all dimensions within the dimensional range, SA-driven MD PSO too, would have the same cost overhead as the SAD PSO. Experimental results indicate that the practical overhead cost is only slightly higher than this.

5.3.2.2 Second SA-driven PSO Approach: aGB Formation by SPSA

The second approach replaces the native FGBF operation with the SPSA to create an *aGB* particle. SPSA is basically applied over the *pbest* position of the *gbest* particle. The *aGB* particle will then guide the swarm instead of *gbest* if it achieves a better fitness score than the (personal best position of) *gbest*. SAD PSO pseudo-code as given in Table 5.10 can then be plugged in between steps 3.3 and 3.4 of *bPSO* pseudo-code.

The extension of the second approach to MD PSO is also quite straightforward. In order to create an *aGB* particle, for all dimensions in the given range (i.e., $(\forall d \in [D_{\min}, D_{\max}])$) SPSA is applied individually over the personal best position of each *gbest*(d) particle and furthermore, the aforementioned competitive selection ensures that $xy_{\text{aGB}}^{d}(t)$, $\forall d \in [D_{\min}, D_{\max}]$ is set to the best of the $xx_{\text{aGB}}^{d}(t + 1)$ and $xy_{\text{aGB}}^{d}(t)$. As a result, SPSA creates one *aGB* particle providing (potential) GB solutions $(xy_{\text{aGB}}^{d}(t + 1)$, $\forall d \in [D_{\min}, D_{\max}])$ for all dimensions in the given dimensional range. The pseudo-code of the second approach as given in

Table 5.10 PSO Plug-in for the second approach

A2) SA-driven PSO Plug-in (ξ, a, c, A, α, γ)
1. Create a new *aGB* particle, $\{x_{\text{aGB}}(t + 1),\ y_{\text{aGB}}(t + 1)\}$
2. Let $k = t$, $\hat{\theta}_k = \hat{y}(t)$ and $L = f$
3. Let s and $c_k = c/k^{\gamma}$
4. Compute $L(\hat{\theta}_k + c_k\Delta_k)$ and $L(\hat{\theta}_k - c_k\Delta_k)$
5. Compute $\hat{g}_k(\hat{\theta}_k)$ using Eq. (5.6)
6. Compute $x_{\text{aGB}}(t) = \hat{\theta}_{k+1}$ using Eq. (2.9)
7. Compute $f(x_{\text{aGB}}(t + 1)) = L(\hat{\theta}_{k+1})$
8. If $(f(x_{\text{aGB}}(t + 1)) < f(y_{\text{aGB}}(t)))$ then $y_{\text{aGB}}(t + 1) = x_{\text{aGB}}(t + 1)$
9. Else $y_{\text{aGB}}(t + 1) = y_{aGB}(t)$
10. If $(f(y_{\text{aGB}}(t + 1)) < f(\hat{y}(t)))$ then $\hat{y}(t) = y_{\text{aGB}}(t + 1)$

Table 5.11 MD PSO Plug-in for the second approach

A2) *SA-driven MD PSO Plug-in* (ξ, *a*, *c*, *A*, α, γ)

1. Create a new *aGB* particle, $\{xx_{aGB}^d(t+1), xy_{aGB}^d(t+1)\}$ for $\forall d \in [D_{min}, D_{max}]$
2. For $\forall d \in [D_{min}, D_{max}]$ do:

    2.1. *Let* $k = t$, $\hat{\theta}_k = x\hat{y}^d(t)$ *and* $L = f$

    2.2. *Let* $a_k = a/(A+k)^{\alpha}$ *and* $c_k = c/k^{\gamma}$

    2.3. *Compute* $L(\hat{\theta}_k + c_k\Delta_k)$ *and* $L(\hat{\theta}_k - c_k\Delta_k)$

    2.4. *Compute* $\hat{g}_k(\hat{\theta}_k)$ *using* Eq. (5.6)

    2.5. *Compute* $xx_{aGB}^d(t+1) = \hat{\theta}_{k+1}V$ *using* Eq. (2.9)

    2.6. If $(f(xx_{aGB}^d(t+1)) < f(xy_{aGB}^d(t)))$ then $xy_{aGB}^d(t+1) = xx_{aGB}^d(t+1)$

    2.7. Else $xy_{aGB}^d(t+1) = xy_{aGB}^d(t)$

    2.8. If $(f(xy_{aGB}^d(t+1)) < f(xy_{gbest(d)}^d(t)))$ then $xy_{gbest(d)}^d(t) = xy_{aGB}^d(t+1)$

3. End For
4. Re-assign *dbest*: dbest = $\arg \min_{d \in [D_{min}, D_{max}]} (f(xy_{gbest(d)}^d(t)))$

Table 5.11 can then be plugged in between steps 3.2 and 3.3 of the MD PSO pseudo-code, given in Table 5.12.

Note that in the second SAD PSO approach, there are three extra fitness evaluations (as opposed to two in the first one) at each iteration. Yet as in the first approach, it is possible to further decrease the cost of SAD PSO by *one* (from three to two fitness evaluations per iteration). Let $\hat{\theta}_k + c_k\Delta_k = \hat{y}(t)$ in step 2 and thus $L(\hat{\theta}_k + c_k\Delta_k)$ is known a priori. Then *aGB* formation follows the same analogy as before and the only difference is that the *aGB* particle is formed not from $\hat{\theta}_k = \hat{y}(t)$ but from $\hat{\theta}_k = \hat{y}(t) - c_k\Delta_k$. However, in this approach a major difference in the computational cost may occur since in each iteration there are inevitably $3(D_{max} - D_{min})$ (or $2(D_{max} - D_{min})$ for lowcost application) fitness evaluations, which can be significant.

5.3.3 Applications to Non-linear Function Minimization

The same seven benchmark functions given in Table 5.3 are used in this section but without the dimensional terms (see their original form in Table 5.12). Recall that *Sphere*, *De Jong*, and *Rosenbrock* are unimodal functions and the rest are multimodal, meaning that they have many local minima. Recall further that on the macroscopic level *Griewank* demonstrates certain similarities with unimodal functions especially when the dimensionality is above 20; however, in low dimensions it bears a significant noise.

Table 5.12 Benchmark functions without dimensional bias

Function	Formula	Initial range	Dimension set: $\{d\}$
Sphere	$F_1(x, d) = \left(\sum_{i=1}^{d} x_i^2 \right)$	$[-150, 75]$	20, 50, 80
De Jong	$F_2(x, d) = \left(\sum_{i=1}^{d} i x_i^4 \right)$	$[-50, 25]$	20, 50, 80
Rosenbrock	$F_3(x, d) = \left(\sum_{i=1}^{d} 100 \left(x_{i+1} - x_i^2 \right)^2 + (x_i - 1)^2 \right)$	$[-50, 25]$	20, 50, 80
Rastrigin	$F_4(x, d) = \left(\sum_{i=1}^{d} 10 + x_i^2 - 10 \cos(2\pi x_i) \right)$	$[-500, 250]$	20, 50, 80
Griewank	$F_5(x, d) = \left(\frac{1}{4,000} \sum_{i=1}^{d} x_i^2 - \prod_{i=1}^{d} \cos\left(\frac{x_i}{\sqrt{i+1}} \right) \right)$	$[-500, 250]$	20, 50, 80
Schwefel	$F_6(x, d) = \left(418.9829\, d + \sum_{i=1}^{d} x_i \sin\left(\sqrt{|x_i|} \right) \right)$	$[-500, 250]$	20, 50, 80
Giunta	$F_7(x, d) = \left(\sum_{i=1}^{d} \sin\left(\frac{16}{15} x_i - 1 \right) + \sin^2\left(\frac{16}{15} x_i - 1 \right) + \frac{1}{50}\sin\left(4\left(\frac{16}{15} x_i - 1 \right) \right) + \frac{268}{1,000} \right)$	$[-500, 250]$	20, 50, 80

Table 5.13 Statistical results from 100 runs over seven benchmark functions

Functions	d	SPSA		bPSO		SAD PSO (A2)		SAD PSO (A1)	
		μ	σ	μ	σ	μ	σ	μ	σ
Sphere	20	0	0	0	0	0	0	0	0
	50	0	0	0	0	0	0	0	0
	80	0	0	135.272	276.185	0	0	0	0
De Jong	20	0.013	0.0275	0	0	0	0	0	0
	50	0.0218	0.03	9.0445	26.9962	0.0075	0.0091	0.2189	0.6491
	80	0.418	0.267	998.737	832.1993	0.2584	0.4706	13546.02	4,305.04
Rosenbrock	20	1.14422	0.2692	1.26462	0.4382	1.29941	0.4658	0.4089	0.2130
	50	3.5942	0.7485	15.9053	5.21491	12.35141	2.67731	2.5472	0.3696
	80	5.3928	0.7961	170.9547	231.9113	28.1527	5.1699	5.2919	0.8177
Rastrigin	20	204.9169	51.2863	0.0429	0.0383	0.0383	0.0369	0.0326	0.0300
	50	513.3888	75.7015	0.0528	0.0688	0.0381	0.0436	0.0353	0.0503
	80	832.9218	102.1792	0.7943	0.9517	0.2363	0.6552	0.1240	0.1694
Griewank	20	0	0	0	0	0	0	0	0
	50	$1.0631e + 007$	$3.3726e + 006$	50.7317	191.1558	0	0	3074.02	13.989
	80	$2.825e + 007$	$5.7896e + 006$	24.978	23.257	20.733	24.160	378.210	137,410
Schwefel	20	0.3584	0.0794	1.7474	0.3915	0.3076	0.0758	0.3991	0.0796
	50	0.8906	0.1006	10.2027	2.2145	0.8278	0.1093	0.9791	0.1232
	80	1.4352	0.1465	21.8269	5.1809	1.3633	0.1402	1.5528	0.1544
Giunta	20	42,743	667.2494	495.0777	245.1220	445.1360	264.1160	445.1356	249.5412
	50	10,724	1,027.6	4,257	713.1723	3,938.9	626.9194	3,916.2	758.3290
	80	17,283	1,247.9	9,873.6	1,313	8,838.2	1,357	8,454.2	1,285.3

Both approaches of the proposed SAD PSO along with the "low cost" application are tested over seven benchmark functions and compared with the *bPSO* and standalone SPSA application. The results are shown in Table 5.13. The same termination criteria as the combination of the maximum number of iterations allowed (iterNo = 10,000) and the cut-off error ($\varepsilon_C = 10^{-5}$) were used. Three dimensions (20, 50, and 80) for the sample functions are used in order to test the performance of each technique. PSO (*bPSO* and SAD PSO) used a swarm size, $S = 40$ and w was linearly decreased from 0.9 to 0.2. Also the values for A, α, γ, a, and c were set as recommended to 60, 0.602, 0.101, 1, and 1, for all functions. No parameter tuning was done on purpose for SPSA since it may not be feasible for many practical applications, particularly the ones where the underlying fitness surface is unknown. In order to make a fair comparison among SPSA, *bPSO*, and SAD-PSO, the number of evaluations is kept equal (so $S = 38$ and $S = 37$ are used for both SAD PSO approaches and the number of evaluations is set to $40 \times 10,000 = 4e+5$ for SPSA). For each function and dimension, 100 runs are performed and the first- and second-order statistics (mean, μ and standard deviation, σ) of the fitness scores are reported in Table 5.13 whilst the best statistics are highlighted. During each run, the operation terminates when the fitness score drops below the cut-off error and it is assumed that the global minimum of the function is reached, henceforth; the score is set to 0. Therefore, an average score $\mu = 0$ means that the method converges to the global minimum at every run.

As the entire statistics in the right side of Table 5.13 indicate, either SAD PSO approach achieves an equal or superior average performance statistics over all functions regardless of the dimension, modality, and without any exception. In other words, SAD PSO works equal or better than the best of *bPSO* and SPSA— even though either of them might have a quite poor performance for a particular function. Note especially that if SPSA performs well enough (meaning that the setting of the critical parameters, e.g., a and c is appropriate), then a significant performance improvement can be achieved by SAD PSO, i.e., see for instance De Jong, Rosenbrock, and Schwefel. On the other hand, if SPSA does not perform well, even much worse than any other technique, SAD PSO still outperforms *bPSO* to a certain degree, e.g., see Giunta and particularly Griewank for $d = 50$ where SAD PSO can still converge to the global optimum ($\mu = 0$) although SPSA performance is rather low. This supports the aforementioned claim, i.e., the PSO update for *gbest* is so poor that even an underperforming SPSA implementation can still improve the overall performance significantly. Note that the opposite is also true, that is, SAD PSO, which internally runs SPSA for *gbest* achieves better performance than SPSA alone.

Based on the results in Table 5.13, we can perform comparative evaluations with some of the promising PSO variants such as [1, 30–32] where similar experiments are performed over some or all of these benchmark functions. For example in Angeline [32], a tournament selection mechanism is formed among particles and the method is applied over four functions (*Sphere, Rosenbrock, Rastrigin,* and *Griewank*). Although the method is applied over a reduced positional range, ±15, and at low dimensions (10, 20, and 30), the mean scores were in the range {0.3, 1,194}. As a result, both better and worse performances than *bPSO*, depending on the function, were reported. In Esquivel and Coello Coello [30], *bPSO* and two PSO variants, *GCPSO* and mutation-extended PSO over three neighborhood topologies are applied to multimodal functions, Rastrigin, Schwefel, and Griewank. Although the dimension is rather low (30), none of the topologies over any PSO variant converged to the global minimum and the mean scores were reported in the range of {0.0014, 4,762}. In Riget and Vesterstrom [1], a diversity guided PSO variant, *ARPSO*, along with two competing methods, *bPSO* and *GA* were applied over the multimodal functions, Rastrigin, Rosenbrock, and Griewank at dimensions, 20, 50, and 100. The experimental showed that none of the three methods converged to the global minimum except ARPSO for Rastrigin at dimension 20. *ARPSO* performed better than *bPSO* and *GA* for Rastrigin and Rosenbrock but worse for Griewank. The CPSO proposed in Bergh and Engelbrecht [24] was applied to five functions including Sphere, Rastrigin, Rosenbrock, and Griewank. The dimension of all functions is fixed to 30 and in this dimension, CPSO performed better than *bPSO* in 80 % of the experiments. Finally in Richards and Ventura [31], dynamic sociometries via *ring* and *star* were introduced among the swarm particles and the performance of various combinations of swarm size and sociometry over six functions (the ones used in this section except Schwefel) was reported. Although the tests were performed over comparatively reduced positional ranges and at a low dimension (30), the experimental results indicate that none of the sociometry and swarm size combinations converged to the global minimum of multimodal functions except for some cases of the Griewank function.

The statistical comparison between low-cost mode and the original (full cost) is reported in Table 5.14. The statistics in the table indicate that both modes within both approaches usually obtain a similar performance but occasionally a significant gap is visible. For instance, lowcost mode achieves a significantly better performance within the second SAD PSO approach for De Jong and Griewank functions at $d = 80$. The opposite is true for Schwefel particularly at $d = 20$.

In order to verify if the results are *statistically* significant, statistical significance test is next applied between each SAD PSO approach and each technique (*bPSO* and SPSA) using the statistical data given in Table 5.13. Let H_0 be the *null hypothesis*, which states that there is no difference between the proposed and competing techniques (i.e., the statistical results occur by chance). We shall then define two common threshold values for *P*, 5 % and 1 %. If the *P* value, which is the probability of observing such a large difference (or larger) between the statistics, is less than either threshold, then we can reject H_0 with the corresponding confidence level. To accomplish this, the standard *t* test was performed and the t values were computed between the pair of competing methods. Recall that the formula for the *t* test is as follows:

Table 5.14 Statistical results between full-cost and low-cost modes from 100 runs over seven benchmark functions

Functions	d	Full cost mode				Low cost mode			
		SAD PSO (A2)		SAD PSO (A1)		SAD PSO (A2)		SAD PSO (A1)	
		μ	σ	μ	σ	μ	σ	μ	σ
Sphere	20	0	0	0	0	0	0	0	0
	50	0	0	0	0	0	0	0	0
	80	0	0	0	0	0	0	0	0
De Jong	20	0	0	0	0	0	0	0	0
	50	0.0075	0.0091	0.2189	0.6491	0.0073	0.0036	23.7977	48.2012
	80	0.2584	0.4706	13546.02	4305.04	0.0326	0.0290	14136	4,578.8
Rosenbrock	20	1.29941	0.4658	0.4089	0.2130	1.1412	0.4031	0.3124	0.2035
	50	12.35141	2.67731	2.5472	0.3696	9.2063	2.2657	2.5864	0.8232
	80	28.1527	5.1699	5.2919	0.8177	24.0142	4.9823	12.9923	2.8497
Rastrigin	20	0.0383	0.0369	0.0326	0.0300	0.0263	0.0634	0.0383	0.0327
	50	0.0381	0.0436	0.0353	0.0503	0.0066	0.0083	0.0062	0.0082
	80	0.2363	0.6552	0.1240	0.1694	0.0053	0.0086	0.0043	0.0065
Griewank	20	0	0	0	0	0	0	0	0
	50	0	0	3074.02	13989	0.0018	0.0125	3033.1	16588
	80	20733	24,160	378,210	137,410	143,5.794	1,247.7	342,230	114,280
Schwefel	20	0.3076	0.0758	0.3991	0.0796	0.7538	0.2103	2.1327	0.5960
	50	0.8278	0.1093	0.9791	0.1232	1.2744	0.2282	6.4797	1.0569
	80	1.3633	0.1402	1.5528	0.1544	1.6965	0.2165	9.5252	1.6241
Giunta	20	445.1160	264.1160	445.1356	249.5412	375.6510	242.7301	387.8901	249.8883
	50	3938.9	626.9194	3,916.2	758.3290	3,282	890.3426	3,688.3	908.4803
	80	8,838.2	1,357	8,454.2	1,285.3	7,922.1	1,320.2	12,132	1,579.4

Table 5.15 t test results for statistical significance analysis for both SPSA approaches, A1 and A2

Functions	d	Pair of competing methods			
		bPSO (A2)	*bPSO* (A1)	SPSA (A2)	SPSA (A1)
Sphere	20	0	0	0	0
	50	0	0	0	0
	80	4.90	4.90	0	0
De Jong	20	0	0	4.73	4.73
	50	3.35	3.27	4.56	3.03
	80	12.00	28.62	2.95	31.46
Rosenbrock	20	0.54	17.56	2.88	21.42
	50	6.06	25.55	31.50	12.54
	80	6.16	7.14	43.51	0.88 (*)
Rastrigin	20	0.86	2.12	39.95	39.95
	50	1.80	2.05	67.81	67.81
	80	4.83	6.93	81.49	81.50
Griewank	20	0	0	0	0
	50	2.65	2.16	31.52	31.51
	80	1.27 (*)	25.35	48.76	48.13
Schwefel	20	36.11	33.75	4.63	3.62
	50	42.28	41.59	4.23	5.56
	80	39.48	39.12	3.55	5.53
Giunta	20	1.39	1.43	589.42	593.75
	50	3.35	3.27	56.37	53.31
	80	5.48	7.73	45.81	49.28

Table 5.16 t Table presenting degrees of freedom versus probability

Degrees of freedom	P: Probability			
	0.1	0.05	0.01	0.001
100	1.29	1.66	2.364	3.174
∞	1.282	1.645	2.325	3.090

$$t = \frac{\mu_1 - \mu_2}{\sqrt{\frac{(n_1-1)\sigma_1^2+(n_2-1)\sigma_2^2}{n_1+n_2-2}\left(\frac{n_1+n_2}{n_1 n_2}\right)}} \tag{5.7}$$

where $n_1 = n_2 = 100$ is the number of runs. Using the first- and second-order statistics presented in Table 5.13, the overall t test values are computed and enlisted in Table 5.15. In those entries with 0 value, both methods have a zero mean and zero variance, indicating convergence to the global optimum. In such cases, H_0 cannot be rejected. In those nonzero entries, t test values corresponding to the best approach are highlighted. In the t tests the *degrees of freedom* is $n_1 + n_2 - 2 = 198$. Table 5.16 presents two corresponding entries of t test values required to reject H_0 at several levels of confidence (one-tailed test). Accordingly,

H_0 can be rejected and hence all results are statistically significant beyond the confidence level of 0.01 except the two entries shown with a (*) in Table 5.15. Note that the majority of the results are statistically significant beyond the 0.001 level of confidence (e.g., the likelihood to occur by chance is less than 1 in 1,000 times).

5.4 Summary and Conclusions

This chapter focused on a major drawback of the PSO algorithm: the poor *gbest* update. This can be a severe problem, which may cause premature convergence to local optima since *gbest* as the common term in the update equation of all particles, is the primary *guide* of the swarm. Therefore, a solution for the social problem in PSO is the main goal of this chapter, i.e., "*Who will guide the guide?*" which resembles the rhetoric question posed by Plato in his famous work on government: "Who will guard the guards?" (*Quis custodiet ipsos custodes?*). At first the focus is drawn on improving the global convergence of PSO and in turn, MD PSO, as its convergence performance is still limited to the same level as PSO, which suffers from the lack of diversity among particles. This leads to a premature convergence to local optima especially when multimodal problems are optimized in high dimensions. Realizing that the main problem lies in fact in the inability of using the available diversity among the vector components of swarm particles' positions, the FGBF technique adapted in this section addresses this problem by collecting the best components and fractionally creating an artificial global best, *aGB*, particle that has the potential to be a better "guide" then the swarm's native *gbest* particle. When used with FGBF, MD PSO exhibits such an impressive speed gain that their mutual performance surpasses *bPSO* by several magnitudes. Experimental results over nonlinear function minimization show that except in few minority cases, the convergence to the global minimum at the target dimension is achieved within fewer than 1,000 iterations on the average, mostly only within few hundreds or even less. Yet, the major improvement occurs in the convergence accuracy. MD PSO with FGBF finds the global minimum at the target dimension for all runs over all functions without any exception. This is a substantial achievement in the area of PSO-based nonlinear function minimization.

FGBF was then tested in another challenging domain, namely optimization in dynamic environments. In order to make comparative evaluations with other techniques in the literature, FGBF with multiswarms is then applied over a conventional benchmark system, the Moving Peak Benchmark, MPB. The results over MPB with common settings (i.e., *Scenario 2*) clearly indicate the superior performance of FGBF with multiswarms over other PSO-based methods. To make the benchmark more generic for real-world applications where the optimum dimension may be unknown too, MPB is extended to a multidimensional system in which there is a certain amount of dependency among dimensions. Note that without such dependency embedded, the benchmark would be just a bunch of independent MPBs in different dimensions, and thus a distinct and independent optimization process would be sufficient for each dimension. The optimization algorithm combining FGBF and multiswarms with MD PSO exhibits both global convergence ability and an impressive speed gain, so that their mutual performance surpasses *bPSO* by several magnitudes. The experiments conducted over the extended MPB approve that MD PSO with multiswarms and FGBF always finds and tracks the optimum dimension where the global peak resides. On both (conventional and extended) MPBs, the proposed techniques generally find and track the global peak, yet they may occasionally converge to a near-optimum peak, particularly if the height difference happens to be insignificant.

However, FGBF is not an entirely generic method, which makes it hard or even impossible to perform if the individual dimensional scores cannot be computed. To remedy this drawback, SPSA is adopted to guide the *gbest* particle toward the "right" direction with the gradient estimate of the underlying surface whilst avoiding local traps due to its stochastic nature. In that, the SAD PSO has an identical process with the basic PSO, because the guidance is only provided to *gbest* particle—not the whole swarm. In SAD PSO, two approaches were introduced, where SPSA is explicitly used. The first approach replaces the PSO update of *gbest* with SPSA, whereas in the second an *aGB* particle, is created, which can replace *gbest* if it has to a better fitness score than gbest itself. Both SAD PSO approaches were tested over seven nonlinear functions and the experimental results demonstrated that they achieved a superior performance over all functions regardless of the dimension, modality. Especially if the setting of the critical parameters, e.g., a and c is appropriate, then a significant performance gain can be achieved by SAD PSO. If not, SAD PSO still outperforms *bPSO*. This shows that SPSA, even without proper parameter setting still performs *better* than the PSO's native *gbest* update. Furthermore, the complexity overhead in SAD PSO is negligible, i.e., only two (or three in the second approach) extra fitness evaluations per iteration (one less in the lowcost mode) are needed. The experimental results show that the low-cost mode does not cause a noticeable performance loss; on the contrary, it may occasionally perform even better.

Finally, both approaches are also integrated into MD PSO, which defines a new particle formation and integrates the ability of dimensional navigation into the core of the PSO process. Recall that such flexibility negates the requirement of setting the dimension in advance, since swarm particles can now converge to the global solution at the optimum dimension simultaneously.

5.5 Programming Remarks and Software Packages

As described in the related sections of the previous chapters, the test-bed application, **PSO_MDlib**, is designed to implement MD PSO operations for the purpose of multidimensional nonlinear function minimization and dynamic system (MPB) optimization. The skipped operations that are plugged into the **template**

<class T, class X> bool CPSO_MD<T,X>::Perform() function, FGBF is first explained in the following section. We shall then explain the implementation details of MPB and the application of MD PSO with FGBF over it. We shall explain the implementation details of both SA-driven and MD PSO with FGBF over the dynamic data clustering application in the next chapter.

5.5.1 FGBF Operation in PSO_MDlib Application

In the previous chapters, the standalone MD PSO operation implemented within the **template <class T, class X> bool CPSO_MD<T,X>::Perform**() function was explained. To enable the FGBF operation, the MD PSO object should be created with the mode setting "FGBF", **CPSO_MD<CSolSpace<COneClass>,COne-Class> *pPSO = new CPSO_MD<CSolSpace<COneClass>,COneClass> (_psoDef._noAgent, _psoDef._maxNoIter, _psoDef._eCutOff, _psoDef._mode)**.

Table 5.17 presents the implementation of FGBF pseudo-code given in Table 5.2. The function **m_fpFindGBDim(m_pPA, m_noP)** determines (flags) the best dimensional components of each particle and constitutes the array $a[j] =$

$\arg \min\limits_{a \in [1, S] j \in [1, D_{\max}]} (f(a, j))$ given in Table 5.2. Therefore, steps 1 and 2 in the table

are implemented in the member function **m_fpFindGBDim(m_pPA, m_noP)**. The step 3.1, that is the construction of the *aGB* particle, is implemented within the first for loop, **for(a=0;a<m_noP;a++)**, and note that the initial *aGB* particle is represented by the object, **xyb_i**, which is filled by the best dimensional components of the flagged particles (by the boolean array, **bInGB[ ]**). Then this new *aGB* particle first competes against the personal best position of the previous *aGB* particle and if it beats it, then the personal best position of the new *aGB* particle will be taken from this new *aGB* particle's position; otherwise, the previous *aGB* particle resides (the new one is not taken into account at all). Therefore, steps 3.2, 3.3, and 3.4 are implemented within the second for loop, **for (int cur_d=m_xd-Min; cur_d<m_xdMax; cur_d++)**, and finally the best overall position is replaced in the personal best position of the final *gbest* particle, **m_pPA[m_g-best[cur_d-m_xdMin]]->GetPBPos(cur_d)**, for dimension **cur_d**. Finally, the step 5, the update of the *dbest*, is implemented within the last for loop, which basically searches for the best overall fitness score for the (final) *gbest* particles in all dimensions. Whichever, dimension leads to the minimum score will thus be assigned as the new *dbest* for this iteration.

Table 5.17 Implementation of FGBF pseudo-code given in Table 5.2

```
if(m_mode == FGBF)
{ // Apply FGBF..
        m_fpFindGBDim(m_pPA, m_noP); // Determine the potential GB dimensions first..
        for( a=0;a<m_noP;a++)
        { // Create the global best particle from the best partial dimensions..
                T* xx = m_pPA[a]->GetPos(); // Current position of the particle a in cur. dim..
                X *pCC = xx->GetPos(); // all dimensions (pixel pos.) of this particle..
                int nDim = xx->GetDim(); // and the total dim.
                bool *bInGB = xx->GetGBDim(); // The boolean array per dim..
                for(int c=0; c<nDim; c++) // for all potential dimensions..
                if(bInGB[c] )
                    xyb_i->Replace(&pCC[c],c); // Assing the min. score dimension into the GB solution..
        } // for..

        for (int cur_d=m_xdMin; cur_d<m_xdMax; cur_d++)
        {// Now Check the Fractional GB score for this dim..
        double score = m_fpGetScore(xyb_i, cur_d); // Get score for fractional dim. sol for dim=d..

        if(score < score_min[cur_d-m_xdMin])
        { // if a better score is obtained than before..
                *xyb[cur_d-m_xdMin] = xyb_i; // The global best sol. for dim=d so far..
                score_min[cur_d-m_xdMin] = score; // The best score for dim=d so far..
                if(cur_d == m_dbest) pInfo->m_gbest = m_noP; // FGB is the true gbest
            } // if..

        if( score_min[cur_d-m_xdMin] < (score = m_pPA[m_gbest[cur_d-m_xdMin]]->GetPBScore(cur_d)))
        { // If this is better than the GB sol. of the swarm in dim=d..
        // then update all dimensions GB to fractional GB: xyb
        T *xyb_a = m_pPA[m_gbest[cur_d-m_xdMin]]->GetPBPos(cur_d); // The GB agent's solution for dim=d..
        *xyb_a = xyb[cur_d-m_xdMin]; // update the agent's GB in this dimension to fractional best..
        m_pPA[m_gbest[cur_d-m_xdMin]]->SetPBScore(score_min[cur_d-m_xdMin],cur_d);//Set best score..
        } // if..
        else{ // No the agents achieved better GB than the fractional, so copy back..
        score_min[cur_d-m_xdMin] = m_pPA[m_gbest[cur_d-m_xdMin]]->GetPBScore(cur_d);
        } // else..
        } // for..

        // Find/update dbest if changed..
        double best_score = score_min[0];
        int best_dim = m_xdMin;
        for(int dim = 1; dim<m_xdMax-m_xdMin; dim++)
        if(score_min[dim] < best_score)
        { // This is the best score achieved either from GB Agent or Fractional set-up..
                best_score = score_min[dim];
                best_dim = dim+m_xdMin;
        } // if..
        m_dbest = best_dim; // new (updated) dbest..
} // if FGBF..
```

5.5.2 MD PSO with FGBF Application Over MPB

PSO_MDlib when configured for MPB optimization during the compile time by simply defining "MOVING_PEAKS_BENCHMARK". When defined, another entry point **main**() function (at the bottom) will be compiled instead for MDPSO with FGBF optimization over MPB environment, which is implemented in the source file *movpeaks.cpp* (and *movpeaks.h*). The MD PSO with FGBF operation is almost identical as in the nonlinear function minimization; except the fitness function (**MPB**()) and the setting of the change period by calling **pPSO-**

Table 5.18 The environmental change signaling from the main MD PSO function

```
if(m_changeFreq && iter && iter %   m_changeFreq == 0)
{ // update each particle's history, e.g. personal best values..
    m_fpGetScore(NULL, -1); // Signal the change..
    for(int dim = 0; dim<m_xdMax-m_xdMin; dim++)
        score_min[dim] = 1e+9; // re-init..
    for(int d = m_xdMin; d<m_xdMax; d++)
    { // for all dimensions in the range..
        for (int a=0;a<m_noP';a++)
        { // for all particles..
            double score_old = m_pPA[a]->GetPBScore(d); // the old score..
            double score_new = m_fpGetScore(m_pPA[a]->GetPBPos(d), d); // and the new one..
            if(score_old != score_new)
            m_pPA[a]->SetPBScore(score_new, d); // Set the Pbest score..

        } // for..
    } // for..

} // if..
```

Table 5.19 MD PSO with FGBF implementation for MPB

```
        init_peaks(); TracePeaks ();
        setChangeFreq((CHANGE_FREQ * (1+_psoDef._noAgent)));
        set_output_file( MPBfp );

        double min_x = 0, max_x = 0; // init..
        int dim = 0; // init..
        getMPBparams( dim, min_x, max_x ); // MBP params..
        double dv = (max_x-min_x)/4; // vel..
        pPosD = (double*) calloc( dim, sizeof( double ) ); // init..

CPSO_MD<CSolSpace<COneClass>,COneClass> *pPSO = new CPSO_MD<CSolSpace<COneClass>,COneClass>(
_psoDef._noAgent, _psoDef._maxNoIter, _psoDef._eCutOff, _psoDef._mode);

        pPSO->Init( dim, dim + 1, 0, 0, min_x, max_x, -dv, dv );
        pPSO->SetFitnessFn(MPB);
        pPSO->SetChangeFreq(CHANGE_FREQ);
        pPSO->SetDistFn(GetDist);
        pPSO->SetFindGBDimFn(FindGBDim);

        pPSO->Perform();

        int dbest = 0, a=0;
        double score = 1000;
        CSolSpace<COneClass> *pSol = pPSO->GetBestPos(dbest, a, score);

        fprintf( MPBfp, "%d %lf\n", pPSOdebug->m_noIter, score ); // log it..
        printf("%d\titer=%d, score=%5.2lf, dim=%d\n", run, pPSOdebug->m_noIter, score, dbest);
        delete pPSO;
        free( pPosD );free_peaks();
```

>SetChangeFreq(CHANGE_FREQ). This will trigger the MPB resetting within the **template <class T, class X> bool CPSO_MD<T,X>::Perform**() function at the beginning of the main for loop of the MD PSO run, as shown in Table 5.18. When the change period, **m_changeFreq**, is a nonzero value, then the signaling is sent to the MPB code by calling **m_fpGetScore(NULL, -1)**, which will call **change_peaks**() function within the fitness function, **MPB**(). In this case the MPB will go through an environmental change, the peaks will shift and no fitness evaluation will be performed within the function. Besides this, MD PSO initialization and the fitness function are of course different. It can also be performed several times (runs) and at each run, the C code given in Table 5.19 is executed. The first three function calls (**init_peaks**(); **TracePeaks**(); **setChangeFreq**()) are performed for MPB initialization. The fitness function pointer of the benchmark (MPB) is then given to the MD PSO object as, **pPSO->SetFitnessFn(MPB)**. Note that the dimension (fixed) and the positional ranges, where the MD PSO is performed, are set by the call **getMPBparams(dim, min_x, max_x)** and the default values are set as: **dim = 5, and max_x = -min_x = 50**. The rest of the MPB's default parameters are all set at the beginning of the *movpeaks.cpp* source file.

The first three function calls (**init_peaks**(); **TracePeaks**(); **setChangeFreq**()) are performed for MPB initialization. The fitness function pointer (MPB) is then given to the MD PSO object, **pPSO** by calling **pPSO->SetFitnessFn(MPB)**. As mentioned earlier, besides this fitness function and the setting of the nonzero

Table 5.20 The fitness function **MPB**()

```
double MPB(CSolSpace<COneClass> *pPos, int dim)
{
        if(dim < 0 || pPos == NULL)
        { // Change is signalled..
                change_peaks();
                return -1;
        } // if..

        float coord = 0;
        COneClass *pVal = pPos->GetPos();
        for (int i=0;i<dim;i++) pPosD[ i ] = pVal[i].m_x ;

        double result  = eval_movpeaks (pPosD); // the current global_max - height..
        double *pDimScore = get_dim_score(); // Dim. score array..
        for (i=0;i<dim;i++) pVal[i].m_bScore = pDimScore[i];

        double cur_error = get_current_error(); // fitness error = global_peak - cur_height..
        int evals = get_number_of_evals(); // returns the number of evaluations so far
        double offline_error = get_offline_error();  // returns offline performance
        ...
        return result;
}
```

change period, the rest of the MD PSO with FGBF operation is identical with the nonlinear function minimization.

Finally, the fitness function **MPB**() is given in Table 5.20. Note that the first **if**() statement checks for the signal for environmental change, that is sent from the native MD PSO function, **Perform**(). If signal is sent (e.g., by assigning **dim = -1** to) then the function **change_peaks**() changes the MPB environment and the function returns abruptly without fitness evaluation. Otherwise, the position stored in **pPos** proposed by a particle will be evaluated by the function **eval_movpeaks**(), which returns the difference of the current height from the global peak height. The other function calls are for debug purposes.

References

1. J. Riget, J.S. Vesterstrom, A diversity-guided particle swarm optimizer—The ARPSO, Technical report, Department of Computer Science, University of Aarhus, 2002
2. G.-J. Qi, X.-S. Hua, Y. Rui, J. Tang, H.-J. Zhang, Image classification with Kernelized spatial-context. IEEE Trans. Multimedia **12**(4), 278–287 (2010). doi:10.1109/TMM.2010.2046270
3. F. Van den Bergh, A.P. Engelbrecht, A new locally convergent particle swarm optimizer, in *Proceedings of the IEEE International Conference on Systems, Man, and Cybernetics*, (2002), pp. 96–101

4. A. Abraham, S. Das, S. Roy, Swarm intelligence algorithms for data clustering. in *Soft Computing for Knowledge Discovery and Data Mining book*, Part IV, (2007), pp. 279–313, Oct 25 2007
5. T.M. Blackwell, Particle swarm optimization in dynamic environments. *Evolutionary Computation in Dynamic and Uncertain Environments*, Studies in Computational Intelligence, vol. 51 (Springer, Berlin, 2007) pp. 29–49
6. Y.-P. Chen, W.-C. Peng, M.-C. Jian, Particle swarm optimization with recombination and dynamic linkage discovery. IEEE Trans. Syst. Man Cybern. Part B **37**(6), 1460–1470 (2007)
7. K.M. Christopher, K.D. Seppi, The Kalman swarm. A new approach to particle motion in swarm optimization, in *Proceedings of the Genetic and Evolutionary Computation Conference, GECCO*, (2004), pp. 140–150
8. L.-Y. Chuang, H.W. Chang, C.J. Tu, C.H. Yang, Improved binary PSO for feature selection using gene expression data. Comput. Biol. Chem. **32**(1), 29–38 (2008)
9. R. Eberhart, P. Simpson, R. Dobbins, *Computational Intelligence,PC Tools* (Academic, Boston, 1996)
10. H. Higashi, H. Iba, Particle Swarm Optimization with Gaussian Mutation, in *Proceedings of the IEEE swarm intelligence symposium*, (2003), pp. 72–79
11. S. Janson, M. Middendorf, A hierarchical particle swarm optimizer and its adaptive variant. IEEE Trans. Syst. Man Cybern. Part B **35**(6), 1272–1282 (2005)
12. B. Kaewkamnerdpong, P.J. Bentley, Perceptive particle swarm optimization: an investigation, in *Proceedings of IEEE Swarm Intelligence Symposium*, (California, 2005), pp. 169–176, 8–10 June 2005
13. U. Kressel, Pairwise Classification and support vector machines. in *Advances in Kernel Methods—Support Vector Learning* (1999)
14. J.J. Liang, A.K. Qin, Comprehensive learning particle swarm optimizer for global optimization of multimodal functions. IEEE Trans. Evol. Comput. **10**(3), 281–295 (2006)
15. Y. Lin, B. Bhanu, Evolutionary feature synthesis for object recognition. IEEE Trans. Man Cybern. Part C **35**(2), 156–171 (2005)
16. M. Løvberg, T. Krink, Extending particle swarm optimizers with self-organized criticality, in *Proceedings of the IEEE Congress on Evolutionary Computation*, vol. 2 (2002), pp. 1588–1593
17. W. McCulloch, W. Pitts, A logical calculus of the ideas immanent in nervous activity. Bull. Math. Biophys. **7**, 115–133 (1943)
18. J. Pan, W.J. Tompkins, A real-time QRS detection algorithm. IEEE Trans. Biomed. Eng. **32**(3), 230–236 (1985)
19. T. Peram, K. Veeramachaneni, C.K. Mohan, Fitness-distance-ratio based particle swarm optimization, in *Proceedings of the IEEE Swarm Intelligence Symposium*, (IEEE Press, 2003) pp. 174–181
20. K. Price, R.M. Storn, J.A. Lampinen, Differential Evolution: A Practical Approach to Global Optimization (Springer, Berlin, 2005). ISBN 978-3-540-20950-8
21. A.C. Ratnaweera, S.K. Halgamuge, H.C. Watson, Particle swarm optimiser with time varying acceleration coefficients, in *Proceedings of the International Conference on Soft Computing and Intelligent Systems*, (2002), pp. 240–255
22. Y. Shi, R.C. Eberhart, A modified particle swarm optimizer, in *Proceedings of the IEEE Congress on Evolutionary Computation*, (1998), pp. 69–73
23. Y. Shi, R.C. Eberhart, Fuzzy adaptive particle swarm optimization. in *Proceedings of the IEEE Congress on Evolutionary Computation*, (IEEE Press, 2001), vol. 1, pp. 101–106
24. F. van den Bergh, A.P. Engelbrecht, A cooperative approach to particle swarm optimization. IEEE Trans. Evol. Comput. **3**, 225–239 (2004)
25. X. Xie, W. Zhang, Z. Yang, A dissipative particle swarm optimization, in *Proceedings of the IEEE Congress on Evolutionary Computation*, vol. 2 (2002), pp. 1456–1461
26. X. Xie, W. Zhang, Z. Yang, Adaptive particle swarm optimization on individual level, in *Proceedings of the Sixth International Conference on Signal Processing*, vol. 2 (2002), pp. 1215–1218

27. X. Xie, W. Zhang, Z. Yang, Hybrid particle swarm optimizer with mass extinction, in *Proceedings of the International Conference on Communication, Circuits and Systems*, vol. 2 (2002), pp. 1170–1173

28. W.-J. Zhang, X.-F. Xie, DEPSO: Hybrid particle swarm with differential evolution operator, in *Proceedings of the IEEE International Conference on System, Man, and Cybernetics*, vol. 4 (2003), pp. 3816–3821

29. P.I. Angeline, Evolutionary optimization versus particle swarm optimization: Philosophy and performance differences. in *Evolutionary Programming VII, Conference EP'98*, Springer Verlag, Lecture Notes in Computer Science No. 1447, (California, USA, 1998). pp. 410–601

30. S.C. Esquivel, C.A. Coello Coello, On the use of particle swarm optimization with multimodal functions, in *Proceedings of 1106 the IEEE Congress on Evolutionary Computation*, (IEEE Press, 2003), pp. 1130–1136

31. M. Richards, D. Ventura, Dynamic sociometry in particle swarm optimization, in *Proceedings of the Sixth International Conference on Computational Intelligence and Natural Computing*, (North Carolina, 2003) pp. 1557–1560

32. P.J. Angeline, Using Selection to Improve Particle Swarm Optimization, in *Proceedings of the IEEE Congress on Evolutionary Computation*, (IEEE Press, 1998), pp. 84–89

33. J. Branke, Moving peaks benchmark (2008), http://www.aifb.unikarlsruhe.de/~jbr/MovPeaks/. Accessed 26 June 2008

34. T.M. Blackwell, J. Branke, Multi-Swarm Optimization in Dynamic Environments. *Applications of Evolutionary Computation*, vol. 3005 (Springer, Berlin, 2004), pp. 489–500

35. T.M. Blackwell, J. Branke, Multiswarms, exclusion, and anti-convergence in dynamic environments. IEEE Trans. Evol. Comput. **10**(4), 51–58 (2004)

36. X. Li, J. Branke, T. Blackwell, Particle swarm with speciation and adaptation in a dynamic environment, in *Proceedings of Genetic and Evolutionary Computation Conference*, (Seattle Washington, 2006), pp. 51–58

37. R. Mendes, A. Mohais, DynDE: a differential evolution for dynamic optimization problems. *IEEE Congress on Evolutionary Computation*, (2005) pp. 2808–2815

38. I. Moser, T. Hendtlass, A simple and efficient multi-component algorithm for solving dynamic function optimisation problems. *IEEE Congress on Evolutionary Computation*, (2007), pp. 252–259

39. M.A. Styblinski, T.-S. Tang, Experiments in nonconvex optimization: stochastic approximation with function smoothing and simulated annealing. Neural Netw. **3**(4), 467–483 (1990)

40. H.J. Kushner, G.G. Yin, *Stochastic Approximation Algorithms and Applications* (Springer, New York, 1997)

41. S.B. Gelfand, S.K. Mitter, Recursive stochastic algorithms for global optimization. SIAM J. Control Optim. **29**(5), 999–1018 (1991)

42. D.C. Chin, A more efficient global optimization algorithm based on Styblinski and Tang. *Neural Networks*, (1994), pp. 573–574

43. J.C. Spall, Multivariate stochastic approximation using a simultaneous perturbation gradient approximation. *IEEE Transactions on Automatic Control*, **37**(3), 332–341 (1992)

44. J.C. Spall, Implementation of the simultaneous perturbation algorithm for stochastic optimization. IEEE Trans. Aerosp. Electron. Syst. **34**, 817–823 (1998)

45. J.L. Maryak, D.C. Chin, Global random optimization by simultaneous perturbation stochastic approximation, in *Proceedings of the 33rd Conference on Winter Simulation*, (Washington, DC, 2001), pp. 307–312

46. Y. Maeda, T. Kuratani, Simultaneous Perturbation Particle Swarm Optimization. *IEEE Congress on Evolutionary Computation, CEC'06*, (2006) pp. 672–676

Chapter 6
Dynamic Data Clustering

> *Data are a precious thing and will last longer than the systems themselves.*
>
> Tim Berners-Lee

Clustering in the most basic terms, is the collection of patterns, which are usually represented as vectors or points in a multi-dimensional data space, into groups (clusters) based on similarity or proximity. Such an organization is useful in pattern analysis, classification, machine learning, information retrieval, spatial segmentation, and many other application domains. Cluster validity analysis is the assessment of the clustering method's output using a specific criterion for optimality, i.e., the so-called clustering validity index (CVI). Therefore, the optimality of any clustering method can only be assessed with respect to the CVI, which is defined over a specific data (feature) representation with a proper distance (similarity) metric. What characterizes a clustering method further depends on its *scalability* over the dimensions of the data and solution spaces, i.e., whether or not it can perform well enough on a large dataset; say with a million patterns and having large number of clusters (e.g., $\gg 10$). In the former case, the complexity of the method may raise an infeasibility problem and the latter case shows the degree of its immunity against the well-known phenomenon, "the curse of dimensionality". Even humans, who perform quite well in 2D and perhaps in 3D, have difficulties of interpreting data in higher dimensions. Nevertheless, most real problems involve clustering in high dimensions, in that; the data distribution can hardly be modeled by some ideal structures such as *hyperspheres*.

Given a CVI, clustering is a multi-modal problem especially in high dimensions, which contains many sub-optimum solutions such as over- and under-clustering. Therefore, well-known deterministic methods such as *K-means*, Max–Min [1, 2], FCM [2, 3], SOM [2], etc., are susceptible to get trapped to the closest local optimum since they are all greedy descent methods, which start from a random point in the solution space and perform a *localized* search. This fact eventually turns the focus on stochastic Evolutionary Algorithms (EAs) [4] such as Genetic Algorithms (GAs) [5], Genetic Programming (GP) [6], Evolution Strategies (ES), [7], and Evolutionary Programming (EP), [8], all of which are motivated by the natural evolution process and thus make use of evolutionary operators. The common point of all is that EAs are in population-based nature and can perform a *globalized* search. So they may avoid becoming trapped in a local optimum and find the optimum solution; however, this is never guaranteed. Many works in the literature

S. Kiranyaz et al., *Multidimensional Particle Swarm Optimization for Machine Learning and Pattern Recognition*, Adaptation, Learning, and Optimization 15, DOI: 10.1007/978-3-642-37846-1_6, © Springer-Verlag Berlin Heidelberg 2014

have shown that EA-based methods outperform their deterministic counterparts, [5, 9, 10]. However, EAs suffer from the sensitivity of high parameter dependence (e.g., crossover and mutation probabilities, internal parameters, etc.), in that one has to tune some or all parameters to suit the application in hand. They further present a high complexity, which makes them applicable only over limited datasets. As the most popular EA method, GAs in general use binary string representations and this may create the ambiguity of finding an appropriate quantization level for data representation. Several researchers have shown that PSO exhibits a better clustering performance than the aforementioned techniques [11–13]; however, when the problem is multi-modal, PSO may also become trapped in local optima [14] due to the premature convergence problem especially when the search space is of high dimensions [13]. Furthermore, PSO has so far been applied to simple clustering problems [11–13], where the data space is limited and usually in low dimensions and the number of clusters (hence the solution space dimension) is kept reasonably low (e.g., <10). Moreover, all clustering methods mentioned earlier are static in nature, that is, the number of clusters has to be specified a priori. This is also true for PSO since in its basic form it can only be applied to a search space with a fixed dimension.

In this chapter, we shall address data clustering as an optimization problem by utilizing techniques for finding the optimal number of clusters in a (fixed) multi-dimensional data or feature space. As detailed in the next section, in order to accomplish *dynamic* clustering where the optimum number of clusters is also determined within the process, we shall utilize FGBF in conjunction with the MD PSO, both of which are adapted to the data clustering problem in hand.

6.1 Dynamic Data Clustering via MD PSO with FGBF

6.1.1 The Theory

Based on the discussion in Sect. 3.4.2, it is obvious that the clustering problem requires the determination of the solution space dimension (i.e., number of clusters, K) and an effective mechanism to avoid local optima traps (both dimensionally and spatially) particularly in complex clustering schemes in high dimensions (e.g., $K > 10$). The former requirement justifies the use of the MD PSO technique while the latter calls for FGBF. At time t, the particle a in the swarm, $\xi = \{x_1, \ldots, x_a, \ldots, x_S\}$, has the positional component formed as, $xx_a^{xd_a(t)}(t) = \{c_{a,1}, \ldots, c_{a,j}, \ldots, c_{a,xd_a(t)}\} \Rightarrow xx_{a,j}^{xd_a(t)}(t) = c_{a,j}$ meaning that it represents a potential solution (i.e., the cluster centroids) for the $xd_a(t)$ number of clusters while jth component being the jth cluster centroid. Apart from the regular limits such as (spatial) velocity, $V_{\max}$, dimensional velocity, $VD_{\max}$, and dimension range $D_{\min} \leq xd_a(t) \leq D_{\max}$, the N dimensional data space is also limited with some practical spatial range, i.e., $X_{\min} < xx_a^{xd_a(t)}(t) < X_{\max}$. In case this range is exceeded even for a single dimension j, $xx_{a,j}^{xd_a(t)}(t)$, then all positional components of the particle for the respective dimension $xd_a(t)$ are initialized randomly within the range (i.e., refer to step 1.3.1 in MD PSO pseudo-code) and this further contributes to the overall diversity. The following validity index is used to obtain computational simplicity with minimal or no parameter dependency,

$$f\left(xx_a^{xd_a(t)}, Z\right) = Q_e\left(xx_a^{xd_a(t)}\right)(xd_a(t))^{\alpha} \text{ where}$$

$$Q_e\left(xx_a^{xd_a(t)}\right) = \frac{1}{xd_a(t)} \sum_{j=1}^{xd_a(t)} \frac{\sum_{\forall z_p \in xx_{a,j}^{xd_a(t)}} \left\| xx_{a,j}^{xd_a(t)} - z_p \right\|}{\left\| xx_a^{xd_a(t)} \right\|} \tag{6.1}$$

where Q_e is the quantization error (or the average intra-cluster distance) as the *Compactness* term and $(xd_a(t))^{\alpha}$ is the *Separation* term, by simply penalizing higher cluster numbers with an exponential, $\alpha > 0$. Using $\alpha = 1$, the validity index yields the simplest form (i.e., only the nominator of Q_e) and becomes entirely parameter-free.

On the other hand, (hard) clustering has some constraints. Let $C_j = \{xx_{a,j}^{xd_a(t)}(t)\} = \{c_{a,j}\}$ be the set of data points assigned to a (potential) cluster centroid $xx_{a,j}^{xd_a(t)}(t)$ for a particle a at time t. The clusters C_j, $\forall j \in [1, xd_a(t)]$ should maintain the following constraints:

1. Each data point should be assigned to one cluster set, i.e., $\bigcup_{j=1}^{xd_a(t)} C_j = Z$

2. Each cluster should contain at least one data point, i.e., $C_j \neq \{\phi\}$, $\forall j \in [1, xd_a(t)]$

3. Two clusters should have no common data points, i.e., $C_i \cap C_j = \{\phi\}$, $,i \neq j$ and $\forall i,j \in [1, xd_a(t)]$

In order to satisfy the 1st and 3rd hard clustering constraints, before computing the clustering fitness score via the validity index function in (6.1), all data points are first assigned to the *closest* centroid. Yet there is no guarantee for the fulfillment of the 2nd constraint since $xx_a^{xd_a(t)}(t)$ is set (updated) by the internal dynamics of the MD PSO process and hence any dimensional component (i.e., a potential cluster candidate), $xx_{a,j}^{xd_a(t)}(t)$, can be in an abundant position (i.e., no closest data point exists). To avoid this, a high penalty is set for the fitness score of the particle, i.e., $f(xx_a^{xd_a(t)}, Z) \approx \infty$, if $\{xx_{a,j}^{xd_a(t)}\} = \{\phi\}$ for any j.

The major outlines so far given are sufficient for the standalone application of the MD PSO technique for a dynamic clustering application; however, the FGBF operation presents further difficulties since for the aGB creation the selection of the best or the most promising dimensions (i.e., the cluster centroids) among all dimensions of swarm particles is not straightforward. Recall that in step 2 of the

FGBF pseudo-code, the index array of such particles yielding the minimum $f(a, j)$ for the jth dimension, can be found as, $a[j] = \arg \min\limits_{a \in [1, S] j \in [1, D_{max}]} (f(a, j))$. This was straightforward for the nonlinear function minimization where each dimension of the solution space is distinct and corresponds to an individual dimension of the data space. However, in the clustering application, any (potential) cluster centroid of each particle, $xx_{a,j}^{xd_a(t)}(t)$, is updated independently and can be any arbitrary point in N dimensional data space. Furthermore, data points assigned to the jth dimension of a particle a, $(\forall z_p \in xx_{a,j}^{xd_a(t)}(t))$, also depend on the distribution of the other dimensions (centroids), i.e., the "closest" data points are assigned to the jth centroid only because the other centroids happen to be at a farther location. Inserting this particular dimension (centroid) into another particle (say aGB, in case selected), might create an entirely different assignment (or cluster) including the possibility of having no data points assigned to it and thus violating the 2nd clustering constraint. To avoid this problem, a new approach is adopted for step 2 to obtain $a[j]$. At each iteration, a subset among all dimensions of swarm particles is first formed by verifying the following: a dimension of any particle is selected into this subset if and only if there is at least one data point that is closest to it. Henceforth, the creation of the aGB particle within this verified subset ensures that the 2nd clustering constraint will (always) be satisfied. Figure 6.1 illustrates the formation of the subset on a sample data distribution with 4 clusters. Note that in the figure, all dimensions of the entire swarm particles are shown as '+' but the red ones belonging to the subset have at least one (or more) data points closest whereas the blue ones have none and hence they are discarded.

Once the subset centroids are selected, then the objective is to compose $a[j]$ with the most promising D_{max} centroids selected from the subset in such a way that each

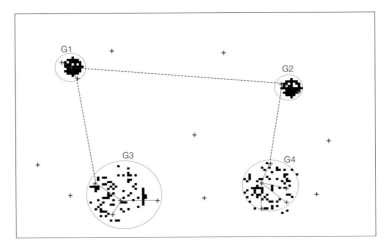

Fig. 6.1 The formation of the centroid subset in a sample clustering example. The black dots represent data points over 2D space and each colored '+' represents one centroid (*dimension*) of a swarm particle

dimensional component of the aGB particle with K dimensions $(xx^K_{aGB,j}(t), \forall j \in [1, K])$, which is formed from $a[j]$ (see step 3.1 of FGBF in MD PSO pseudo-code) can represent one of the true clusters, i.e., being in a close vicinity of its centroid. To accomplish this, only such D_{max} dimensions that fulfill the two clustering criteria, *Compactness* and *Separation* are selected and then stored in $a[j]$. To achieve well-separated clusters and to avoid the selection of more than one centroid representing the same cluster, *spatially* close centroids are first grouped using a Minimum Spanning Tree (MST) [15] and then a certain number of centroid groups, say $d \in [D_{min}, D_{max}]$, can be obtained simply by breaking $(d - 1)$ longest MST branches. From each group, one centroid, which provides the highest *Compactness* score (i.e., minimum dimensional fitness score, $f(a,j)$) is then selected and inserted into $a[j]$ as the jth dimensional component. During the computation of the validity index $f(xx_a^{xd_a(t)}, Z)$ in (6.1), $f(a, j)$ can simply be set as the jth term of the summation in Q_e expression, such as,

$$f(a, j) = \frac{\sum_{\forall z_p \in xx_{a,j}^{xd_a(t)}} \left\| xx_{a,j}^{xd_a(t)} - z_p \right\|}{\left\| xx_a^{xd_a(t)} \right\|} \tag{6.2}$$

In Fig. 6.1, a sample MST is formed using 14 subset centroids as the nodes and 13 branches are shown as the red lines connecting the closest nodes (in a minimum span). Breaking the 3 longest branches (shown as the dashed lines) thus reveals the 4 groups (G1, ..., G4) among which one centroid yielding the minimum $f(a,j)$ can then be selected as an individual dimension of the aGB particle with 4 dimensional components (i.e., $d = K = 4$, $xx^K_{aGB,j}(t), \forall j \in [1, K]$).

6.1.2 Results on 2D Synthetic Datasets

In order to test the clustering performance of the standalone MD PSO, we used the same 15 synthetic data spaces as shown in Fig. 3.6, and to make the evaluation independent from the choice of the parameters, we simply used Q_e in Eq. (6.1) as the CVI function. Note that this is just a naïve selection and any other suitable CVI function can also be selected since it is a black box implementation for MD PSO. We also use the same PSO parameters and settings given in Sect. 3.4.2, and recall that the clustering performance degraded significantly for complex datasets or datasets with large number of clusters (e.g., >10).

As stated earlier, MD PSO with FGBF, besides its speed improvement, has its primary contribution over the accuracy of the clustering, i.e., converging to the true number of clusters, K, and correct localization of the centroids. As typical results shown in Fig. 6.2, MD PSO with FGBF meets the expectations on clustering accuracy, but occasionally results in a slightly higher number of clusters

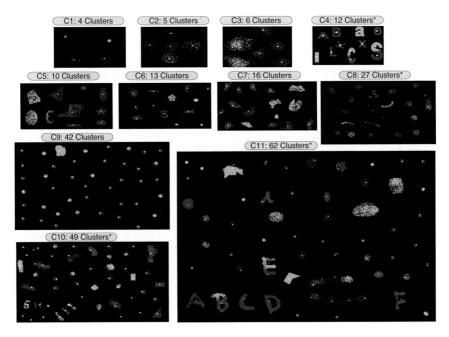

Fig. 6.2 Typical clustering results via MD PSO with FGBF. Over-clustered samples are indicated with *

(over-clustering). This is due to the use of a simple but quite impure validity index in (6.1) as the fitness function and for some complex clustering schemes it may, therefore, yield its minimum score at a slightly higher number of clusters. A sample clustering operation validating this fact is shown in Fig. 6.3. Note that the true number of clusters is 10, which is eventually reached at the beginning of the operation, yet the minimum score achieved with $K = 10(\sim 750)$ remains higher than the one with $K = 11(\sim 610)$ and than the final outcome, $K = 12(\sim 570)$ too. The main reason for this is that the validity index in (6.1) over long (and loose) clusters such as 'C' and 'S' in the figure, yields a much higher fitness score with one centroid than two or perhaps more and therefore, over all data spaces with such long and loose clusters (e.g., $C4$, $C8$, $C10$, and $C11$), the proposed method yields a slight over-clustering but never under-clustering. Improving the validity index or adapting a more sophisticated one such as Dunn's index [16] or many others, might improve the clustering accuracy.

An important observation worth mentioning is that clustering complexity (more specifically modality) affects the proposed methods' mutual performance much more than the total cluster number (dimension). For instance, MD PSO with FGBF clustering (with $S = 640$) over data space $C9$ can immediately determine the true cluster number and the accurate location of the centroids with a slight offset (see Fig. 6.4), whereas this takes around 1,900 iterations for $C8$. Figure 6.5 shows time instances where aGB (with index number 640) becomes the GB particle. It

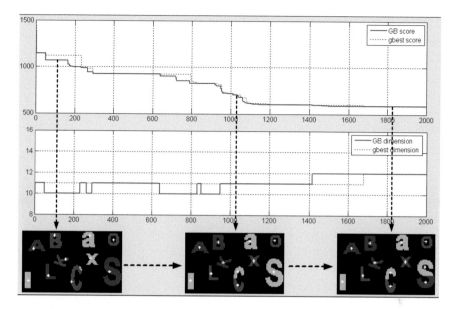

Fig. 6.3 Fitness score (*top*) and dimension (*bottom*) plots vs. iteration number for a MD PSO with FGBF clustering operation over *C*4. 3 clustering snapshots at iterations 105, 1,050, and 1,850, are presented below

immediately (at the 1st iteration) provides a "near optimum" GB solution with 43 clusters and then the MD PSO process (at the 38th iteration) eventually finds the global optimum with 42 clusters (i.e., see the 1st snapshot in Fig. 6.4). Afterward the ongoing MD PSO process corrects the slight positional offset of the cluster centroids (e.g., compare 1st and 2nd snapshots in Fig. 6.4). So when the clusters are compact, uniformly distributed, and have similar shape, density, and size, thus yielding the simplest form, it becomes quite straightforward for FGBF to select the 'most promising' dimensions with a greater accuracy. As the complexity (modality) increases, different centroid assignments and clustering combinations have to be assessed to converge toward the global optimum, which eventually becomes a slow and tedious process.

Recall from the earlier discussion (Sect. 5.1.4) about the application of the proposed methods over nonlinear function minimization (both standalone MD PSO and MD PSO with FGBF), a certain speed improvement occurs in terms of reduction in iteration number and a better fitness score is achieved, when using a larger swarm. However, the computational complexity (per iteration) also increases since the number of evaluations (fitness computations) is proportional to the number of particles. The same trade-off also exists for clustering application and a significantly higher computational complexity of the mutual application of the proposed methods can occur due to the spatial MST grouping for the selection of the well-separated centroids. As explained in the previous section, MST is the essence of choosing the "most promising" dimensions (centroids) so as to form

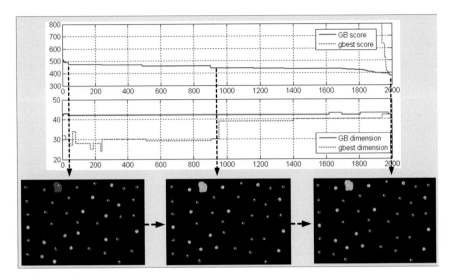

Fig. 6.4 Fitness score (*top*) and dimension (*bottom*) plots vs. iteration number for a MD PSO with FGBF clustering operation over *C*9. 3 clustering snapshots at iterations 40, 950, and 1,999, are presented below

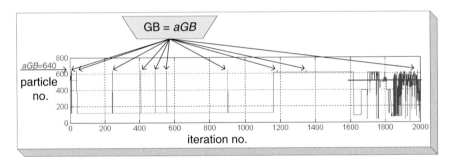

Fig. 6.5 Particle index plot for the MD PSO with FGBF clustering operation shown in Fig. 6.4

the best possible *aGB* particle. However, it is a costly operation $(O(N_{SS}^2))$ where N_{SS} is the subset size, which is formed by those dimensions (potential centroids) having at least one data item closest to it. Therefore, N_{SS} tends to increase if a larger swarm size is used and/or MD PSO with FGBF clustering is performed over large and highly complex data spaces.

Table 6.1 presents average processing times per iteration over all sample data spaces and using 4 different swarm sizes. All experiments are performed on a computer with P-IV 3 GHz CPU and 1 GB RAM. Note that the processing times tend to increase in general when data spaces get larger but the real factor is the complexity. The processing for a highly complex data structure, such as *C*10, may require several times more computations than a simpler but comparable-size data

Table 6.1 Processing time (in msec) per iteration for MD PSO with FGBF clustering using 4 different swarm sizes. Number of data items is presented in parenthesis with the sample data space

S	C1 (238)	C2 (408)	C3 (1,441)	C4 (1,268)	C5 (3,241)	C6 (1,314)	C7 (3,071)	C8 (5,907)	C9 (2,192)	C10 (3,257)	C11 (12,486)
80	19.7	57	140	688	864.1	231.5	690.8	1,734.5	847.7	4,418.2	8,405.1
160	31.7	104.9	357.3	1,641	1,351.3	465.5	2,716.3	3,842.8	1,699.4	13,693.7	26,608.6
320	62.3	222.9	1,748.8	4,542.5	3,463.9	1,007	3,845.7	7,372.7	4,444.5	55,280.6	62,641.9
640	153.4	512	3,389.4	17,046.4	8,210.1	4,004.5	11,398.6	23,669.8	14,828.2	159,642.3	212,884.6

space, such as $C5$. Therefore, on such highly complex data spaces, the swarm size should be kept low, e.g., $80 \leq S \leq 160$, for the sake of a reasonable processing time.

6.1.3 Summary and Conclusions

In this section we presented a robust dynamic data clustering technique based on MD PSO with FGBF. Note that although the ability of determining the optimum dimension where the global solution exists is gained with MD PSO, its convergence performance is still limited to the same level as $bPSO$, which suffers from the lack of diversity among particles. This leads to a premature convergence to local optima especially when multi-modal problems are optimized at high dimensions. Ideally the (clustering) complexity can be thought of as synonymous to (function) modality, i.e., speed and accuracy performances of both methods drastically degrade with increasing complexity. Needless to say that the true number of clusters has to be set in advance for $bPSO$ whereas MD PSO finds it as a part of the optimization process and hence exhibits a slower convergence speed compared to $bPSO$. When MD PSO is enhanced with FGBF, a significant speed improvement is achieved and such cooperation provides accurate clustering results over complex data spaces. Since the clustering performance also depends on the validity index used, occasional over-clustering can be encountered where we have shown that such results indeed correspond to the global minimum of the validity index function used. As a result, the true number of clusters and accurate centroid localization are achieved at the expense of increased computational complexity due to the usage of MST.

6.2 Dominant Color Extraction

6.2.1 Motivation

Dominant Color (DC) extraction is basically a dynamic color quantization process, which seeks for such prominent color centers that minimize the quantization error. To this end, studying human color perception and similarity measurement in the color domain becomes crucial and there is a wealth of research performed in this field. For example in [17], Broek et al. focused on the utilization of color categorization (called focal colors) for content-based image retrieval (CBIR) purposes and introduced a new color matching method, which takes human cognitive capabilities into account. They exploited the fact that humans tend to think and perceive colors *only* in 11 basic categories (black, white, red, green, yellow, blue, brown, purple, pink, orange, and gray). In [18], Mojsilovic et al. performed a series of psychophysical experiments analyzing how humans perceive and measure similarity in the domain of color patterns. One observation worth mentioning here is that the human eye cannot perceive a large number of colors at the same time, nor it is able to distinguish similar (close) colors well. Based on this, they showed that at the coarsest level of judgment, the human visual system (HVS) primarily uses dominant colors (i.e., few prominent colors in the scenery) to judge similarity.

Many existing DC extraction techniques, particularly the ones widely used in CBIR systems such as MPEG-7 Dominant Color Descriptor (DCD), have severe drawbacks and thus show a limited performance. The main reason for this is because most of them are designed based on some heuristics or naïve rules that are *not* formed with respect to what humans or more specifically the human visual system (HVS) finds "relevant" in color similarity. Therefore, it is of decisive importance that human color perception is considered while modeling and describing any color composition of an image. In other words, when a particular color descriptor is designed entirely based on HVS and color perception rules, further discrimination power and hence certain improvements in retrieval performance can be achieved. For this reason in this section we shall present a *fuzzy* model to achieve a perceptual distance metric over HSV (or HSL) color space, which provides means of modeling color in a way HVS does. In this way the discrimination between distinct colors is further enhanced, which in turn improves the clustering (and DC extraction) performance.

In order to solve the problems of static quantization in color histograms, various DC descriptors, e.g., [19–23], have been developed using dynamic quantization with respect to the image color content. DCs, if extracted properly according to the aforementioned color perception rules, can indeed represent the prominent colors in any image. They have a global representation, which is compact and accurate; and they are also computationally efficient. MPEG-7 DC descriptor (DCD) is adopted as in [20] where the method is designed with respect to HVS color perceptual rules. For instance, HVS is more sensitive to changes in smooth regions than in detailed regions. Thus colors are quantized more coarsely in the detailed regions while smooth regions have more importance. To exploit this fact, a smoothness weight $(w(p))$ is assigned to each pixel (p) based on the variance in a local window. Afterward, the *General Lloyd Algorithm* (*GLA*, also referred to as *Linde–Buzo–Gray* and it is equivalent to the well-known *K-means* clustering method [2]) is used for color quantization. For a color cluster C_i, its centroid c_i is calculated by,

$$c_i = \frac{\sum w(p)x(p)}{\sum w(p)}, \quad x(p) \in C_i \tag{6.3}$$

and the initial clusters for *GLA* is determined by using a weighted distortion measure, defined as,

$$D_i = \sum w(p)\|x(p) - c_i\|^2, \quad x(p) \in C_i \tag{6.4}$$

This is used to determine which clusters to split until either a maximum number of clusters (DCs), $N_{DC}^{\max}$, is achieved or a maximum allowed distortion criterion, ε_D, is reached. Hence, pixels with smaller weights (detailed regions) are assigned fewer clusters so that the number of color clusters in the detailed regions where the likelihood of outliers' presence is high, is therefore suppressed. As the final step, an agglomerative clustering (AC) is performed on the cluster centroids to further merge similar color clusters so that there is only one cluster (DC) hosting all similar color components in the image. A similarity threshold T_s is assigned to the maximum color distance possible between two similar colors in a certain color domain (CIE-Luv, CIE_Lab, etc.). Another merging criterion is the color area, that is, any cluster should have a minimum amount of coverage area, T_A, so as to be assigned as a DC; otherwise, it will be merged with the closest color cluster since it is just an outlier. Another important issue is the choice of the color space since a proper color clustering scheme for DC extraction tightly relies on the metric. Therefore, a perceptually uniform color space should be used and the most common ones are CIE-Luv and CIE-Lab, which are designed in such a way that color distances perceived by HVS are also equal in L_2 (Euclidean) distance in these spaces. For CIE-Luv, a typical value for T_S is between 10 and 25, T_A is between 1–5 % and $\varepsilon_D < 0.05$ [23].

Particularly for dominant color extraction, the optimal (true) number of DCs in an image is also unknown and should thus be determined within the clustering process, in an optimized way and without critical parameter dependency. MPEF-7 DCD, as a modified K-means algorithm, does not address these requirements at all. Therefore, the reason for the use of MD PSO with FGBF clustering is obvious. Furthermore, humans tend to think and describe color the way they perceive it. Therefore, in order to achieve a color (dis-) similarity metric taking HVS into account, HSV (or HSL), which is a perceptual color space and provides means of modeling color in a way HVS does, is used in the presented technique for extracting dominant colors. Note that in a typical image with 24-bit RGB representation, there can be several thousands of distinct colors, most of which cannot be perceived by HVS. Therefore, to reduce the computational complexity of RGB to HSV color transformation and particularly to speed up the dynamic clustering process via MD PSO and FGBF, a pre-processing step, which creates a limited color palette in RGB color domain, is first performed. In this way such a massive, yet unperceivable amount of colors in RGB domain can be reduced to a reasonable number, e.g., $256 < n < 512$. To this end, we used the Median Cut method [24] because it is fast (i.e., $O(n)$) and for such a value of n, it yields an image which can hardly be (color-wise) distinguished from the original. Only the RGB color components in the color palette are then transformed into HSV (or HSL) color space over which the dynamic clustering technique is applied to extract the dominant colors, as explained next.

6.2.2 Fuzzy Model over HSV-HSL Color Domains

Let $c_1 = \{h_1, s_1, v_1\}$ and $c_2 = \{h_2, s_2, v_2\}$ be two colors in HSV domain. Assume for the sake of simplicity that the hue is between 0 and 360° and both s and v are *unit* normalized. The normalized *Euclidean* distance between s and v can be defined as,

$$\|c_1 - c_2\|^2 = (v_1 - v_2)^2 + (s_1 \cos(h_1) - s_2 \cos(h_2))^2 \\ + (s_1 \sin(h_1) - s_2 \sin(h_2))^2 \tag{6.5}$$

During the dynamic clustering process by MD PSO and FGBF, the problem of using this equation for computing a color distance between a candidate color centroid, $xx_{a,j}^d(t)$, $\forall j \in [1, d]$ and a color in the palette, $z_p \in xx_a^d(t)$, as in Eq. (6.1) is that it has a limited discrimination power between distinct colors, as it basically yields arbitrary fractional numbers despite the fact that HVS finds "no similarity" in between. Therefore, instead of using this typical distance metric for all color pairs, we adopt a perceptual approach in order to improve discrimination between different colors. Recall from the earlier discussion that humans can recognize and distinguish 8 to 12 colors. Recall that in [17], the authors exploited the fact that humans tend to think and perceive colors only in 11 basic categories. Hence above a certain hue difference between two colors, it is obvious that they become entirely different for HVS, e.g., yellow and green are as different as yellow and blue or cyan or black or purple, etc. So if the hue difference is above a certain limit, a maximum difference should be used (i.e., 1.0). We have selected an upper limit by considering distinct colors number as only 8, therefore, the perceptual threshold is, $\Delta T_H = 360/8 = 45°$. In practice, however, even a lower hue threshold can also be used; because, two colors for instance with 40° of hue difference can hardly have any similarity—yet 45° present a safe margin leaving any subjectivity out.

We then use a *fuzzy* color model for further discrimination. As shown in Fig. 6.6, for a fixed hue, e.g., red for HSV and green for HSL, a typical saturation (S) versus Value (V) or Lightness (L) plot can be partitioned into 5 regions: White (W), Black (B), Gray (G), Color (C), and Fuzzy (F), which is a transition area among others. W, B, and G are the areas where there is absolutely no color (hue) component whereas in F, there is a hint of a color presence with a known hue but perhaps not fully saturated. In C, the color described by its hue, is fully perceivable with a varying saturation and value. It is a fact that the borders among color regions are highly subjective and this is the sole reason to use a large Fuzzy region, so as to address this subjectivity in color perception and thus to contain the error. This is the reason why there is no need for drawing precise boundaries of F (even if possible) or the boundaries between $W \leftrightarrow G$ and $B \leftrightarrow G$ because between two colors, *say* one in C and one in F, or both in C or both in F, the same distance metric shall anyway be applied (as in Eq. 6.5) provided that they have hue differences less than ΔT_H thus presenting some degree of color similarity. This is not a condition in other cases where at least one color is from either of the "no color"

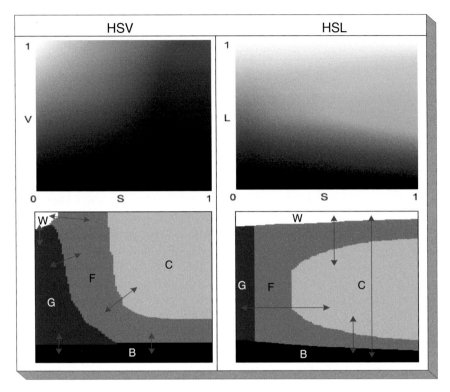

Fig. 6.6 Fuzzy model for distance computation in HSV and HSL color domains (*best viewed in color*)

areas. For instance, between $W \leftrightarrow G$ and $B \leftrightarrow G$, the distance should only be computed over V (or L) components because they have no *perceivable* color components. The boundaries are only important to distinguish areas such as C, W, and B (and between $C \leftrightarrow G$) where there is no similarity among them. Therefore, as shown in the HSL map on the right with blue arrows, if two colors, despite the fact that they have similar hues (i.e., $\Delta H < \Delta T_H$), happen to be in such regions, the maximum distance (1.0) shall be applied rather than computing Eq. (6.5).

6.2.3 DC Extraction Results

We have made comparative evaluations against MPEG-7 DCD over a sample database with 110 images, which are selected from *Corel* database in such a way that their prominent colors (DCs) are easy to be recognized by a simple visual inspection. We used the typical internal PSO parameters (c_1, c_2 and w) as in [25]. Unless otherwise stated, in all experiments in this section, the two critical PSO parameters, swarm size (S) and number of iterations, (*iterNo*) are set as 100 and

500, respectively. Their effects over the DC extraction are then examined. The dimension (search) range for DC extraction is set as $D_{min} = 2$, $D_{max} = 25$. This setting is in harmony with the maximum number of DCs set by the MPEG-7 DCD, i.e., $N_{DC}^{max} = 25$. Finally the size of the initial color palette created by the Median Cut method is set as 256.

6.2.3.1 Comparative Evaluations Against MPEG-7 DCD

In order to demonstrate the strict parameter dependency of MPEG-7 DCD, we have varied only two parameters, T_A and T_S while keeping the others fixed, i.e., $N_{DC}^{max} = 25$, and $\varepsilon_D = 0.01$. Experiments are performed with three sets of parameters: $P1 : T_A = 1\ \%$, $T_S = 15$, $P2 : T_A = 1\ \%$, $T_S = 25$, and $P3 : T_A = 5\ \%$, $T_S = 25$. The number of DCs (per image) plots obtained from the 110 images in the sample database using each parameter set can be seen in Fig. 6.7. It is evident from the figure that the number of DCs is strictly dependent on the parameters used and can vary significantly, i.e., between 2 and 25.

Figures 6.8 and 6.9 show some visual examples from the sample database. In both figures, the first column shows the output of the Median-Cut algorithm with 256 (maximum) colors, which is almost identical to the original image. The second and the rest of the three columns show the back-projected images using the DCs extracted from the presented technique and MPEG-7 DCD with those three parameter sets, respectively. Note that the parts, where DC centroids cannot be accurately localized or missed completely by MPEG-7 DCD, are pointed with (yellow) arrows. There is an ambiguity for deciding which parameter set yield the best visual performance although it would have naturally been expected from the first set, $P1 : T_A = 1\ \%$, $T_S = 15$, where the highest number of DCs are extracted (see the *red* plot in Fig. 6.7), but it is evident that $P2$ and $P3$ can also yield "comparable or better" results; however, it is a highly subjective matter.

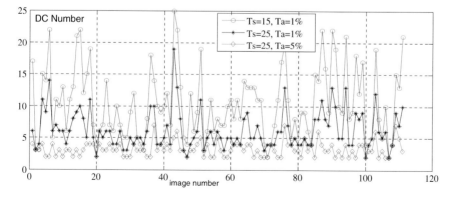

Fig. 6.7 Number of DC plot from three MPEG-7 DCDs with different parameter set over the sample database

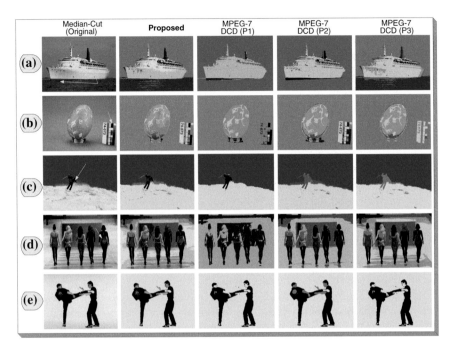

Fig. 6.8 The DC extraction results over 5 images from the sample database (*best viewed in color*)

According to the results, one straightforward conclusion is that not only the number of DCs significantly varies but DC centroids, as well, change drastically depending on the parameter values used. On the other hand, it is obvious that the best DC extraction performance is achieved by the presented technique, where none of the prominent colors are missed or mislocated while the "true" number of DCs is extracted. However, we do not in any waw claim that the presented technique achieves the minimum quantization error (or the mean square error, MSE), due to two reasons. First, the optimization technique is applied over a regularization (fitness) function where the quantization error minimization (i.e., minimum *Compactness*) is only one part of it. The other part, implying maximum *Separation*, presents a constraint so that minimum MSE has to be achieved using the least number of clusters (DCs). The second and the main reason is that the computation of MSE is typically performed in RGB color space, using the *Euclidean* metric. Recall that the presented DC extraction is performed over HSV (or HSL) color domain, which is discontinuous and requires nonlinear transformations, and using a *fuzzy* distance metric with respect to the HVS perceptual rules for enhancing the discrimination power. Therefore, the optimization in this domain using such a *fuzzy* metric obviously cannot ensure a minimum MSE in RGB domain. Besides that, several studies show that MSE is *not* an appropriate metric for visual (or perceptual) quality (e.g., [26]) and thus we hereby avoid using it as a performance measure.

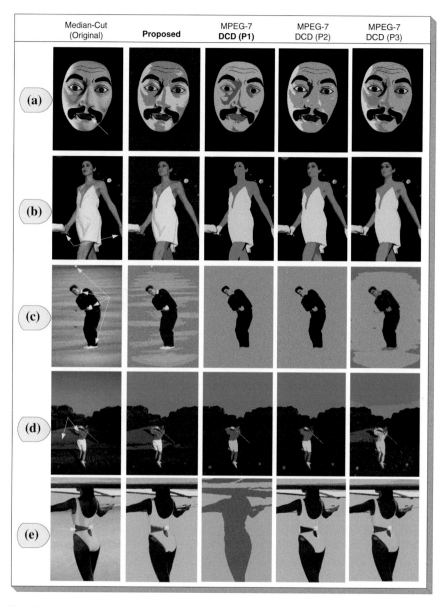

Fig. 6.9 The DC extraction results over 5 images from the sample database (*best viewed in color*)

6.2.3.2 Robustness and Parameter Insensitivity

Due to its stochastic nature, there is a concern about robustness (defined here as repeatability) of the results. In this section, we perform several experiments to examine whether or not the results are consistent in regard to accuracy of the DC centroids and their numbers. Repeatability would be a critical problem for deterministic methods such as K-means, Min–Max, etc., if the initial color (cluster) centroids are randomly chosen, as the original algorithm suggests. Eventually such methods would create a different clustering scheme each time they are performed since they are bound to get trapped to *the nearest* local optimum from the initial position. The solution to this problem induced by MPEG-7 DCD method is to change the random initialization part to a fixed (deterministic) initial assignment to the existing data points so that the outcome, DC centroids, and the number of DCs extracted, will be the same each time the algorithm is performed over a particular image with the same parameters. This would also be a practical option for the presented technique, i.e., fixing the initialization stage and using a constant *seed* for the random number generator that MD PSO uses. However, as a global optimization method, we shall demonstrate that MD PSO with FGBF can most of the time converge to (near-) optimal solutions, meaning that, the number of DCs and their centroids extracted from the presented dynamic clustering technique shall be consistent and perceptually intact. Furthermore, in order to show that significant variations for two major parameters, *iterNo* and S, do not cause drastic changes on the DC extraction, we will use three parameter sets: $P1$: $S = 50$, *iterNo* $= 500$, $P2$: $S = 50$, *iterNo* $= 1,000$ and $P3$: $S = 100$, *iterNo* $= 1,000$. With each parameter set, we run the presented technique (with random initialization and random seeds) 100 times over two images. The DC number histograms per image and per parameter set are as shown in Fig. 6.10. In the first image (left), it is certain that the number of DCs is either 2 or 3, as one might argue the yellowish color of the dot texture over the object can be counted as a DC or not. For the image on the right, it is rather difficult to decide the exact number of DCs, since apart from blue, the remaining 5 colors, red, pink, yellow, green, and brown have certain *shades*.

It is, first of all, obvious that the presented technique is parameter invariant since in both cases, the significant parameter variations, (particularly from $P1$ to $P3$ where both *iterNo* and S are doubled) only cause a *slight* difference over the histograms. A high degree of robustness (repeatability) is also achieved since *all* runs in the first image yielded either 2 or 3 DCs, as desired and >95 % of the runs in the second image, the number of DCs is in the range 16 ± 1. Among the back-projected images, it is evident that quite similar/almost identical DCs are anyway extracted even though they have different number of DCs (e.g., see the two with 14 and 18 DCs). As a result of the perceptual model used, the number of DCs can slightly vary, somewhat reflecting the *subjectivity* in HVS color perception, but similar DCs are extracted by the presented technique regardless of the parameter set used.

6.2.3.3 Computational Complexity Analysis

The computational complexity of the DC extraction technique depends on two distinct processes. First is the pre-processing stage which creates a limited color palette in RGB color domain using the Median Cut method and the following RGB

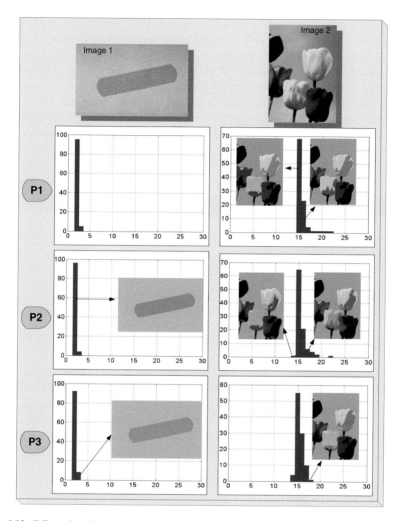

Fig. 6.10 DC number histograms of 2 sample images using 3 parameter sets. Some typical back-projected images with their DC number pointed are shown within the histogram plots (*best viewed in color*)

to HSV color transformation. Recall that the Median Cut is a fast method (i.e., $O(n)$), which has the same computational complexity as K-means. The following color transformation has an insignificant processing time since it is only applied to a reduced number of colors. As the dynamic clustering technique based on MD PSO with FGBF is stochastic in nature, a precise computational complexity analysis is not feasible; however, there are certain attributes, which proportionally affect the complexity such as swarm size (S), the total number of iteration (*IterNo*) and the dimension of the data space, (n). Moreover, the complexity of the validity index used has a direct impact over the total computational cost since for each particle (and at each iteration) it is used to compute the fitness of that particle. This is the main reason of using such a simple (and parameter independent) validity index as in Eq. (6.1). In that, the presented fuzzy color model makes the computational cost primarily dependent on the color structure of the image because the normalized *Euclidean* distance that is given in Eq. (6.5) and is used within the validity index function is obviously quite costly; however, recall that it may not be used at all for such color pairs that do not show any perceptual color similarity. This further contributes to the infeasibility of performing an accurate computational complexity analysis for the presented technique. For instance, it takes on the average of 4.3 and 17.2 s to extract the DCs for the images 1 and 2 shown in Fig. 6.10, respectively. Nevertheless, as any other EA the DC extraction based on MD PSO with FGBF is slow in nature and may require indefinite amount of iterations to converge to the global solution.

6.2.4 Summary and Conclusions

In this section, the dynamic clustering technique based on MD PSO with FGBF is applied for extracting *"true"* number of dominant colors in an image. In order to improve the discrimination among different colors, a *fuzzy* model over HSV (or HSL) color space is then presented so as to achieve such a distance metric that reflects HVS perception of color (dis-) similarity. The DC extraction experiments using MPEG-7 DCD have shown that the method, although a part of the MPEG-7 standard, is highly dependent on the parameters. Moreover, since it is entirely based on K-means clustering method, it can create artificial colors and/or misses some important DCs due to its convergence to local optima, thus yielding critical over- and under-clustering. Consequently, a mixture of different colors, and hence artificial DCs or DCs with shifted centroids, may eventually occur. This may also cause severe degradations over color textures since the regular textural pattern cannot be preserved if the true DC centroids are missed or shifted. Using a simple CVI, we have successfully addressed these problems and a superior DC extraction is achieved with ground-truth DCs. The optimum number of DCs can slightly vary on some images, but the number of DCs on such images is hardly definitive, rather subjective and thus in such cases the dynamic clustering based on a stochastic optimization technique can converge to some near-optimal solutions. The technique presented in this section shows a high level of robustness for parameter insensitivity and hence the main idea is that instead of struggling to fine tune several parameters to improve performance, which is not straightforward—if possible at all, the focus can now be drawn to designing better validity index functions or improving the ones for the purpose of higher DC extraction performance in terms of perceptual quality.

6.3 Dynamic Data Clustering via SA-Driven MD PSO

The theory behind the SA-driven PSO and its multi-dimensional extension, SA-driven MD PSO is well explained in the previous chapter. Recall that unlike the FGBF method, the main advantage of this global convergence technique is its generic nature—the applicability to any problem without any need of adaptation or tuning. Therefore, the clustering application of SA-driven (MD) PSO requires no changes for both Simultaneously Perturbed Stochastic Approximation (SPSA) approaches, and the (MD) PSO particles are encoded in the same way as was explained in Sect. 6.1. In this section we shall focus on the application of SA-driven MD PSO in dynamic clustering.

6.3.1 SA-Driven MD PSO-Based Dynamic Clustering in 2D Datasets

In order to test each SA-driven MD PSO approach over clustering, we created 8 synthetic data spaces as shown in Fig. 6.11 where white dots (pixels) represent data points. For illustration purposes each data space is formed in 2D; however, clusters are formed with different shapes, densities, sizes, and inter-cluster distances to test the robustness of clustering application of the proposed approaches against such variations. Furthermore, recall that the number of clusters determines the (true) dimension of the solution space in a PSO application and hence it is also kept varying among data spaces to test the converging accuracy to the true (solution space) dimension. As a result, significantly varying complexity levels are established among all data spaces to perform a general-purpose evaluation of each approach.

Unless stated otherwise, the maximum number of iterations is set to 10,000 as before; however, the use of cut-off error as a termination criterion is avoided since

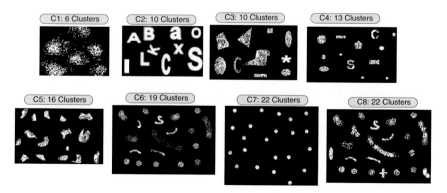

Fig. 6.11 2D synthetic data spaces carrying different clustering schemes

it is not feasible to set a unique ε_C value for all clustering schemes. The positional range can now be set simply as the natural boundaries of the 2D data space. For MD PSO, we used the swarm size, $S = 200$ and for both SA-driven approaches, a reduced number is used in order to ensure the same number of evaluation among all competing techniques. w is linearly decreased from 0.75 to 0.2 and we again used the recommended values for A, α and γ as 60, 0.602, and 0.101, whereas a and c are set to 0.4 and 10, respectively. For each dataset, 20 clustering runs are performed and the 1st and 2nd order statistics (mean, μ and standard deviation, σ) of the fitness scores and *dbest* values converged are presented in Table 6.2.

According to the statistics in Table 6.2, similar comments can be made as in the PSO application on nonlinear function minimization, i.e., either SA-driven approach achieves a superior performance over all data spaces regardless of the number of clusters and cluster complexity (modality) without any exception. The superiority hereby is visible on the average fitness scores achieved as well as the proximity of the average *dbest* statistics to the optimal dimension. Note that d in the table is the optimal dimension, which may be different than the true number of clusters due to the validity index function used.

Some further important conclusions can be drawn from the statistical results in Table 6.2. First of all, the performance gap tends to increase as the cluster number (dimension of the solution space) rises. For instance all methods have fitness scores in a close vicinity for the data space $C1$ while both SA-driven MD PSO approaches perform significantly better for $C7$. Note, however, that the performance gap for $C8$ is not as high as in $C7$, indicating SPSA parameters are not appropriate for $C8$ (as a consequence of fixed SPSA parameter setting). On the other hand, in some particular clustering runs, the difference in the average fitness scores in Table 6.2 does not correspond to the actual improvement in the clustering quality. Take for instance the two clustering runs over $C1$ and $C2$ in Fig. 6.12, where some clustering instances with the corresponding fitness scores are shown. The first (left-most) instances in both rows are from severely erroneous clustering operation although only a mere difference in fitness scores occurs with the instances in the second column, which have significantly less clustering errors. One the other hand, the proximity of the average *dbest* statistics to the optimal dimension may be another alternative for the evaluation of the clustering performance; however, it is likely that two runs, one with severely under- and another with over-clustering, may have an average *dbest* that is quite close to the optimal dimension. Therefore, the standard deviation should play an important role in the evaluation and in this aspect; one can see from the statistical results in Table 6.2 that the second SA-driven MD PSO approach ($A2$) in particular achieves the best performance (i.e., converging to the true number of clusters and correct localization of the centroids) while the performance of the standalone MD PSO is the poorest.

For visual evaluation, Fig. 6.13 presents the *worst* and the *best* clustering results of the two competing techniques, standalone versus SA-driven MD PSO, based on the highest (worst) and lowest (best) fitness scores achieved among the 20 runs. The clustering results of the best performing SA-driven MD PSO approach, as highlighted in Table 6.2, are shown while excluding $C1$ since results

Table 6.2 Statistical results from 20 runs over 8 2D data spaces

Clusters	No.	d	MD PSO				SA-driven (A2)				SA-driven (A1)			
			Score		dbest		Score		dbest		Score		dbest	
			μ	σ	μ	σ	μ	σ	μ	σ	μ	σ	μ	σ
C1	6	6	1,456.5	108.07	6.4	0.78	1,455.2	103.43	6.3	0.67	1,473.8	109	6.2	1.15
C2	10	12	1,243.2	72.12	10.95	2.28	1,158.3	44.13	12.65	2.08	1,170.8	64.88	11.65	1.56
C3	10	11	3,833.7	215.48	10.4	3.23	3,799.7	163.5	11.3	2.57	3,884.8	194.03	11.55	2.66
C4	13	14	1,894.5	321.3	20.2	3.55	1,649.8	243.38	19.75	2.88	1,676.2	295.8	19.6	2.32
C5	16	17	5,756	1,439.8	19	7.96	5,120.4	1,076.3	22.85	4.17	4,118.3	330.31	21.8	2.87
C6	19	28	21,533	4,220.8	19.95	10.16	18,323	1,687.6	26.45	2.41	20,016	3,382	22.3	6.97
C7	22	22	3,243	1,133.3	21.95	2.8	2,748.2	871.1	23	2.51	2,380.5	1,059.2	22.55	2.8
C8	22	25	6,508.85	1,014	17.25	10.44	6,045.1	412.78	26.45	3.01	5,870.25	788.6	23.5	5.55

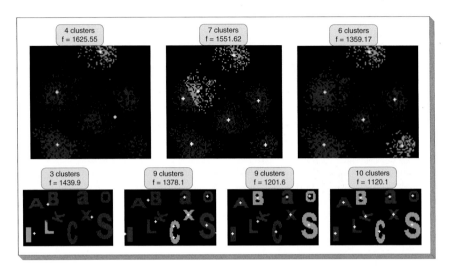

Fig. 6.12 Some clustering runs with the corresponding fitness scores (*f*)

of all techniques are quite close for this data space due to its simplicity. Note first of all that the results of the (standalone) MD PSO deteriorate severely as the complexity and/or the number of clusters increases. Particularly in the *worst* results, the critical errors such as under-clustering often occur with dislocated cluster centroids.

For instance 4 out of 20 runs for *C*6 result in severe under-clustering with 3 clusters, similar to the one shown in the figure whereas this goes up to 10 out of 20 runs for *C*8. Although the clusters are the simplest in shape and in density for *C*7, due to the high solution space dimension (e.g., number of clusters = 22), even the *best* MD PSO run is not immune to under-clustering errors. In some of the *worst* SA-driven MD PSO runs too, a few under-clusterings do occur; however, they are minority cases in general and definitely not as severe as in MD PSO runs. It is quite evident from the *worst* and the *best* results in the figure that SA-driven MD PSO achieves a significantly superior clustering quality and usually converges to a close vicinity of the global optimum solution.

6.3.2 Summary and Conclusions

In this section, SA-driven MD PSO is applied to the dynamic data clustering problem and tested over 8 synthetic data spaces in 2D with ground truth clusters. The statistical results obtained from the clustering runs approve the superiority of SA-driven MD PSO in terms of global convergence. As in SA-driven PSO application for nonlinear function minimization, we have applied a *fixed* set of SPSA parameters and hence we can make the same conclusion as before about the

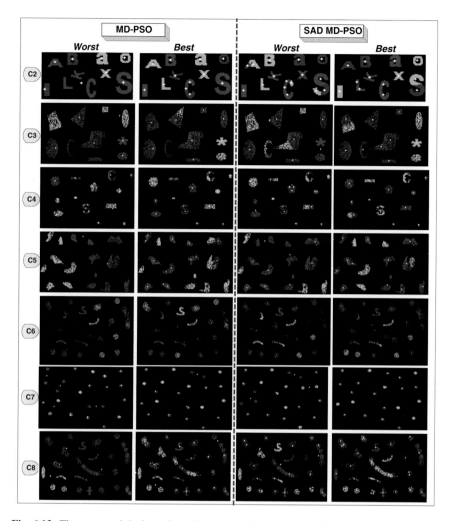

Fig. 6.13 The worst and the best clustering results using standalone (*left*) and SA-driven (*right*) MD PSO

effect of the SPSA parameters over the performance. Furthermore, we have noticed that the performance gap widens especially when the clustering complexity increases since the performance of the standalone MD PSO operation, without proper guidance, severely deteriorates. One observation worth mentioning is that the second approach on SA-driven MD PSO has a significant overhead cost, which is anyway balanced by using a reduced number of particles in the experiments; therefore, the low-cost mode should be used with a limited dimensional range for those applications with high computational complexity.

6.4 Programming Remarks and Software Packages

As the basics described in Sect. 4.4.2, the major MD PSO test-bed application is
PSOTestApp where several MD PSO applications, mostly based on data (or
feature) clustering, are implemented. In this section, we shall describe the programming details for performing MD PSO-based dynamic clustering applications:
(1) 2D data clustering with FGBF, (2) 3D dynamic color quantization, and (3) 2D
data clustering with SA-driven MD PSO. These operations that are plugged into
the **template <class T, class X> bool CPSO_MD <T,X>::Perform()** function,
the first clustering application performed in **CPSOcluster** class is explained in the
following section. We shall then explain the implementation details the second
application performed in **CPSOcolorQ** class. Finally, the third application, which
is also performed in **CPSOcluster** class will be explained. Note that the interface
CPSOcluster class was described in Sect. 4.4.2, therefore the focus is mainly
drawn on the FGBF plug-in function for both MD PSO modes, FGBF and SA-
driven.

6.4.1 FGBF Operation in 2D Clustering

MD PSO application for 2D clustering operations over synthetic datasets is briefly
explained in Sect. 4.4.2. Recall that there are two sample CVI functions implemented: **ValidityIndex()** and **ValidityIndex2()**. Recall further that the data space
(**class X**) is implemented by the **CPixel2D** class declared in the **Pixel2D.h** file.
This class basically contains all the arithmetic operations that can be performed
over 2D pixels with coordinates (**m_x** and **m_y**). Table 6.3 presents the initialization steps performed in the **CPSOcluster::PSOThread()** function and presents
the rest of the code that initiates MD PSO (with FGBF) clustering operation. The
positional and velocity ranges (**psoVelMin, psoVelMax, psoPosMin, psoPos-
Max**) are then set according to the frame dimensions. Then the MD PSO object
(**m_pPSO**) is initialized with these settings. Finally, the call, **m_pPSO->Per-
form()**, will start the MD PSO clustering operation with the user setting MD PSO
mode as FGBF. The main function in **PSOTestApp**, **CPSO_MD<T,X>::Per-
form()**, is identical to the one in **PSO_MDlib** application except the plug-in FGBF
operations for clustering and feature synthesis. This plugin is called in the following **if()** statement,

```
if(m_mode == FGBF)
{
    If (m_bFS == true) FGBF_FSFn(xyb, xyb_i, score_min);//FGBF for Syn-
    thesis Function…
        else FGBF_CLFn(xyb, xyb_i, score_min);//FGBF for CLustering Function…
}//if FGBF…
```

Table 6.3 MD PSO initialization in the function **CPSOcluster::PSOThread()**

```
...
      // SET 3 Fn. ptrs for Fitness, Dist (for MST) and GBDim..
      m_pPSO->SetFitnessFn(ValidityIndex2);
      m_pPSO->SetDistFn(GetDist);
      m_pPSO->SetFindGBDimFn(FindGBDim);
      m_pPSO->SetSAparam(m_saParam); // Set SA params..

      CPixel2D psoVelMin, psoVelMax, psoPosMin, psoPosMax;
      psoVelMin.m_x = m_psoParam._xvMin; psoVelMin.m_y=m_psoParam._xvMin;
      psoVelMax.m_x = m_psoParam._xvMax; psoVelMax.m_y=m_psoParam._xvMax;
      psoPosMin.m_x = 0; psoPosMin.m_y=0;
      psoPosMax.m_x = Xs-1; psoPosMax.m_y=Ys-1;

      m_pPSO->Init(m_psoParam._dMin, m_psoParam._dMax, m_psoParam._vdMin, m_psoParam._vdMax, pso-
PosMin, psoPosMax, psoVelMin, psoVelMax);

      int time = timeGetTime();
      m_pPSO->Perform(); // Apply MD PSO for clustering..
      time -= timeGetTime();
      DbgOut("\n\n*** Time (msec) per iter.= %5.2lf\n\n", (double) time / m_psoParam._maxNoIter);

      GetResults(fp); // Get Results on Image and Show them on UI..
...
```

In this section, we shall focus on the FGBF function, **FGBF_CLFn()**, for any (2D, 3D, or N-D) clustering operation. The other function, **FGBF_FSFn()**, for feature syntheses will be covered in Chap. 10.

Recall from the FGBF application over clustering that in order to achieve well-separated clusters and to avoid the selection of more than one centroid representing the same cluster, *spatially* close centroids are first grouped using a MST and then a certain number of centroid groups, say $d \in [D_{\min}, D_{\max}]$, can be obtained simply by breaking $(d-1)$ longest MST branches. From each group, one centroid, which provides the highest *Compactness* score [i.e., minimum dimensional fitness score, $f(a,j)$ as given in Eq. (6.2) is then selected and inserted into $a[j]$ as the jth dimensional component. Table 6.4 presents the actual code accomplishing the first two steps: (1) By calling **m_fpFindGBDim(m_pPA, m_noP)**, a subset among all potential centroids (dimensions of swarm particles) is first formed by verifying the following: a dimension of any particle is selected into this subset if and only if there is at least one data point that is closest to it, and (2) Formation of the MST object by grouping the *spatially* close centroids (dimensions).

Table 6.5 presents how the first step is performed. In the first loop, it is resetting the Boolean array that is present in each swarm particle, which actually holds the $a[j]$ array. Then it determines which particle has a dimensional component (a potential centroid) in its current dimension that is closest to one of the data points (a white pixel in **CPixel2D *pPix** structure). The closest centroid is then selected as the candidate dimension (one of the red '+' in Fig. 6.1), all of which are then used to form the MST within the **FGBF_CLFn()** function. Note that in the *for-loop* over all swarm particles, the dimensions selected within the **fpFindGBDim()** function of each particle are then appended into the MST object by **pMSTp->AppendItem(&pCC[c])**. In this way the MST is formed from the selected dimensions (centroids) as illustrated in Fig. 6.1.

Table 6.4 MST formation in the function **CPSO_MD<T,X>::FGBF_CLFn()**

```
template <class T, class X> void CPSO_MD<T,X>::FGBF_CLFn(T** xyb, T* xyb_i, double* score_min)
{
        CMSTGlobal<X> *pMSTp = new CMSTGlobal<X> (m_fpGetDist, 0); // Create cell MST obj..
        m_fpFindGBDim(m_pPA, m_noP); // Determine the potential GB dimensions first..

        bool bCreated = false;
        for( int a=0;a<m_noP;a++)
        { // Create the MST with the potential cluster centroids..
                T* xx = m_pPA[a]->GetPos(); // Current position of the particle a in cur. dim..
                X *pCC = xx->GetPos(); // all dimensions (pixel pos.) of this particle
                int nDim = xx->GetDim(); // and the total dim.
                bool *bInGB = xx->GetGBDim(); // The boolean array per dim..
                for(int c=0; c<nDim; c++) // for all potential cluster candidates..
                        if(bInGB[c]))
                        {
                                if(!bCreated) {
                                        pMSTp->Create(&pCC[c]);
                                        bCreated = true; // Append first item into MST..
                                } else pMSTp->AppendItem(&pCC[c]) ; // Append the rest..
                        } // if..
        } // for..
...
```

Once the MST is formed, then its longest branches are iteratively broken to form the group of centroids. Table 6.6 presents the code breaking the longest MST branches within a loop starting from 1 to D_{max}. Recall that the best centroid candidate selected from each group will be used to form the corresponding dimensional component of the *aGB* particle. Each (broken) group is saved within the MST array, **pMSTarr[]**, and in the second for loop at the bottom, there is a search for the longest branch among all MSTs stored in this array. Once found, the MST with the (next) longest branch is then broken into two siblings and each is added into the array.

Finally, as presented in Table 6.7 the *aGB* particle (**xyb[]**) is formed by choosing the best dimensional components (the potential centroids with the minimum $f(a,j)$) where the individual dimensional scores are computed within the CVI function for all particles and stored in **m_bScore** member of **CPixel2D** class. Note that the **if**() statement checks whether the *aGB* formation is within the dimensional range, $d \in \{D_{min}, D_{max}\}$. If so, then within a *for* loop of all MST groups, the best dimensional component is found and assigned to the corresponding dimension of the *aGB* particle. Once the *aGB* particle is formed (for that dimension), then recall that it has to compete with the best of previous *aGB* and *gbest* particles. If it surpasses, only then it will be the GB particle of the swarm for that dimension. This is then repeated for each dimensions in the range, $d \in \{D_{min}, D_{max}\}$, and the GB particle is formed according to the competition result.

Table 6.8 shows the implementation of the CVI function that is given in Eq. (6.1). Recall that the entire 2D dataset (white pixels) is stored in the link list **s_pPixelQ**. In the first *for* loop each of them is assigned to the closest potential cluster centroids (stored in the current position of the particle in dimension **_nDim**) and stored in an array of link lists **pClusterQ[cm]**. In this way we can

Table 6.5 MST formation in the function **CPSO_MD<T,X>::FGBF_CLFn()**

```
void CPSOcluster::FindGBDim(CParticle<CSolSpace<CPixel2D>,CPixel2D>** pPA, int noP)
{
        // First Reset all candidacy of all dimensions..
        for (int a=0;a<noP;a++)
        {
            CSolSpace<CPixel2D>* pPos = pPA[a]->GetPos(); // Current position of the particle a in cur. dim..
            bool *blnGB = pPos->GetGBDim(); // The boolean array per dim..
            memset(blnGB, 0, pPos->GetDim()*sizeof(bool)); // Reset all..
        } // for..

        int min_c, min_a;

        // Step 1: Compute the avg. intra-cluster distance..
        s_pPixelQ->Reset();
        for(CPixel2D *pPix = NULL; (pPix = s_pPixelQ->Next()); )
        { // For all pixels in the array..
            float min_dist2 = 1e+9; // init..
            for (int a=0;a<noP;a++)
            { // for all particles..
                CSolSpace<CPixel2D>* pPos = pPA[a]->GetPos(); // Current position of the particle a in cur. dim..
                CPixel2D *pCC = pPos->GetPos(); // all dimensions (pixel pos.) of this particle
                int nDim = pPos->GetDim();
                for(int c=0; c<nDim; c++)
                { // for all potential cluster candidates..
                    float  dist2  =  ((pCC[c].m_x-pPix->m_x)*(pCC[c].m_x-pPix->m_x))  +  ((pCC[c].m_y-pPix-
>m_y)*(pCC[c].m_y-pPix->m_y));
                    if(dist2<min_dist2)
                    {
                        min_dist2=dist2; // assign the point to the closest cluster center and compute its dist..
                        min_c = c; // in this dimension (pix. pos.)..
                        min_a = a; // for this particle..
                    } // if..
                } // for..
            } // for..
            pPA[min_a]->GetPos()->SetGPDim(min_c); // This dim. is granted for candidacy for GB formation..
        } // for..
}
```

know the group (cluster) of pixels represented with the **cm**th cluster centroid. The second *for* loop then evaluates each cluster centroid by computing the quantization error, Q_e if and only if it has at least one or more data points assigned to it. Otherwise, the entire set of potential centroids proposed by the particle in that dimension will be discarded due to the violation of the second clustering constraint by assigning its fitness value to a very large value (i.e., $1e + 9$) imitating a practical infinity value. As mentioned earlier, individual dimensional scores are also computed and stored in the **m_bScore** member, that will then be used to perform FGBF. Finally, the Q_e computed for a particle position is multiplied by the *Separation* term, $(xd_a(t))^\alpha$, where $\alpha = 3/2$ in this CVI function and returned as the fitness (CVI) score of the current position of the particle.

6.4.2 DC Extraction in **PSOTestApp** *Application*

In order to perform DC extraction over natural images, the second action, namely "2. 3D Color Quantization over color images" should be selected in the GUI of the

Table 6.6 Formation of the centroid groups by breaking the MST iteratively

```
int noC = MIN(noItem, m_xdMax-1); // No of clusters from MST..
CMSTGlobal<X> **pMSTarr = new CMSTGlobal<X>* [noC]; // Create the clusters array..
CMSTGlobal<X>* pMST1, *pMST2, *pMST; // child MSTs..
pMSTp->BreakLongestBranch(pMST1, pMST2); // First  break parent into 2 children..

pMSTarr[0] = pMST1; pMSTarr[1] = pMST2; // initial insertion into array..
int cur_d = 2; // Number of MSTs so far present..
CQueueList < CBranch<X> >* pBranchQ = pMSTp->GetBranchQ();
int noBranch= pBranchQ->GetItemNo();

for (int m=1; m<noC; m++)
{
        If (cur_d >= m_xdMin) // We start to evalute after min. number of dimensions..
        {
           ...
           ...
        } // if..

        if(cur_d == noC) break;

        // LAST: find the MST with the next longest branch
        int index = 0;
        CBranch<X> *pLongBranch = pBranchQ->GetItem(noBranch-m); // Get the next longest one..
        for (int i=0; i<cur_d; i++)
        {
                If (pMSTarr[i]->IsBranch( pLongBranch ) )
                {       // next longest branch belongs to this MST?
                        pMST = pMSTarr[i]; // This MST is going to break into 2..
                        index = i; // keep the index in array..
                        break; // OK..
                } // if..
        } // for..

        CMSTGlobal<X>* pMSTtemp = NULL;
        pMST->BreakLongestBranch(pMSTtemp, pMSTarr[cur_d++]);
        delete pMST; // Local clean up..
        pMSTarr[index] = pMSTtemp; // the new one is replaced instead the one broken..
} // for..
```

PSOTestApp application as shown in Fig. 4.4 and then one or more images should be selected as input. Then the main dialog object from **CPSOtestAppDlg** class will use the object from **CPSOcolorQ** class in the **CPSOtestAppDlg::OnDopso()** function to extract the DCs (by the command: **m_PSOc.ApplyPSO(m_pXImList, m_psoParam, m_saParam)**). **CPSOcolorQ** class is quite similar to the **CPSO-cluster** having the same CVI function, and an identical code for MD PSO and FGBF operations. Recall that DC extraction is nothing but a dynamic clustering operation in 3D color space. Therefore, the few differences lie in the initialization, formation of the color dataset. Moreover, the template class <X> is now implemented within **CColor** class, which has three color components, **m_c1, m_c2, m_c3** and the weight of the color, **m_weight**. As shown in Table 6.9, during the initialization of the DC extraction operation in the beginning of the **CPSOcolorQ::PSOThread()** function, for the RBG frame buffer of each input image, there is a pre-processing step, which performs the median-cut method to create a color palette of number of colors, **MAX_COLORS** that is a constant value initially set

Table 6.7 Formation of the *aGB* particle

```
if(cur_d >= m_xdMin) // We start to evaluate after min. number of dimensions..
{
// Now there are "cur_d" number of MSTs available, get the best pos.s in each cluster..
        for (int c=0; c< cur_d; c++)
        { // for all clusters so far created..
                X* pCC = pMSTarr[c]->GetItem(0); // Get the first sol. pos.
                X* pCCmin = pCC; // init..
                double min = pCC->m_bScore; // init..
                for(int j=1; j<pMSTarr[c]->GetItemNo(); j++)
                {
                        X* pCCNext = pMSTarr[c]->GetItem(j); // get the next pos. in this cluster..
                        if(pCCNext->m_bScore < min)
                        {
                                min = pCCNext->m_bScore; // update the min. score..
                                pCCmin = pCCNext; // The min. score so far in this cluster..
                        } // if..
                } // for..
                xyb_i->Replace(pCCmin,c); // Assing the min. score dimension into the GB solution..
        } // for..

        // Now Check the Fractional GB score for this dim..
        double score = m_fpGetScore(xyb_i, cur_d); // Get the score for fractional dim. sol for dim=d..

        if(score < score_min[cur_d-m_xdMin])
        { // if a better score is obtained than before..
                *xyb[cur_d-m_xdMin] = xyb_i; // The global best sol. for dim=d so far..
                score_min[cur_d-m_xdMin] = score; // The best score for dim=d so far..
        } // if..

        if( score_min[cur_d-m_xdMin] < (score = m_pPA[m_gbest[cur_d-m_xdMin]]->GetPBScore(cur_d)))
        { // If this is better than the GB sol. of the swarm in dim=d..
                // then update all dimensions GB to fractional GB: xyb
                T *xyb_a = m_pPA[m_gbest[cur_d-m_xdMin]]->GetPBPos(cur_d); // The GB agent's solution for dim=d..
                *xyb_a = xyb[cur_d-m_xdMin]; // update the agent's GB in this dimension to fractional best..
                m_pPA[m_gbest[cur_d-m_xdMin]]->SetPBScore(score_min[cur_d-m_xdMin], cur_d); // Set the best score..
        } // if..
        else { // No the agents achieved better GB than the fractional, so copy back..
                score_min[cur_d-m_xdMin] = m_pPA[m_gbest[cur_d-m_xdMin]]->GetPBScore(cur_d);
        } // else..

} // if..
```

to 256. Over the this color palette, the dynamic clustering operation by MD PSO with FGBF will extract the optimum number of DCs with respect to the CVI function, **CPSOcolorQ::ValidityIndex2()**. It can be performed in one of the four color spaces: RGB, LUV, HSV, and HSL by setting **m_usedColorSpace** to either of the **CS_RGB, CS_LUV, CS_HSV, CS_HSL** variables in the constructor of the class, **CPSOcolorQ**. If the color space selected is not RGB, then the color palette is also converted to the selected color space and stored in the **CColor** array **s_ppColorA**. As explained in Sect. 1.2, each color space has, first of all, a distinct color distance function, i.e., HSV and HSL uses the distance metric given in Eq. (6.5), while RGB and LUV use a plain Euclidean function. Each color distance metric is performed in a distinct static function, which is stored in the function pointer **s_fpDistance**. In this way, both the **CPSOcolorQ:: ValidityIndex2()** and **CPSOcolorQ:: FindGBDim()** functions can use this static function pointer to

Table 6.8 The CVI function, **CPSOcluster::ValidityIndex2**

```
double CPSOcluster::ValidityIndex2(CSolSpace<CPixel2D> *pPos, int _nDim)
{
        CPixel2D *pCC = pPos->GetPos();
        int nDim = (pPos->GetDim() < _nDim) ? pPos->GetDim() : _nDim;
        CQueueList<CPixel2D>*   pClusterQ = new CQueueList<CPixel2D> [nDim];

        // Step 1: Compute the avg. intra-cluster distance..
        s_pPixelQ->Reset();
        for(CPixel2D *pPix = NULL; (pPix = s_pPixelQ->Next()); )
        { // For all pixels in the array..
                float min_dist2 = (pCC[0].m_x-pPix->m_x)*(pCC[0].m_x-pPix->m_x) +
                                  (pCC[0].m_y-pPix->m_y)*(pCC[0].m_y-pPix->m_y); // init..
                int cm=0; // init..
                for(int c=1; c<nDim; c++)
                { // for all potential cluster candidates..
                        float dist2 = ((pCC[c].m_x-pPix->m_x)*(pCC[c].m_x-pPix->m_x)) +
                                      ((pCC[c].m_y-pPix->m_y)*(pCC[c].m_y-pPix->m_y));
                        if(dist2<min_dist2)
                        {
                             min_dist2=dist2; // assign the point to the closest cluster center and compute its dist..
                             cm = c; // min. so far..
                        }
                } // for..
                pClusterQ[cm].Insert(pPix); // this pixel belongs to cm.th cluster..
        } // for..

        float intra_avg_dist2 = .0;
        for(int c=0; c<nDim; c++)
        {
                int noC = pClusterQ[c].GetItemNo();
                if( noC == 0)
                {
                        intra_avg_dist2 = 1e+9; // NoGo for such clustering scheme..
                        pCC[c].m_bScore = 1e+9; // NoGo for this dimension score within this solution cand..
                        continue; // Just check the other dimensions..
                } // if..
                float dist_c = .0;
                for(CPixel2D *pPix = NULL; (pPix = pClusterQ[c].Remove());) // For all pixels in the array..
                        dist_c += ((pCC[c].m_x-pPix->m_x)*(pCC[c].m_x-pPix->m_x)) +
                                  ((pCC[c].m_y-pPix->m_y)*(pCC[c].m_y-pPix->m_y));
                dist_c /= noC; // dist. per item..
                pCC[c].m_bScore = dist_c; // Set the dimension score within this solution cand..
                intra_avg_dist2 += dist_c; // the total sum..
        } // for..
        delete[] pClusterQ; // clean up..

        return (sqrt(nDim))*intra_avg_dist2;
}
```

compute the distance between two colors regardless from the choice of the color space. Furthermore, the positional range setting for MD PSO swarm particles vary with respect to the choice of the color space. For example, $\{X_{min}, X_{max}\} = \{\{0, 0, 0\}, \{255, 255, 255\}\}$ for RGB color space. As mentioned earlier, apart from such initialization details, the rest of the operation is identical the 2D dynamic clustering application explained in the previous section. Once the DCs are extracted, they are back projected to an output image in the **CPSOcolorQ:: GetResults()** function as some resultant images are shown in Figs. 6.8 and 6.9.

Table 6.9 Initialization of DC extraction in **CPSOcolorQ::PSOThread()** function

```
void CPSOcolorQ::PSOThread()
{
char fbase[511];
m_theImage = CreateImage(NULL); // Create an image decoder/encoder obj..

m_pFLQ->Reset();
for(m_pFL = NULL; (m_pFL = m_pFLQ->Next()); )
{ // For all the image files, apply PSO clustering..

        // Step 1: Decode & Get the RGB frame buffer ready for PSO indexing..
        sprintf(image_name, "%s\\%s.%s", m_pFL->m_dir, m_pFL->m_title, m_pFL->m_ext);
        sprintf(fbase, "%sCLUSTERS\\", m_pFL->m_dir); CreateDirectory(fbase, NULL);
        DecodeImage(m_theImage, image_name);
        sprintf(image_name, "%sCLUSTERS\\%s_doco", m_pFL->m_dir, m_pFL->m_title);
        DbgOut("Process %s ...\n", image_name);

        memset( s_ppColorA, 0, MAX_COLORS * sizeof(CColor*) );

        m_pOriginalRGB = GetExtImage(m_theImage, &m_width, &m_height); // 32bit RGB = [B, G, R, Alpha];
        m_pIndexImage  = (unsigned int*)calloc( m_width*m_height, sizeof(unsigned int) );

        int numColors3 = PreProcessing(); // MCA + Back Projection..
        int numColors = numColors3/3;  // size of the color pallette..

        CColor   psoVelMin, psoVelMax, psoPosMin, psoPosMax; // PSO param. for velocity eq.n..

        // Color Conversion to the color domain where MD PSO Color Quant. will be applied..
        if (m_usedColorSpace == CS_RGB)
        {
          ...
          ...
}
```

6.4.3 SA-DRIVEN Operation in PSOTestApp Application

Both SA-driven approaches, A1 and A2, are implemented within the **template
<class T, class X> bool CPSO_MD<T,X>::Perform()** function as two separate
plug-ins, and the code for 2D data clustering (or any other MD PSO application) is
identical due to the fact that both approaches are generic. The first SA-driven plug-
in that implements the second SA-driven approach (A2), is called in the following
if() statement,

if(m_mode == SAD)
{//A2: Create aGB particle and compete with gbest at each dimension...

  SADFn(xyb, score_min, iter);
}//if SAD...

and the plug-in function, **SADFn()**, is given in Table 6.10.

Recall that this approach is also an *aGB* formation, similar to the one given in
Table 6.7 while the underlying method is now SPSA, *not* FGBF, and thus it can be
used for any MD-PSO application. The new *aGB* particle is formed on each
dimension by applying SPSA over the personal best position of the *gbest* particle

Table 6.10 The plug-in function **SADFn()** for the second SA-driven approach, A2

```
template <class T, class X> void CPSO_MD<T,X>::SADFn(T** xyb, double* score_min, int iter )
{
        for (int cur_d=m_xdMin; cur_d<m_xdMax; cur_d++)
        {
                float ak = m_sap.a / pow(m_sap.A + iter+1, m_sap.alpha);
                float ck = m_sap.c / pow(iter+1, m_sap.gamma);

                X* delta = BernoulliX(cur_d); // Delta..
                for(int d=0; d<cur_d; ++d) delta[d] *= ck; // scale with ck..

                T *xyb_a = m_pPA[m_gbest[cur_d-m_xdMin]]->GetPBPos(cur_d); // Theta: The GB agent's PB pos.

                *xyb[cur_d-m_xdMin] = xyb_a ; // = Theta

#ifdef LOWCOST
                double yplus = m_pPA[m_gbest[cur_d-m_xdMin]]->GetPBScore(cur_d); // This is Theta + ck.delta ..
                *xyb[cur_d-m_xdMin] -= (delta); // = Theta
                *xyb[cur_d-m_xdMin] = (delta); // = Theta - ck.delta
                double yminus = m_fpGetScore(xyb[cur_d-m_xdMin], cur_d); // Get the score for theta - ck.delta..
                *xyb[cur_d-m_xdMin] += (delta); // = Theta..

#else

                *xyb[cur_d-m_xdMin] += (delta); // = Theta + ck.delta
                double yplus = m_fpGetScore(xyb[cur_d-m_xdMin], cur_d); // Get the score for theta + ck.delta..

                *xyb[cur_d-m_xdMin] = xyb_a ; // = Theta
                *xyb[cur_d-m_xdMin] -= (delta); // = Theta - ck.delta
                double yminus = m_fpGetScore(xyb[cur_d-m_xdMin], cur_d); // Get the score for theta - ck.delta..
                *xyb[cur_d-m_xdMin] = xyb_a ; // = Theta

#endif

                for(d=0; d<cur_d; ++d) delta[d] *= (ak*(yplus-yminus)/(2*ck*ck)); // ak*ghat...
                *xyb[cur_d-m_xdMin] -= (delta); // = Theta - ak*ghat ..

                xyb[cur_d-m_xdMin]->CheckLimit(); // to be sure that soln is not out of border..

                // Now Check the Fractional GB score for this dim..
                double score = m_fpGetScore(xyb[cur_d-m_xdMin], cur_d); // Get the score of the SAD position..

                if(score < score_min[cur_d-m_xdMin])
                { // if a better score is obtained than before..
                        ...
                        ...
```

(margin annotations: `for dim=cur_d..` appears beside the line with `T *xyb_a`)

on that dimension, **cur_d**: **m_pPA[m_gbest[cur_d-m_xdMin]]->GetPB-Pos(cur_d)**, basically implementing the pseudo-code given in Table 5.11. There is a low cost mode, which is applied if the "LOWCOST" definition is made. Once the new *aGB* particle for that dimension is created, its fitness function is computed and then compared with the personal best position of *gbest* particle in that dimension. This comparison is also identical to the FGBF operation given in Table 6.7 and thus excluded in the table above.

The first SA-driven approach is directly plugged into the MD PSO code as shown in Table. When this SAD mode is chosen (**m_mode = SADv1**), as the

Table 6.11 The plug-in for the first SA-driven approach, A1

```
if(m_mode == SADv1 && a == m_gbest[cur_dim_a-m_xdMin])
{            // Apply SA to the gbest particles of each dimension = PB sol. of gbest particles..

       float ak = m_sap.a / pow(m_sap.A + iter+1, m_sap.alpha);
       float ck = m_sap.c / pow(iter+1, m_sap.gamma);
       X* delta = BernoulliX(cur_dim_a); // Delta..
       for(int d=0; d<cur_dim_a; ++d) delta[d] *= ck; // scale with 2*ck..

       *xyb[cur_dim_a-m_xdMin] = xx ; // = Theta

#ifdef LOWCOST
       double yplus = m_pPA[a]->GetCurScore(); // This is Theta + ck.delta ..

       *xyb[cur_dim_a-m_xdMin] -= (delta); // = Theta
       *xyb[cur_dim_a-m_xdMin] -= (delta); // = Theta - ck.delta
       double yminus = m_fpGetScore(xyb[cur_dim_a-m_xdMin], cur_dim_a); // Get the score for theta - ck.delta..
       *xyb[cur_dim_a-m_xdMin] += (delta); // = Theta

#else
       *xyb[cur_dim_a-m_xdMin] -= (delta); // = Theta - ck.delta
       double yminus = m_fpGetScore(xyb[cur_dim_a-m_xdMin], cur_dim_a); // Get the score for theta - ck.delta..
       *xyb[cur_dim_a-m_xdMin] = xx ; // = Theta
       *xyb[cur_dim_a-m_xdMin] += (delta); // = Theta + ck.delta
       double yplus = m_fpGetScore(xyb[cur_dim_a-m_xdMin], cur_dim_a); // Get the score for theta - ck.delta..

       *xyb[cur_dim_a-m_xdMin] = xx ; // = Theta
#endif

       for(d=0; d<cur_dim_a; ++d) delta[d] *= (ak*(yplus-yminus)/(2*ck*ck)); // ak*ghat...
       *xyb[cur_dim_a-m_xdMin] -= (delta); // = Theta - ak*ghat ..
       *xx = xyb[cur_dim_a-m_xdMin]; // current estimate from theta - ck.delta
}
```

pseudo-code given in Table, SPSA will only update the position of the native *gbest* particle, **m_gbest[cur_dim_a-m_xdMin]** on its current dimension **cur_dim**. As in the second SA-driven approach, there is a low-cost mode applied with the same definition "LOWCOST" (Table 6.11).

References

1. J.T. Tou, R.C. Gonzalez, *Pattern Recognition Principles* (Addison-Wesley, London, 1974)
2. A.K. Jain, M.N. Murthy, P.J. Flynn, Data clustering: A review. ACM Comput. Rev. **31**(3), 264–323 (1999)
3. J.C. Bezdek, *Pattern Recognition with Fuzzy Objective Function Algorithms* (Plenum, New York, 1981)
4. A. Antoniou, W.-S. Lu, *Practical Optimization, Algorithms and Engineering Applications* (Springer, USA, 2007)
5. D. Goldberg, *Genetic Algorithms in Search, Optimization and Machine Learning* (Addison-Wesley, Reading, MA, 1989), pp. 1–25
6. J. Koza, *Genetic Programming: On the Programming of Computers by Means of Natural Selection* (MIT Press, Cambridge, MA, 1992)

7. T. Back, F. Kursawe, Evolutionary Algorithms for Fuzzy Logic: A Brief Overview, in *Proceedings of Fuzzy Logic and Soft Computing, World Scientific*, Singapore, Nov 1995, pp. 3–10
8. U.M. Fayyad, G.P. Shapire, P. Smyth, R. Uthurusamy, *Advances in Knowledge Discovery and Data Mining* (MIT Press, Cambridge, MA, 1996)
9. A.P. Engelbrecht, *Fundamentals of Computational Swarm Intelligence* (Wiley, Chichester, 2005)
10. T. Ince, S. Kiranyaz, J. Pulkkinen, M. Gabbouj, Evaluation of global and local training techniques over feed-forward neural network architecture spaces for computer-aided medical diagnosis. Expert Syst. Appl. **37**(12), 8450–8461 (2010). doi:10.1016/j.eswa.2010.05.033. (Article ID 4730)
11. M.G. Omran, A. Salman, A.P. Engelbrecht, Dynamic clustering using particle swarm optimization with application in image segmentation. Pattern Anal. Appl. **8**, 332–344 (2006)
12. M.G. Omran, A. Salman, A.P. Engelbrecht, *Particle Swarm Optimization for Pattern Recognition and Image Processing* (Springer, Berlin, 2006)
13. F. Van den Bergh, An Analysis of Particle Swarm Optimizers, PhD thesis, Department of Computer Science, University of Pretoria, Pretoria, South Africa, 2002
14. J. Riget, J.S. Vesterstrom, A Diversity-Guided Particle Swarm Optimizer—The ARPSO, Technical report, Department of Computer Science, University of Aarhus, 2002
15. J. R. Kruskal, On the Shortest Spanning Subtree of a Graph and the Traveling Salesman Problem, in *Proceedings of AMS*, vol. 7, no. 1, 1956
16. J.C. Dunn, Well separated clusters and optimal fuzzy partitions. J. Cyber. **4**, 95–104 (1974)
17. E.L. Van den Broek, P.M.F. Kisters, L.G. Vuurpijl, The Utilization of Human Color Categorization for Content-Based Image Retrieval, in *Proceedings of Human Vision and Electronic Imaging IX*, San José, CA (SPIE, 5292), 2004, pp. 351–362
18. A. Mojsilovic, J. Kovacevic, J. Hu, R.J. Safranek, K. Ganapathy, Matching and retrieval based on the vocabulary and grammar of color patterns. IEEE Trans. Image Process. **9**(1), 38–54 (2000)
19. A. Abraham, S. Das, S. Roy, Swarm Intelligence Algorithms for Data Clustering, in *Proceedings of Soft Computing for Knowledge Discovery and Data Mining Book, Part IV*, Oct. 25, 2007, pp. 279–313
20. Y. Deng, C. Kenney, M. S. Moore, B.S. Manjunath, Peer Group Filtering and Perceptual Color Image Quantization, in *Proceedings of IEEE International Symposium on Circuits and Systems, ISCAS*, vol. 4, 1999, pp. 21–24
21. J. Fauqueur, N. Boujemaa, Region-Based Image Retrieval: Fast Coarse Segmentation and Fine Color Description, in *Proceedings of IEEE Int. Conf. on Image Processing (ICIP'2002)*, (Rochester, USA, 2002), pp. 609–612
22. A. Mojsilovic, J. Hu, E. Soljanin, Extraction of perceptually important colors and similarity measurement for image matching, retrieval and analysis. IEEE Trans. Image Process. **11**, 1238–1248 (2002)
23. B.S. Manjunath, J.-R. Ohm, V.V. Vasudevan, A.Yamada, Color and texture descriptors. IEEE Trans. On Circuits and Systems for VideoTechnology. **11**, 703–715 (2001)
24. A. Kruger, Median-cut color quantization. Dr Dobb's J. Softw. Tools Prof. Program. **19**(10), 46–55 (1994)
25. Y. Shi, R.C. Eberhart, A Modified Particle Swarm Optimizer, in *Proceedings of the IEEE Congress on Evolutionary Computation*, 1998, pp. 69–73
26. Z. Wang, A.C. Bovik, H.R. Sheikh, E.P. Simoncelli, Image quality assessment: From error visibility to structural similarity. IEEE Trans. Image Process. **13**(4), 600–612 (2004)

Chapter 7
Evolutionary Artificial Neural Networks

If you just have a single problem to solve, then fine, go ahead and use a neural network. But if you want to do science and understand how to choose architectures, or how to go to a new problem, you have to understand what different architectures can and cannot do.

Marvin Minsky

Artificial neural networks (ANNs) are known as "universal approximators" and "computational models" with particular characteristics such as the ability to learn or adapt, to organize or to generalize data. Because of their automatic (self-adaptive) process and capability to learn complex, nonlinear surfaces, ANN classifiers have become a popular choice for many machine intelligence and pattern recognition applications. In this chapter, we shall present a technique for automatic design of Artificial Neural Networks (ANNs) by evolving to the optimal network configuration(s) within an architecture space (AS), which is a family of ANNs. The AS can be formed according to the problem in hand encapsulating indefinite number of network configurations. The evolutionary search technique is entirely based on multidimensional Particle Swarm Optimization (MD PSO). With a proper encoding of the network configurations and parameters into particles, MD PSO can then seek positional optimum in the error space and dimensional optimum in the AS. The optimum dimension converged at the end of a MD PSO process corresponds to a unique ANN configuration where the network parameters (connections, weights, and biases) can then be resolved from the positional optimum reached on that dimension. In addition to this, the proposed technique generates a ranked list of network configurations, from the best to the worst. This is indeed a crucial piece of information, indicating what potential configurations can be alternatives to the best one, and which configurations should not be used at all for a particular problem. In this chapter, the architecture space is defined over feed-forward, fully connected ANNs so as to use the conventional techniques such as back-propagation and some other evolutionary methods in this field. We shall then apply the evolutionary ANNs over the most challenging synthetic problems to test its optimality on evolving networks and over the benchmark problems to test its generalization capability as well as to make comparative evaluations with the several competing techniques. We shall demonstrate that MD PSO evolves to optimum or near-optimum networks in general and has a superior generalization capability. In addition, MD PSO naturally favors a low-dimension solution when it exhibits a competitive performance with a high dimension counterpart and such a

S. Kiranyaz et al., *Multidimensional Particle Swarm Optimization for Machine Learning*   187
and Pattern Recognition, Adaptation, Learning, and Optimization 15,
DOI: 10.1007/978-3-642-37846-1_7, © Springer-Verlag Berlin Heidelberg 2014

native tendency eventually steers the evolution process toward the compact network configurations in the architecture space instead of more complex ones, as long as optimality prevails.

7.1 Search for the Optimal Artificial Neural Networks: An Overview

In the fields of machine learning and artificial intelligence, the evolutionary search mimics the process of natural evolution for finding the optimal solution to complex high dimensional, multimodal problems. In other words, it is basically a search process using an evolutionary algorithm (EA) to determine the best possible (optimal or near-optimal) design among a large collection of potential solutions according to a given cost function. Up to date, designing a (near) optimal network architecture is made by a human expert and requires a tedious trial and error process. Specifically, determining the optimal number of hidden layers and the optimal number of neurons in each hidden layer is the most critical task. For instance, an ANN with no or too few hidden layers may not differentiate among complex patterns, and instead may lead to only a linear estimation of such—possibly nonlinear—problem. In contrast, if an ANN has too many nodes/layers, it might be affected severely by noise in data due to over-fitting, which eventually leads to a poor generalization. Furthermore, proper training of complex networks is often a time-consuming task. The optimum number of hidden nodes/layers might depend on the input/output vector sizes, the amount of training and test data, and more importantly the characteristics of the problem, e.g., its dimensionality, nonlinearity, dynamic nature, etc.

The era of ANNs started with the simplified neurons proposed by McCulloch and Pitts in 1943 [1], and particularly after 1980s, ANNs have widely been applied to many areas, most of which used feed-forward ANNs trained with the back-propagation (BP) algorithm. As detailed in Sect. 3.4.3, BP has the advantage of performing directed search, that is, weights are always updated in such a way as to minimize the error. However, there are several aspects, which make the algorithm not guaranteed to be universally useful. The most troublesome is its strict dependency on the learning rate parameter, which, if not set properly, can either lead to oscillation or an indefinitely long training time. Network paralysis [2] might also occur, i.e., as the ANN trains, the weights tend to be quite large values and the training process can come to a virtual standstill. Furthermore, BP eventually slows down by an order of magnitude for every extra (hidden) layer added to ANN [2]. Above all; BP is only a gradient descent algorithm applied on the error space, which can be complex and may contain many deceiving local minima (multi-modal). Therefore, BP gets most likely trapped into a local minimum, making it entirely dependent on the initial settings. There are many BP variants and extensions which try to address some or all of these problems, see, e.g., [3–5], yet all share one major drawback, that is, the ANN architecture has to be fixed in advance and the question of which specific ANN structure should be used for a particular problem still remains unsolved.

Several remedies can be found in the literature, some of which are briefly described next. Let N_I, N_H and N_O be the number of neurons in the input, the hidden, and the output layers of a 2-layers feed-forward ANN. Jadid and Fairbain in [6] proposed an upper bound on N_H such as $N_H \leq N_{TR}/(R + N_I + N_O)$ where N_{TR} is the number of training patterns. Masters [7] suggested that ANN architecture should resemble to a pyramid with $N_H \approx \sqrt{N_I + N_O}$. Hecht-Nielsen [8] proved that $N_H \leq N_I + 1$ by using Kolmogrov theorem. Such boundaries may only give an idea about the architecture range that should be applied in general but many problems with high nonlinearity and dynamic nature may require models that far exceed these bounds. Therefore, these limits can only serve as a rule of thumb which should further be analyzed.

Designing the optimal network architecture can be thought as a search process within the AS containing all potential and feasible architectures. Some attempts are found in the literature such as the research on constructive and pruning algorithms, [9–12]. The former methods initially assume a minimal ANN and insert nodes and links as warranted while the latter proceeds to the opposite way, i.e., starting with a large network, superfluous components are pruned. However, Angeline et al. in [13] pointed out that "*Such structural hill climbing methods are susceptible to becoming trapped at structural local minima.*" The reasoning behind this is clarified by Miller et al. in [14], stating that the AS is non-differentiable, complex, deceptive, and multimodal. Therefore, those constructive and pruning algorithms eventually face similar problems in the AS as BP does in the error (weight) space. This makes EA [15] such as genetic algorithm (GA) [16], genetic programming (GP) [17], evolutionary strategies (ES), [18], and evolutionary programming (EP), [19], more promising candidates for both training and evolving the ANNs. Furthermore, they have the advantage of being applicable to any type of ANNs, including feed-forward ANNs, with any type of activation functions.

GAs are a popular form of EAs that rely mainly on reproduction heuristic, *crossover,* and random *mutations.* When used for training ANNs (with a fixed architecture), many researchers [14, 20–23] reported that GAs can outperform BP in terms of both accuracy and speed, especially for large networks. However, as stated by Angeline et al. in [13] and Yao and Liu in [24], GAs are not well suited for evolving networks. For instance, the evolution process by GA suffers from the permutation problem [25], indicating that two identical ANNs may have different representations. This makes the evolution process quite inefficient in producing fit offsprings. Most GAs use binary string representations for connection weights and architectures. This creates many problems, one of which is the representation precision of quantized weights. If weights are coarsely quantized, training might be infeasible since the required accuracy for a proper weight representation cannot be obtained. On the other hand, if too many bits are used (fine quantization), binary strings may be unfeasibly long, especially for large ANNs, and this makes the evolution process too slow or even impractical. Another problem in this binary representation is that network components belonging to the same or neighboring neurons may be placed far apart in the binary string. Due to crossover operation, the interactions among such components might be lost and hence the evolution speed is drastically reduced.

EP-based Evolutionary ANNs (ENNs) in [13, 24] have been proposed to address the aforementioned problems of GAs. The main distinction of EPs from GAs is that they do not use the problematic *crossover* operation, instead make the commitment to *mutation* as the sole operator for searching over the weight and ASs. For instance the so-called EPNet, proposed in [24] uses 5 different mutations: hybrid training, node and connection deletions, node and connection additions. It starts with an initial population of M random networks, partially trains each network for some epochs, selects the network with the best-rank performance as the parent network and if it is improved beyond some threshold, further training is performed to obtain an offspring network, which replaces its parent and the process continues until a desired performance criterion is achieved. Over several benchmark problems, EPNet was shown to discover compact ANNs, which exhibits comparable performance with the other GA-based evolutionary techniques. However, it is not shown that EPNet creates optimal or near-optimal architectures. One potential drawback is the fact that the best network is selected based only on partial training since the winner network as a result of such initial (limited) BP training may not lead to the optimal network at the end. Therefore, this process may eliminate potential networks which may be the optimal or near-optimal ones if only a proper training would have been performed. Another major drawback is the algorithm's dependence on BP as the primary training method which suffers from the aforementioned problems. Its complex structure can only be suitable or perhaps feasible for applications where computational complexity is not a crucial factor and or the training problem is not too complex, as stated in [24], "However, EPNet might take a long time to find a solution to a large parity problem. Some of the runs did not finish within the user-specified maximum number of generations." Finally, as a hybrid algorithm it uses around 15 user-defined parameters/thresholds some of which are set with respect to the problem. This obviously creates a limitation in a generic and modular application domain.

7.2 Evolutionary Neural Networks by MD PSO

7.2.1 PSO for Artificial Neural Networks: The Early Attempts

Recall that PSO, which has obvious ties with the EA family, lies somewhere in between GA and EP. Yet unlike GA, PSO has no complicated evolutionary operators such as *crossover, selection,* and *mutation* and it is highly dependent on stochastic processes. PSO has been successfully applied for training feed-forward [26–29] and recurrent ANNs [30, 31] and several works on this field have shown that it can achieve a superior learning ability to the traditional BP method in terms of accuracy and speed. Only few researchers have investigated the use of PSO for evolutionary design of ANNs or to be precise, the fully connected feed-forward ANNs, the multilayer perceptrons (MLPs) only with single hidden layer. In [26, 29], the PSO–PSO algorithm and its slightly modified variant, PSO–PSO: weight decay (PSO–PSO:WD) have been proposed. Both techniques use an inner PSO to train the weights and an outer one to determine the (optimal) number of hidden nodes. Both methods perform worse classification performance than EP and GA-based Evolutionary Neural Networks (ENNs) on three benchmark problems from *Proben1* dataset [32]. Recently, Yu et al. proposed an improved PSO technique, the so-called IPSONet [28], which achieved a comparable performance in terms of average classification error rate over the same dataset in *Proben1*. All potential network architectures have been encoded into the particles of a single PSO operation, which evaluates their weights and architecture simultaneously. However, such an *all-in-one* encoding scheme makes the dimension of particles too high and thus the method can be applied to only single hidden layer MLPs with a limited number of (hidden) nodes (i.e., maximum 7 was used in [28]). Furthermore, it turns out to be a hybrid technique, which uses GA operators such as *mutation* and *crossover* in order to alleviate the stagnation problem of PSO on such high dimensions.

7.2.2 MD PSO-Based Evolutionary Neural Networks

The major drawback of many PSO variants including the basic method is that they can only be applied to a search space with a fixed dimension. However, in the field of ANNs as well as in many of the optimization problems (e.g., clustering, spatial segmentation, function optimization, etc.), the optimum dimension where the optimum solution lies, is also unknown and should thus be determined. By a proper adaptation, MD PSO can be utilized for designing (near-) optimal ANNs. In this section, the focus is particularly drawn on automatic design of feed-forward ANNs and the search is carried out over all possible network configurations within the specified AS. Therefore, no assumption is made about the number of (hidden) layers and in fact none of the network properties (e.g., feed-forward, differentiable activation function, etc.) is an inherent constraint of the proposed scheme. All network configurations in the AS are enumerated into a dimensional hash table with a proper hash function, which ranks the networks with respect to their complexity. That is, it associates a higher hash index to a network with a higher complexity. MD PSO can then use each index as a unique dimension of the search space where particles can make inter-dimensional navigations to seek an optimum dimension (*dbest*) and the optimum solution on that dimension, $x\hat{y}^{dbest}$. The optimum dimension found naturally corresponds to a distinct ANN architecture where the network parameters (connections, weights and biases) can be resolved from the positional optimum reached on that dimension. One important advantage of the proposed approach compared to other evolutionary methods is that at the end of a MD PSO evolution process, apart from the best solution achieved on the optimum dimension, 2nd, 3rd, etc., ranked solutions, which corresponds to 2nd, 3rd, etc., best network configurations within the AS are readily available. This further indicates other potential ANN configurations that might be convenient to use for the particular problem encountered. For example the best solution might correspond to a highly complex network whereas an acceptable performance can also be achieved by a much simpler one, say the 3rd best solution provided that it yields an acceptable performance loss. In addition, the worst architecture(s) will also be known, indicating what such ANN configurations should not be used at all for the same problem.

In this section, we shall use MD PSO for evolving fully connected, feed-forward ANNs or the so-called MLPs. The reasoning behind this choice is that MLP is the most widely used type of ANNs and conventional methods such as BP can be used for training. Furthermore, we can perform comparative evaluations against other PSO techniques such as IPSONet, PSO–PSO, and PSO–PSO:WD, all of which are used to automatically design MLPs—however, only with a single hidden layer. This may be a serious drawback especially for complex problems, and it is the main reason for their inferior performance with respect to other EAs based on GA and EP. No such assumptions over network properties are made for MD PSO and the AS can be defined over a wide range of configurations, i.e., say from a single-layer perceptron (SLP) to complex MLPs with many hidden layers.

We shall construct the AS as explained in Sect. 3.4.3.2 and herein we shall recall some of the key-points. A range is defined for the minimum and maximum number of layers, $\{L_{\min}, L_{\max}\}$ and the number of neurons for hidden layer l, $\{N^l_{\min}, N^l_{\max}\}$. The sizes of the input and the output layers are usually determined by the problem at hand and hence are assumed to be fixed. Therefore, the AS can now be defined by two range arrays, $R_{\min} = \{N_I, N^1_{\min}, \ldots, N^{L_{\max}-1}_{\min}, N_O\}$ and $R_{\max} = \{N_I, N^1_{\max}, \ldots, N^{L_{\max}-1}_{\max}, N_O\}$, the minimum and the maximum number of neurons allowed for each layer of a MLP, respectively. The size of both arrays is naturally $L_{\max} + 1$ where the corresponding entries define the range of the lth hidden layer. The size of the input and the output layers, $\{N_I, N_O\}$, is fixed and it is the same for all configurations in the AS for any l-layer MLP, where $L_{\min} \le l \le L_{\max}$. $L_{\min} \ge 1$ and $L_{\max}$ can be set to any value that is suitable for the problem at hand. The hash function then enumerates all potential MLP configurations into hash indices, starting from the simplest MLP with $L_{\min} - 1$ hidden layers, each has the minimum number of neurons given in $R_{\min}$, to the most complex network with $L_{\max} - 1$ hidden layers, each has the maximum number of neurons given in $R_{\max}$. In Sect. 3.4.3.2, the following range arrays are used, $R_{\min} = \{9, 1, 1, 2\}$ and $R_{\max} = \{9, 8, 4, 2\}$, which indicate that $L_{\max} = 3$. If $L_{\min} = 1$ then the hash function enumerates all MLP configurations in the AS as shown in

Table 4. Note that in this example, the input and output layer sizes are 9 and 2, which are eventually fixed for all MLP configurations. The hash function associates the 1st dimension to the simplest possible architecture, i.e., a SLP with only the input and the output layers (9×2). From dimensions 2–9, all configurations are 2-layer MLPs with a hidden layer size varying between 1 and 8 (as specified in the 2nd entries of R_{min} and R_{max}). Similarly, for dimensions 10 and higher, 3-layer MLPs are enumerated where the 1st and the 2nd hidden layer sizes are varied according to the corresponding entries in R_{min} and R_{max}. Finally, the most complex MLP with the maximum number of layers and neurons is associated with the highest dimension, 41. Therefore, all 41 entries in the hash table span the AS with respect to the configuration complexity and this eventually determines the dimensional range of the solution space as $D_{min} = 1$ and $D_{max} = 41$.

At time t, suppose that the particle a in the swarm, $\xi = \{x_1, \ldots, x_a, \ldots, x_S\}$, has the positional component formed as,

$$xx_a^{xd_a(t)}(t) = \left\{ \{w_{jk}^0\}, \{w_{jk}^1\}, \{\theta_k^1\}, \{w_{jk}^2\}, \{\theta_k^2\}, \ldots, \{w_{jk}^{O-1}\}, \{\theta_k^{O-1}\}, \{\theta_k^O\} \right\}$$

where $\left\{ w_{jk}^l \right\}$ and $\left\{ \theta_k^l \right\}$ represent the sets of weights and biases of the layer l. Note that the input layer ($l = 0$) contains only weights whereas the output layer ($l = O$) has only biases. By means of such a direct encoding scheme, the particle a represents all potential network parameters of the MLP architecture at the dimension (hash index) $xd_a(t)$. As mentioned earlier, the dimensional range, $D_{min} \leq xd_a(t) \leq D_{max}$, where MD PSO particles can make inter-dimensional jumps, is determined by the AS defined. Apart from the regular limits such as positional velocity range, $\{V_{min}, V_{max}\}$, dimensional velocity range, $\{VD_{min}, VD_{max}\}$, the data space can also be limited with a practical range, i.e., $X_{min} \leq xxd_a^{xd_a(t)}(t) < X_{max}$. In short, only a few boundary parameters need to be defined in advance for the MD PSO process, as opposed to other GA-based methods, or EPNet, which use several parameters and external techniques (e.g., Simulated Annealing, BP, etc.) in a complex process. Setting *MSE* in Eq. (3.15) as the fitness function enables MD PSO to perform evolutions of both network parameters and architectures within its native process.

7.2.3 Classification Results on Synthetic Problems

In order to test and evaluate the performance of MD PSO for evolving ANNs, experiments are performed over a synthetic dataset and real benchmark dataset. The aim is to test the "optimality" of the networks found with the former dataset, and the "generalization" ability while performing comparative evaluations against several popular techniques with the latter dataset. In order to determine which network architectures are (near-) optimal for a given problem, we apply exhaustive BP training over every network configuration in the given AS. As mentioned earlier, BP is a gradient descent algorithm, which for a single run, it is susceptible to get trapped to the nearest local minimum. However, applying BP a large number of times (e.g., $K = 500$ is used in the experiments) with randomized initial parameters eventually increases the chance of converging to (a close vicinity of) the global minimum in the error space. Note that even though K is high, there is still no guarantee of converging to the global optimum with BP; however, the aim is to obtain the "trend" of best performances achievable with every configuration under equal training conditions. In this way the optimality of the networks evolved by MD PSO can be justified.

Due to the reasoning given earlier, the AS is defined over MLPs (possibly including SLP) with any activation function. The input and output layer sizes are determined by the problem. We use the learning parameter for BP as $\lambda = 0.002$ and the iteration number is 10,000. The default PSO parameters for MD PSO are used, i.e. the swarm size, $S = 200$, and the velocity ranges $V_{max} = -V_{min} = X_{max}/2$, and $VD_{max} = -VD_{min} = D_{max}/2$. The dimensional range is determined by the AS defined and the positional range is set as $X_{max} = -X_{min} = 2$. Unless stated otherwise, these parameters are used in all experiments presented in this section. We use the most typical activation functions: *hyperbolic tangent* ($\tanh(x) = \frac{e^x - e^{-x}}{e^x + e^{-x}}$) or *sigmoid* ($sigm(x) = 1/(1 + e^{-x})$) for problems in this section with bipolar and unipolar inputs.

Many investigations in the literature developing EA-based methods for automatic training or evolving ANNs often present limited testing over quite simple synthetic problems. For example in [33], the *bPSO* training algorithm is tested against BP over function approximation problem of $y = 2x^2 + 1$. In another work, [34], a hybrid PSO-BP algorithm is tested over problems such as 3-bit parity and function approximation of $y = \sin(2x)e^{-x}$. Note that the dynamic nature of these functions is quite stationary and it is thus fairly easy to approximate them by MLPs. EPNet in [24] has been tested over N-bit parity problem where N is kept within $4 \leq N \leq 8$ (due to feasibility problems mentioned earlier). Similarly, on such low N-bit parity problems, even the simplest MLPs can provide as satisfactory solutions as any other. This eventually creates an ambiguity over the decision of "optimality" since no matter how limited the AS is defined; still many networks in it can achieve "near-optimum" solutions. For instance consider the AS defined with $R_{min} = \{3, 1, 1, 1\}$ and $R_{max} = \{3, 8, 4, 1\}$ for the 3-bit parity problem in [34], more than 90 % of the configurations in such a limited AS can achieve an MSE less than 10^{-4}. So in order to show the optimality of the network configurations evolved by MD PSO, we shall first of all, use this "limited" AS (R^1: $R^1_{min} = \{N_I, 1, 1, N_O\}$ and $R^1_{max} = \{N_I, 8, 4, N_O\}$) containing the simplest 1-, 2-, or 3-layer MLPs with $L^1_{min} = 1$ and $L^1_{max} = 3$. The set contains 41 networks similar to configurations in Table 4 with different input and output layer dimensions, i.e., $N_I = 9$ and $N_O = 2$. Furthermore, we have selected the following the most challenging set of problems that can be solved with as few as possible "optimal" configurations: function approximation of $y = \cos(x/2)\sin(8x)$, *10-bit parity* and *two-spiral*. In all experiments in this section, we set $K = 500$ for exhaustive BP

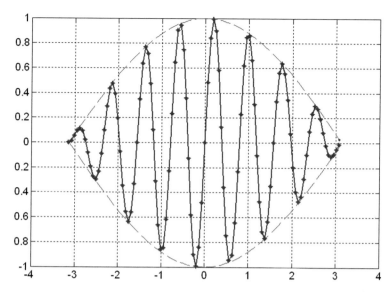

Fig. 7.1 The function $y = \cos(x/2)\sin(8x)$ plot in interval $\{-\pi, \pi\}$ with 100 samples

training and perform 100 MD PSO runs, each of which terminates at the end of 1,000 epochs (iterations).

As shown in Fig. 7.1, the function $y = \cos(x/2)\sin(8x)$ has a highly dynamic nature within the interval $\{-\pi, \pi\}$. 100 samples are taken for training where x coordinate of each sample is fed as input and y is used as the desired (target) output, so all networks are formed with $N_I = N_O = 1$. At the end of each run, the best fitness score (minimum error) achieved, $f(x\hat{y}^{dbest})$, by the particle with the index $gbest(dbest)$ at the optimum dimension $dbest$ is stored. The histogram of $dbest$, which is a hash index indicating a particular network configuration in R^1, eventually provides the crucial information about the (near-) optimal configuration(s).

Figure 7.2 shows the $dbest$ histogram and the error statistics plot from the exhaustive BP training for the function approximation problem. Note that BP can at best perform a linear approximation for most of the simple configurations (hash indices) resulting in MSE ≈ 0.125 and only some 3-layer MLPs (in the form of $1 \times M \times N \times 1$ where $M = 5,6,7,8$ and $N = 2, 3, 4$) yield convergence to near-optimal solutions. Accordingly, from the $dbest$ histogram it is straightforward to see that a high majority (97 %) of the MD PSO runs converges to those near-optimal configurations. Moreover, the remaining solutions with $dbest = 8$ and 9 (indicating 2-layer MLPs in $1 \times 7 \times 1$ and $1 \times 8 \times 1$ forms) achieve a competitive performance with respect to the optimal 3-layer MLPs. Therefore, these particular MD PSO runs are not indeed trapped to local minima; on the contrary, the exhaustive BP training on these networks could not yield a "good" solution. Furthermore, MD PSO evolution in this AS confirmed that 4 out of 41 configurations achieve the minimal MSEs (*mMSEs*), namely $1 \times 7 \times 4 \times 1$ (*mMSE*(40) = 0.0204), $1 \times 7 \times 2 \times 1$ (*mMSE*(23) = 0.0207), $1 \times 7 \times 3 \times 1$ (*mMSE*(33) = 0.0224), and $1 \times 8 \times 1$ (*mMSE*(9) = 0.0292). These are the four local peaks in the histogram indicating that the majority of the runs evolves to these optimum networks whilst the rest converged to similar but near-optimum configurations, which perform only a slightly worse than the optial solutions.

Figure 7.3 presents the fitness (MSE) and the dimension (hash index) plots per iteration for two distinct MD PSO runs where the red curves of both plots belong to the GB particle (the *gbest* particle at each iteration) and the corresponding plots of the blue curve exhibit the behavior of the particular *gbest* particles when the process terminates (i.e., *gbest* particles are 31 and 181 for the left and right plots, respectively). Recall that the main objective of the MD PSO is to find the true dimension where the global optimum resides and at this dimension, its internal process becomes identical with the *bPSO*. The optimum dimensions are 23 and 41 for the left and the right plots. Note that in both experiments, several dimensions became *dbest* before they finally converge to the true optimum. Furthermore, it is interesting to observe the steep descent in MSE of both *gbest* particles (blue curve) after they converge to the optimum dimensions.

Recall that an important advantage of MD PSO process is its ability to provide not only the best network configuration but a ranked list of all possible network configurations in the AS, especially when MD PSO takes a sufficiently long time for evolution. Figure 7.4 illustrates this fact with a typical experiment over $y = \cos(x/2)\sin(8x)$ function approximation problem. The plot on the top is identical to the one in Fig. 7.2, showing the error statistics of the exhaustive BP training

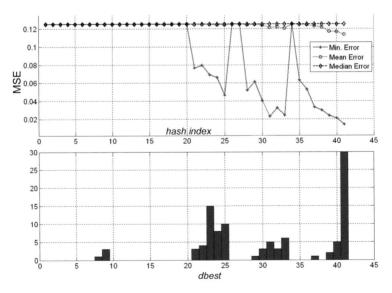

Fig. 7.2 Error statistics from exhaustive BP training (*top*) and *dbest* histogram from 100 MD PSO evolutions (*bottom*) for $y = \cos(x/2)\sin(8x)$ function approximation

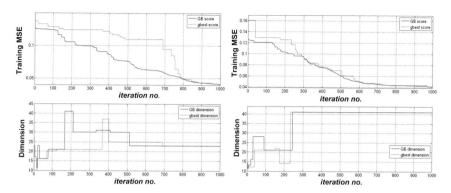

Fig. 7.3 Training MSE (*top*) and dimension (*bottom*) plots vs. iteration number for 17th (*left*) and 93rd (*right*) MD PSO runs

whereas the bottom one shows the minimum MSE (best fitness) achieved per dimension at the end of a MD PSO evolution process with 10,000 iterations (i.e. $\mathrm{mMSE}(d) = f(x\hat{y}^d(10,000))$ $\forall d \in \{1, 41\}$). Note that both curves show a similar behavior for $d > 18$ (e.g. note the peaks at $d = 19–20, 26–27,$ and 34); however, only MD PSO provides a few near-optimal solutions for $d \leq 18$, whereas none of the BP runs managed to escape from local minima (the linear approximation). Most of the optimal and near-optimal configurations can be obtained from this single run, e.g., the ranked list is $\{dbest = 23, 24, 40, 33, 22, 39, 30,...\}$. Additionally, the peaks of this plot reveal the worst network configurations that should not be used for this problem, e.g., $d = 1, 2, 8, 9, 10, 11, 19, 26,$ and 34. Note, however, that this can be a noisy evaluation since there is always the possibility that MD PSO process may also get trapped in a local minimum on these dimensions and/or not sufficient amount of particles visited this dimension due to their natural attraction toward *dbest* (within the MD PSO process), which might be too far away. This is the reason of the erroneous evaluation of dimensions 8 and 9, which should not be on that list.

Figure 7.5 shows the *dbest* histogram and the error statistics plot from the exhaustive BP training for 10-bit parity problem, where $N_I = 10$, $N_O = 1$ are used for all networks. In this problem, BP exhibits a better performance on the majority of the configurations, i.e., $4 \times 0^{-4} \leq \mathrm{mMSE}(d) \leq 10^{-3}$ for $d = 7, 8, 9, 16, 23, 24, 25,$ and $\{29, 41\} - \{34\}$. The 100 MD PSO runs show that there are in fact two optimum dimensions: 30 and 41 (corresponding to 3-layer MLPs in $10 \times 5 \times 3 \times 1$ and $10 \times 8 \times 4 \times 1$ forms), which can achieve minimum MSEs, i.e., $\mathrm{mMSE}(30) < 2 \times 10^{-5}$ and $\mathrm{mMSE}(41) < 10^{-5}$. The majority of MD PSO runs, which is represented in the *dbest* histogram in Fig. 7.5, achieved $\mathrm{MSE}(dbest) < 8 \times 10^{-4}$except the four runs (out of 100) with $\mathrm{MSE}(dbest) > 4 \times 10^{-3}$ for $dbest = 4, 18,$ and 19. These are the minority cases where MD PSO trapped to local minima but the rest evolved to (near-) optimum networks.

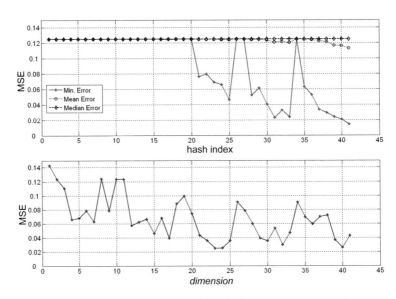

Fig. 7.4 MSE plots from the exhaustive BP training (*top*) and a single run of MD PSO (*bottom*)

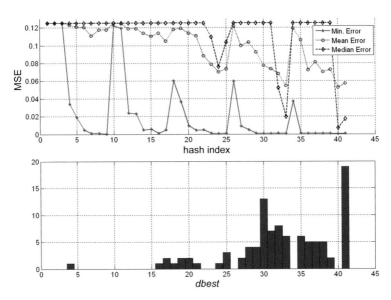

Fig. 7.5 Error statistics from exhaustive BP training (*top*) and *dbest* histogram from 100 MD PSO evolutions (*bottom*) for 10-bit parity problem

The *Two-spirals* problem [35], proposed by Alexis Wieland, is highly nonlinear and promises further interesting properties. For instance, the 2D data exhibits some temporal characteristics where the radius and the angle of the spiral vary with time.

The error space is highly multimodal with many local minima, thus methods, such as BP, encounters severe problems in error reduction [36]. The dataset consists of 194 patterns (2D points), 97 samples in each of the two classes (spirals). The dataset is now used as a benchmark for ANNs by many researchers. Lang and Witbrock in [35] reported that a near-optimum solution could not be obtained with standard BP algorithm over feed-forward ANNs. They tried a special network structure with short-cut links between layers. Similar conclusions are reported by Baum and Lang [36] that the problem is unsolvable with 2-layers MLPs with $2 \times 50 \times 1$ configuration. This, without doubt, is one of the hardest problems in the field of ANNs. Figure 7.6 shows *dbest* histogram and the error statistics plot from the exhaustive BP training for the *two-spirals* problem where $N_I = 2$, $N_O = 1$. It is obvious that none of the configurations yield a sufficiently low error value with BP training and particularly BP can at best perform a linear approximation for most of the configurations (hash indices) resulting MSE ≈ 0.49 and only few 3-layer MLPs (with indices 32, 33, 38 and 41) are able to reduce MSE to 0.3. MD PSO also shows a similar performance with BP, achieving $0.349 \leq \text{mMSE}(dbest) \leq 0.371$ for $dbest = 25, 32, 33, 38$ and 41. These are obviously the best possible MLP configurations to which a high majority of MD PSO runs converged.

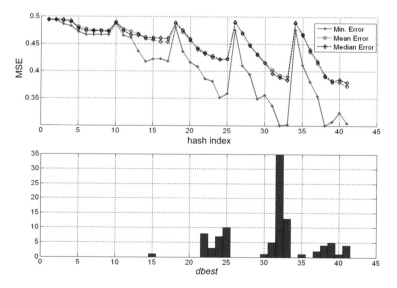

Fig. 7.6 Error (*MSE*) statistics from exhaustive BP training (*top*) and *dbest* histogram from 100 MD PSO evolutions (*bottom*) for the *two-spirals* problem

7.2.4 Classification Results on Medical Diagnosis Problems

In the previous section a set of synthetic problems that are among the hardest and the most complex in the ANN field, has been used in order to test the optimality of MD PSO evolution process, i.e., to see whether or not MD PSO can evolve to the few (near-) optimal configurations present in the limited AS, R^1, which mostly contains shallow MLPs. In this section we shall test the generalization capability of the proposed method and perform comparative evaluations against the most promising, state-of-the-art evolutionary techniques, over a benchmark dataset, which is partitioned into three sets: training, validation and testing. There are several techniques [37] to use training and validation sets individually to prevent over-fitting and thus improve the classification performance in the test data. However, the use of validation set is not needed for EA-based techniques since the latter perform a global search for a solution [38]. Although all competing methods presented in this section use training and validation sets in some way to maximize their classification rate over the test data, we simply combine the validation and training sets to use for training.

From *Proben1* repository [32], we selected three benchmark classification problems, *breast cancer, heart disease* and *diabetes*, which were commonly used in the prior work such as PSO–PSO [29], PSO–PSO:WD [26], IPSONet [28], EPNet [24], GA (basic) [38], and GA (Connection Matrix) [39]. These are medical diagnosis problems, which mainly present the following attributes:

- All of them are real-world problems based on medical data from human patients.
- The input and output attributes are similar to those used by a medical doctor.
- Since medical examples are expensive to get, the training sets are quite limited.

We now briefly describe each classification problem before presenting the performance evaluatio.

1. *Breast Cancer*

The objective of this dataset is to classify breast lumps as either benign or malignant according to microscopic examination of cells that are collected by needle aspiration. There are 699 exemplars of which 458 are benign and 241 are malignant and they are originally partitioned as 350 for training, 175 for validation, and 174 for testing. The dataset consists of 9 input and 2 output attributes. The dataset [32] was created and made public at University of Wisconsin Madison by Dr. William Wolberg.

2. *Diabetes*

This dataset is used to predict diabetes diagnosis among Pima Indians. All patients reported are females of at least 21 years old. There are total of 768 exemplars of which 500 are classified as diabetes negative and 268 as diabetes

positive. The dataset is originally partitioned as 384 for training, 192 for validation, and 192 for testing. It consists of 8 input and 2 output attributes.

3. *Heart Disease*

The initial dataset consists of 920 exemplars with 35 input attributes, some of
which are severely missing. Hence a second dataset is composed using the cleanest
part of the preceding set, which was created at Cleveland Clinic Foundation by Dr.
Robert Detrano. The Cleveland data is called as "heartc" in *Proben1* repository
and contains 303 exemplars but 6 of them still contain missing data and hence
discarded. The rest is partitioned as 149 for training, 74 for validation, and 74 for
testing. There are 13 input and 2 output attributes. The purpose is to predict the
presence of the heart disease according to the input attributes.

The input attributes of all datasets are scaled to between 0 and 1 by a linear
function. Their output attributes are encoded using a 1-of-c representation using
c classes. The *winner-takes-all* methodology is applied so that the output of the
highest activation designates the class. The experimental setup is identical for all
methods and thus fair comparative evaluations can now be made over the classification error rate of the test data. In all experiments in this section we mainly use
R^1 that is specified by the range arrays, $R^1_{min} = \{N_I, 1, 1, N_O\}$ and $R^1_{max} =$
$\{N_I, 8, 4, N_O\}$ containing the simplest 1-, 2-, or 3-layer MLPs where N_I and N_O, are
determined by the number of input and output attributes of the classification
problem. In order to collect some statistics about the results, we perform 100 MD
PSO runs, each using 250 particles ($S = 250$) and terminating at the end of 200
epochs ($E = 200$).

Before presenting the classification results over the test data, there are some
crucial points worth mentioning. First of all, the aforementioned ambiguity over
the decision of "optimality" is witnessed over the (training) datasets, *Diabetes* and
particularly the *Breast Cancer*, as the majority of networks in R^1 can achieve
similar performances. Figure 7.7 demonstrates this fact by two error statistics plots
from the exhaustive BP training ($K = 500$, $\lambda = 0.05$) with 5,000 epochs. Note that
most of the networks trained over both datasets result in minimum MSE values
that are within a narrow range and it is rather difficult to distinguish or separate one
from the other.

Contrary to the two datasets, the *Heart Disease* dataset gives rise to four distinct
sets of network configurations, which can achieve training *mMSEs* below 10^{-2} as
shown in Fig. 7.8. The corresponding indices (dimensions) to these four optimal
sets are located in the lower vicinity of the indices: *dbest* = 9, 25, 32, and 41,
where MD PSO managed to evolve either to them or to those neighboring configurations. Note that the majority of MD PSO runs (>50 %) to evolve the simplest
MLPs with single hidden layer (i.e., from Table 4, *dbest* = 9 is for the MLP
13 × 8 × 2) although BP achieved slightly lower *mMSEs* over other three (near-)
optimal configurations. The main reason is the fact that MD PSO or PSO in
general performs better in low dimensions and recall that premature convergence
problem might also occur when the search space is in high dimensions [40].

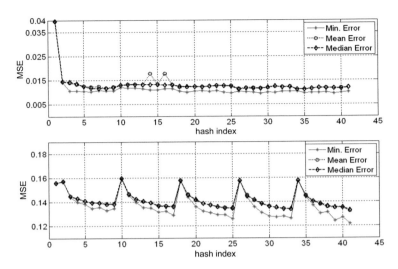

Fig. 7.7 Error statistics from exhaustive BP training over *Breast Cancer* (*top*) and *Diabetes* (*bottom*) datasets

Therefore, MD PSO naturally favors a low-dimension solution when it exhibits a competitive performance with a high dimension counterpart. Such a natural tendency eventually leads the evolution process to compact network configurations in the AS rather than the complex ones, as long as the optimality prevails.

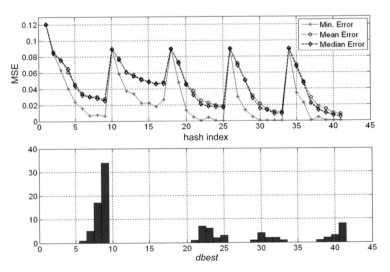

Fig. 7.8 Error statistics from exhaustive BP training (*top*) and *dbest* histogram from 100 MD PSO evolutions (*bottom*) over the *Heart Disease* dataset

Table 7.1 Mean (μ) and standard deviation (σ) of classification error rates (%) over test datasets

Algorithm	Dataset					
	Breast cancer		Diabetes		Heart disease	
	μ	σ	μ	σ	μ	σ
MD PSO	**0.39**	**0.31**	**20.55**	**1.22**	19.53	**1.71**
PSO–PSO	4.83	3.25	24.36	3.57	19.89	2.15
PSO–PSO:WD	4.14	1.51	23.54	3.16	18.1	3.06
IPSONet	1.27	0.57	21.02	1.23	18.14*	3.42
EPNet	1.38	0.94	22.38	1.4	**16.76***	2.03
GA [38]	2.27	0.34	26.23	1.28	20.44	1.97
GA [39]	3.23	1.1	24.56	1.65	23.22	7.87
BP	3.01	1.2	29.62	2.2	24.89	1.71

Table 7.1 presents the classification error rate statistics of MD PSO and other methods from the literature. The error rate in the table refers to the percentage of wrong classification over the test dataset of a benchmark problem. It is straightforward to see that the best classification performance is achieved with the MD PSO technique over the *Diabetes* and *Breast Cancer* datasets. Particularly on the latter set, roughly half of the MD PSO runs resulted in zero (0 %) error rate, meaning that perfect classification is achieved. MD PSO exhibits a competitive performance on the *Heart Disease* dataset. However, note that both IPSONet and EPNet (marked as * in Table 7.1), showing slightly better performances, used a subset of this set (134 for training, 68 for validation, and 68 for testing), excluding 27 entries overall. In [24] this is reasoned as, "27 of these were retained in case of dispute, leaving a final total of 270".

7.2.5 Parameter Sensitivity and Computational Complexity Analysis

First we shall demonstrate the effects of network architecture properties on the classification performance. To do so, a broader AS, R^2, is defined with larger and deeper MLPs with the following range arrays, $R^2 : R^2_{\min} = \{N_I, 6, 6, 3, N_O\}$ and $R^2_{\max} = \{N_I, 12, 10, 5, N_O\}$ using $L_{\min} = 1, L_{\max} = 4$. The second AS has 148 MLP configurations as shown in Table 7.2 where $N_I = 13$ and $N_O = 2$. Table 7.3 presents the classification error rate statistics from both ASs, R^1 and R^2 for a direct comparison. From the table, it is obvious that classification performances with both R^1 and R^2 are quite similar. This is indeed an expected outcome since all classification problems require only the simplest MLPs with a single hidden layer for a (near-) optimal performance. Therefore, no significant performance gain is observed when deeper and more complex MLPs in R^2 are used for these problems and furthermore,

Table 7.2 A sample architecture space with range arrays, $R^2_{\min} = \{13, 6, 6, 3, 2\}$ and $R^2_{\max} = \{13, 12, 10, 5, 2\}$

Dim.	Configuration
1	13×2
2	$13 \times 6 \times 2$
3	$13 \times 7 \times 2$
4	$13 \times 8 \times 2$
5	$13 \times 9 \times 2$
6	$13 \times 10 \times 2$
7	$13 \times 11 \times 2$
8	$13 \times 12 \times 2$
9	$13 \times 6 \times 6 \times 2$
10	$13 \times 7 \times 6 \times 2$
11	$13 \times 8 \times 6 \times 2$
12	$13 \times 9 \times 6 \times 2$
13	$13 \times 10 \times 6 \times 2$
14	$13 \times 11 \times 6 \times 2$
15	$13 \times 12 \times 6 \times 2$
16	$13 \times 6 \times 7 \times 2$
17	$13 \times 7 \times 7 \times 2$
18	$13 \times 8 \times 7 \times 2$
19	$13 \times 9 \times 7 \times 2$
…	…
138	$13 \times 9 \times 9 \times 5 \times 2$
139	$13 \times 10 \times 9 \times 5 \times 2$
140	$13 \times 11 \times 9 \times 5 \times 2$
141	$13 \times 12 \times 9 \times 5 \times 2$
142	$13 \times 6 \times 10 \times 5 \times 2$
143	$13 \times 7 \times 10 \times 5 \times 2$
144	$13 \times 8 \times 10 \times 5 \times 2$
145	$13 \times 9 \times 10 \times 5 \times 2$
146	$13 \times 10 \times 10 \times 5 \times 2$
147	$13 \times 11 \times 10 \times 5 \times 2$
148	$13 \times 12 \times 10 \times 5 \times 2$

Table 7.3 Classification error rate (%) statistics of MD PSO when applied to two architecture spaces

Error rate statistics	Breast cancer		Diabetes		Heart disease	
	R^1	R^2	R^1	R^2	R^1	R^2
μ	**0.39**	0.51	20.55	**20.34**	**19.53**	20.21
σ	**0.31**	0.37	1.22	**1.19**	**1.71**	1.90

MD PSO's evolution process among those complex networks might be degraded by the limited number of iterations (*iterNo* = 200) used for training and high dimensionality/modality of the solution spaces encountered in R^2.

During a MD PSO process, at each iteration and for each particle, first the network parameters are extracted from the particle and input vectors are (forward) propagated to compute the average MSE at the output layer. Therefore, it is not feasible to accomplish a precise computational complexity analysis of the MD PSO evolutionary process since this mainly depends on the networks that the particles converge and in a stochastic process such as PSO this cannot be determined. Furthermore, it also depends on the AS selected because MD PSO can only search for the optimal network within the AS. This will give us a hint that the computational complexity also depends on the problem in hand because particles eventually tend to converge to those near-optimal networks after the initial period of the process. Yet we can count certain attributes which directly affects the complexity, such as the size of the training dataset (T), swarm size (S), and number of epochs to terminate the MD PSO process (E). Since the computational complexity is proportional with the total number of forward propagations performed, then it can be in the order of $O(SET\mu_t)$ where μ_t is an abstract time for the propagation and MSE computation over an average network in the AS. Due to the aforementioned reasons, μ_t cannot be determined a priori and therefore, the abstract time can be defined as the expected time to perform a single forward propagation of an input vector. Moreover the problem naturally determines T, yet the computational complexity can still be controlled by S and E settings.

7.3 Evolutionary RBF Classifiers for Polarimetric SAR Images

In this section, we shall draw the focus on evolutionary radial basis function (RBF) network classifiers that will be evolved to classify terrain data in polarimetric synthetic aperture radar (SAR) images. For the past few decades image and data classification techniques have played an important role in the automatic analysis and interpretation of remote sensing data. Particularly polarimetric SAR data poses a challenging problem in this field due to the complexity of measured information from its multiple polarimetric channels. Recently, a number of applications which use data provided by the SAR systems having fully polarimetric capability have been increasing. Over the past decade, there has been extensive research in the area of the segmentation and classification of polarimetric SAR data. In the literature, the classification algorithms for polarimetric SAR can be divided into three main classes: (1) classification based on physical scattering mechanisms inherent in data [41, 42], (2) classification based on statistical characteristics of data [43, 44], and (3) classification based on image processing techniques [45–47]. Additionally, there has been several works using some combinations of the above classification approaches [41, 43]. While these approaches to the polarimetric SAR classification problem can be based on either supervised or unsupervised methods, their performance and suitability usually depend on applications and the availability of ground truth.

As one of the earlier algorithms, Kong et al. [48] derived a distance measure based on the complex Gaussian distribution and used it for maximum likelihood (ML) classification of single-look complex polarimetric SAR data. Then, Lee et al. [49] used the statistical properties of a fully polarimetric SAR to perform a supervised classification based on complex Wishart distribution. Afterwards, Cloude and Pottier [50] proposed an unsupervised classification algorithm based on their target decomposition theory. Target entropy (H) and target average scattering mechanism (scattering angle, α) calculated from this decomposition have been widely used in polarimetric SAR classification. For multilook data represented in covariance or coherency matrices, Lee et al. [43] proposed a new unsupervised classification method based on a combination of polarimetric target decomposition [50] and the maximum likelihood classifier using the complex Wishart distribution. The unsupervised Wishart classifier has an iterative procedure based on the well-known *K-means* algorithm, and has become a preferred benchmark algorithm due to its computational efficiency and generally good performance. However, this classifier still has some significant drawbacks since it entirely relies on *K-means* for actual clustering, such as it may converge to local optima, the number of clusters should be fixed a priori, its performance is sensitive to the initialization and its convergence depends on several parameters. Recently, a two-stage unsupervised clustering based on the EM algorithm [51] was proposed for classification of polarimetric SAR images. The EM algorithm estimates parameters of the probability distribution functions which represent the elements of a 9-dimensional feature vector, consisting of six magnitudes and three angles of a coherency matrix. Markov random field (MRF) clustering based method exploiting the spatial relation between adjacent pixels in polarimetric SAR images was proposed in [45, 49], a new wavelet-based texture image segmentation algorithm was successfully applied to unsupervised SAR image segmentation problem.

More recently, neural network based approaches [53, 54] for the classification of polarimetric SAR data have been shown to outperform other aforementioned well-known techniques. Compared with other approaches, neural network classifiers have the advantage of adaptability to the data without making a priori assumption of a particular probability distribution. However, their performance depends on the network structure, training data, initialization, and parameters. As discussed earlier, designing an optimal ANN classifier structure and its parameters to maximize the classification accuracy is still a crucial and challenging task. In this section, another feed-forward ANN type, the RBF network classifier which is optimally designed by MD PSO, is employed. For this task, RBFs are purposefully chosen due to their robustness, faster learning capability compared with other feed-forward networks, and superior performance with simpler network architectures. Earlier work on RBF classifiers for polarimetric SAR image classification has demonstrated a potential for performance improvement over conventional techniques [55]. The polarimetric SAR feature vector presented in this section

includes: full covariance matrix, the H/α/A decomposition based features combined with the backscattering power (*Span*), and the gray level co-occurrence
matrix (GLCM)-based texture features as suggested by the results of previous
studies [56, 57]. The performance of the evolutionary RBF network based classifier is evaluated using the fully polarimetric San Francisco Bay and Flevoland
datasets acquired by the NASA/Jet Propulsion Laboratory Airborne SAR (AIR-
SAR) at L-band [57–59]. The classification results measured in terms of confusion
matrix, overall accuracy and classification map are compared with other classifiers.

7.3.1 Polarimetric SAR Data Processing

Polarimetric radars often measure the complex scattering matrix, [S], produced by
a target under study with the objective to infer its physical properties. Assuming
linear horizontal and vertical polarizations for transmitting and receiving, [S] can
be expressed as,

$$S = \begin{bmatrix} S_{hh} & S_{hv} \\ S_{vh} & S_{vv} \end{bmatrix} \tag{7.1}$$

Reciprocity theorem applies in a monostatic system configuration, $S_{hv} = S_{vh}$.
For coherent scatterers only, the decompositions of the measured scattering matrix
[S] can be employed to characterize the scattering mechanisms of such targets.
One way to analyze coherent targets is the Pauli decomposition [43], which
expresses [S] in the so-called Pauli basis $\left\{ [S]_a = \frac{1}{\sqrt{2}} \begin{bmatrix} 1 & 0 \\ 0 & 1 \end{bmatrix}, [S]_b = \frac{1}{\sqrt{2}} \begin{bmatrix} 1 & 0 \\ 0 & -1 \end{bmatrix}, [S]_c = \frac{1}{\sqrt{2}} \begin{bmatrix} 0 & 1 \\ 1 & 0 \end{bmatrix} \right\}$ as,

$$S = \begin{bmatrix} S_{hh} & S_{hv} \\ S_{vh} & S_{vv} \end{bmatrix} = \alpha[S]_a + \beta[S]_b + \gamma[S]_c \tag{7.2}$$

where $\alpha = (S_{hh} + S_{vv})/\sqrt{2}, \beta = (S_{hh} - S_{vv})/\sqrt{2}, \gamma = \sqrt{2}S_{hv}$. Hence by means of
the Pauli decomposition, all polarimetric information in [S] could be represented
in a single RGB image by combining the intensities $|\alpha|^2, |\beta|^2$ and $|\gamma|^2$, which
determine the power scattered by different types of scatterers such as single- or
odd-bounce scattering, double- or even-bounce scattering, and orthogonal polari-
zation returns by volume scattering. There are several other coherent decomposition theorems such as the Krogager decomposition [60] the Cameron
decomposition [61], and SDH (Sphere, Diplane, Helix) decomposition [61] all of
which aim to express the measured scattering matrix by the radar as the combination of scattering responses of coherent scatterers.

Alternatively, the second order polarimetric descriptors of the 3 × 3 average
polarimetric covariance $\langle [C] \rangle$ and the coherency $\langle [T] \rangle$ matrices can be derived

from the scattering matrix and employed to extract physical information from the observed scattering process. The elements of the covariance matrix, [C], can be written in terms of three unique polarimetric components of the complex-valued scattering matrix:

$$
\begin{aligned}
C_{11} &= S_{hh}S_{hh}^*, & C_{21} &= S_{hh}^*S_{hv} \\
C_{22} &= S_{hv}S_{hv}^*, & C_{32} &= S_{hv}^*S_{vv} \\
C_{33} &= S_{vv}S_{vv}^*, & C_{31} &= S_{hh}^*S_{vv}
\end{aligned}
\tag{7.3}
$$

For single-look processed polarimetric SAR data, the three polarimetric components (HH, HV, and VV) have a multivariate complex Gaussian distribution and the complex covariance matrix form has a complex Wishart distribution [49]. Due to the presence of speckle noise and random vector scattering from surface or volume, polarimetric SAR data are often multi-look processed by averaging n neighboring pixels. By using the Pauli-based scattering matrix for a pixel i, $k_i = [S_{hh} + S_{vv}, S_{hh} - S_{vv}, 2S_{hv}]^T / \sqrt{2}$, the multi-look coherency matrix, $\langle [T] \rangle$, can be written as,

$$
\langle T \rangle = \frac{1}{n} \sum_{i=1}^{n} k_i k_i^{*T}
\tag{7.4}
$$

Both coherency $\langle [T] \rangle$ and covariance $\langle [C] \rangle$ are 3×3 Hermitian positive semi-definite matrices, and since they can be converted into one another by a linear transform, both are equivalent representations of the target polarimetric information.

The incoherent target decomposition theorems such as the Freeman decomposition, the Huynen decomposition, and the Cloude-Pottier (or H/α/A) decomposition employ the second order polarimetric representations of PolSAR data (such as covariance matrix or coherency matrix) to characterize distributed scatterers. The H/α/A decomposition [62] is based on eigenanalysis of the polarimetric coherency matrix, $\langle [T] \rangle$:

$$
\langle T \rangle = \lambda_1 e_1 e_1^{*T} + \lambda_2 e_2 e_2^{*T} + \lambda_3 e_3 e_3^{*T}
\tag{7.5}
$$

where $\lambda_1 > \lambda_2 > \lambda_3 \geq 0$ are real eigenvalues, e_1^* implies complex conjugate of e_1 and e_1^T is the transpose of e_1. The corresponding orthonormal eigenvectors e_i (representing three scattering mechanisms) are,

$$
e_i = e^{i\varphi_i} \left[\cos \alpha_i, \sin \alpha_i \cos \beta_i e^{i\delta_i}, \sin \alpha_i \sin \beta_i e^{i\gamma_i} \right]^T
\tag{7.6}
$$

Cloude and Pottier defined entropy H, average of set of four angles $\bar{\alpha}$, $\bar{\beta}$, $\bar{\delta}$, and $\bar{\gamma}$, and anisotropy A for analysis of the physical information related to the scattering characteristics of a medium:

$$
H = - \sum_{i=1}^{3} p_i \log_3 p_i \quad \text{where} \quad p_i = \frac{\lambda_i}{\sum_{i=1}^{3} \lambda_i},
\tag{7.7}
$$

$$\bar{\alpha} = \sum_{i=1}^{3} p_i \alpha_i, \ \bar{\beta} = \sum_{i=1}^{3} p_i \beta_i, \ \bar{\delta} = \sum_{i=1}^{3} p_i \delta_i, \ \bar{\gamma} = \sum_{i=1}^{3} p_i \gamma_i, \qquad (7.8)$$

$$A = \frac{p_2 - p_3}{p_2 + p_3}. \qquad (7.9)$$

For a multi-look coherency matrix, the entropy, $0 \le H \le 1$, represents the randomness of a scattering medium between isotropic scattering ($H = 0$) and fully random scattering ($H = 1$), while the average alpha angle can be related to target average scattering mechanisms from single-bounce (or surface) scattering ($\bar{\alpha} \approx 0$) to dipole (or volume) scattering ($\bar{\alpha} \approx \pi/4$) to double-bounce scattering ($\bar{\alpha} \approx \pi/2$). Due to basis invariance of the target decomposition, H and $\bar{\alpha}$ are roll invariant hence they do not depend on orientation of the target about the radar line of sight. Additionally, information about the target's total backscattered power can be determined by the *span* as,

$$span = \sum_{i=1}^{3} \lambda_i \qquad (7.10)$$

Entropy (H), estimate of the average alpha angle ($\bar{\alpha}$), and *span* calculated by the above noncoherent target decomposition method have been commonly used as polarimetric features of a scatterer in many target classification schemes [43, 63].

7.3.2 SAR Classification Framework

The first operation for SAR classification is naturally the feature extraction. The SAR feature extraction process presented herein utilizes the complete covariance matrix information, the GLCM-based texture features, and the backscattering power (*span*) combined with the $H/\alpha/A$ decomposition [50]. The feature vector from the Cloude–Pottier decomposition includes entropy (H), anisotropy (A), estimates of the set of average angles ($\bar{\alpha}$, $\bar{\beta}$, $\bar{\delta}$, and $\bar{\gamma}$), three real eigenvalues ($\lambda_1, \lambda_2, \lambda_3$), and *span*. As suggested by the previous studies [53, 56] appropriate texture measures for SAR imagery based on the gray level co-occurrence probabilities are included in the feature set to improve its discrimination power and classification accuracy. In this study, *contrast, correlation, energy,* and *homogeneity* features are extracted from normalized GLCMs which are calculated using interpixel distance of 2 and averaging over four possible orientation settings ($\theta = 0°, 45°, 90°, 135°$). To reduce the dimensionality (and redundancy) of input feature space, the principal components transform is applied to these inputs and the most principal components (which contain about 95 % of overall energy in the original feature matrix) are then selected to form a resultant feature vector for each imaged pixel. Dimensionality reduction of input feature information improves efficiency of learning for a neural network classifier due to a smaller number of input nodes (to avoid curse of dimensionality) [64] and reduces computation time. For the purpose of normalizing and scaling the feature vector, each feature dimension is first normalized to have a zero mean and unity standard deviation before principal component analysis (PCA) is applied, and following the PCA outputs are linearly scaled into $[-1, 1]$ interval.

In this section, two distinct training methods for RBF network classifiers, the traditional back-propagation (BP), and particle swarm optimization (PSO) are investigated. The RBF networks and the training algorithm BP are introduced in Sect. 3.4.3.1. For the BP algorithm, RPROP enhancement is used when training RBF networks. The main difference in RPROP is that it modifies the update-values for each parameter according to the sequence of signs of partial derivatives. This only leads to a faster convergence, while the problems of a hill climbing algorithm are not solved. Further details about BP and RPROP can be found in [65, 66], respectively. In order to determine (near-) optimal network architecture for a given problem, we apply exhaustive BP training over every network configuration in the AS defined. For the training based on MD PSO, first the dynamic clustering based on MD PSO is applied to determine the optimal (with respect to minimizing a given cost function for the input–output mapping) number of Gaussian neurons with their correct parameters (centroids and variances). Afterwards, a single run of BP can conveniently be used to compute the remaining network parameters, weights (w), and bias (θ) of the each output layer neuron. Note that once the number of Gaussian networks and their parameters are found, the rest of the RBF network resembles a SLP where BP training results in a unique solution of weights and biases. The overview of the classifier framework for polarimetric SAR images is shown in Fig. 7.9.

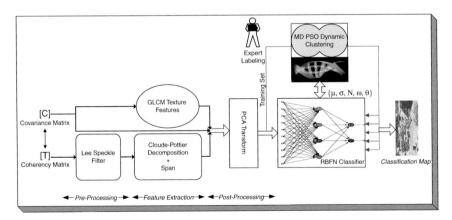

Fig. 7.9 Overview of the evolutionary RBF network classifier design for polarimetric SAR image

7.3.3 Polarimetric SAR Classification Results

In this section, two test images of an urban area (San Francisco Bay, CA) and an agricultural area (Flevoland, the Netherlands), both acquired by the NASA/Jet Propulsion Laboratory's AIRSAR at L-band, were chosen for performance evaluation of the RBF network classifier. Both datasets have been widely used in the polarimetric SAR literature over the past two decades [57–59], and distributed as multi-look processed and publicly available through the polarimetric SAR data processing and educational tool (PolSARpro) by ESA [22]. The original four-look fully polarimetric SAR data of the San Francisco Bay, having a dimension of $900 \times 1{,}024$ pixels, provides good coverage of both natural (sea, mountains, forests, etc.) and man-made targets (buildings, streets, parks, golf course, etc.) with a more complex inner structure. For the purpose of comparing the classification results with the Wishart [43] and the NN-based [53] classifiers, the subarea (Fig. 7.10) with size 600×600 is extracted and used. The aerial photographs for this area which can be used as ground truth are provided by the TerraServer Web site [67]. In this study, no speckle filtering is applied to originally four-look processed covariance matrix data and before GLCM-based texture feature generation to retain the resolution and to preserve the texture information. However, additional averaging, such as using the polarimetry preserving refined Lee filter [68] with 5×5 window, of coherency matrix should be performed prior to the Cloude-Pottier decomposition [50]. For MD PSO based clustering algorithm, the typical internal PSO parameters (c_1, c_2 and w) are used as in [69], also explained in [70]. For all experiments in this section, the two critical PSO parameters, swarm size (S), and number of iterations (*IterNo*), are set as 40 and 1,000, respectively.

Fig. 7.10 Pauli image of 600×600 pixel subarea of San Francisco Bay (*left*) with the 5×5 refined Lee filter used. The training and testing areas for three classes are shown using red rectangles and circles respectively. The aerial photograph for this area (*right*) provided by the U.S. Geological Survey taken on Oct, 1993 can be used as ground-truth

Table 7.4 Summary table of pixel-by-pixel classification results of the RBF-MDPSO classifier over the training and testing area of San Francisco Bay dataset

	Training data			Test data		
	Sea	Urban	Vegetation	Sea	Urban	Vegetation
Sea	14,264	4	0	6,804	0	0
Urban	11	9,422	22	10	6,927	23
Vegetation	10	87	4,496	21	162	6,786

To test the performance of the RBF classifier and compare its classification results, the same training and testing areas for the three classes from the sub San Francisco area (as shown on the Pauli-based decomposition image in Fig. 7.10), the sea (15,810, 6723 pixels, respectively), urban areas (9,362, 6,800), and the vegetated zones (5,064, 6,534), which are manually selected in an earlier study [53], are used. The confusion matrix of the evolutionary RBF classifier on the training and testing areas are given in Table 7.4. The classification accuracy values are averaged over 10 independent runs. From the results, the main drawback of this classifier is the separation of vegetated zones from urban areas. Compared to two other competing techniques, this classifier is able to differentiate better the uniform areas corresponding to main classes of scattering such as the ocean, vegetation, and building areas. In Table 7.5, the overall accuracies in training and testing areas for the RBF classifier trained using the BP and MD PSO algorithms and two competing methods, the Wishart maximum likelihood (WML) classifier [43] and the NN-based classifier [53], are compared. The average accuracies over 10 independent runs for the best configuration of the RBF-BP and RBF-MDPSO classifiers are reported. The RBF classifier trained by the global PSO algorithm is superior to the NN-based, WML, and RBF-BP-based methods with higher accuracies in both training (99.50 %) and testing (98.96 %) areas. Figure 7.11 shows the classification results on the whole subarea image for the RBF-MDPSO based classifier. The classification map of the whole San Francisco Bay image produced by the same classifier is given in Fig. 7.12 for a qualitative (visual) performance evaluation. The evolutionary RBF classifier has the structure of 11 input neurons, 21 Gaussian neurons which the cluster centroids and variance (μ_k and σ_k) are determined by MD PSO-based dynamic clustering the training data, and 3 output neurons.

The classification results in Table 7.5 have been produced by using a high percentage (60 %) of total (training and testing combined) pixels for training. The RBF network classifier is also tested by limiting the percentage of total pixels which were used for classifier training to less than 1 % of the total pixels to be classified. The results over the same testing dataset are shown in Table 7.6. In this case, the classifier trained by the BP or MD PSO algorithms performed still at a high level, achieving accuracies over 95 and 98 %, respectively. Generally, a relatively smaller training dataset can avoid over-fitting and improve generalization performance of a classifier over larger datasets.

Table 7.5 Overall performance comparison (in percent) for San Francisco Bay dataset

Method	Training area	Testing area
RBF-BP	98.00	95.70
WML [43]	97.23	96.16
NN [53]	99.42	98.64
RBF-MDPSO	**99.50**	**98.96**

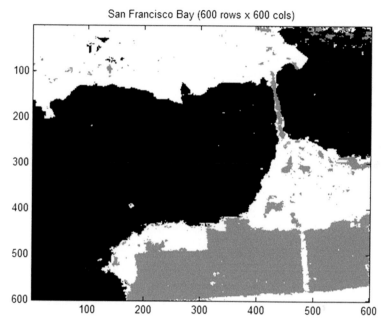

Fig. 7.11 The classification results of the RBF-MDPSO classifier over the 600 × 600 sub-image of San Francisco Bay (*black denotes sea, gray urban areas, white vegetated zones*)

In order to test robustness of the RBF network classifier trained by the MD PSO-based dynamic clustering, 20 independent runs are performed over the San Francisco area image and the resulting cluster number histogram is plotted in Fig. 7.13. Additionally, the plots of a typical run showing the fitness score and dimension versus number of iterations for MD PSO operation are presented in the left side of Fig. 7.13. Based on overall clustering results, it is found that the number of clusters (the optimal number of Gaussian neurons) and their centroids extracted from the MD PSO-based dynamic clustering are generally consistent, indicating the technique is robust (or repeatable).

Next, the evolutionary RBF classifier with this feature set has been applied to the polarimetric image of the Flevoland site, an agricultural area (consists of primarily crop fields and forested areas) in The Netherlands. This original four-look fully polarimetric SAR data has a dimension of 750 × 1024 pixels with 11

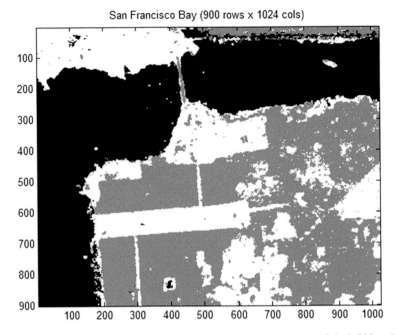

Fig. 7.12 The classification results of the RBF-MDPSO technique for the original (900 × 1024) San Francisco Bay image (*black denotes sea, gray urban areas, white vegetated zones*)

Table 7.6 Overall performance (in percent) using smaller training set (<%1 of total pixels) for San Francisco Bay dataset

Method	Training area	Testing area
RBF-BP	100	95.60
RBF-MDPSO	100	**98.54**

identified crop classes {stem beans, potatoes, lucerne, wheat, peas, sugar beet, rape seed, grass, forest, bare soil, and water}. The available ground truth for 11 classes can be found in [71]. To compare classification results the same 11 training and testing sets are used with those of the NN-based [53], wavelet-based [59], and ECHO [72] classifiers. In Table 7.7, the overall accuracies in training and testing areas of the Flevoland dataset for the RBF classifier trained using the BP and MD PSO algorithms and three state-of-the-art methods, the ECHO [71], wavelet-based [59], and NN-based [53] classifiers, are shown. The overall classification accuracies of the RBF-based classifier framework are quite high. The percentage of correctly classified training and testing pixels in the Flevoland L-band image for the evolutionary (MD PSO) RBF method are given in Table 7.8. Figure 7.14 shows the classification results of the proposed evolutionary RBF classifier for the Flevoland image.

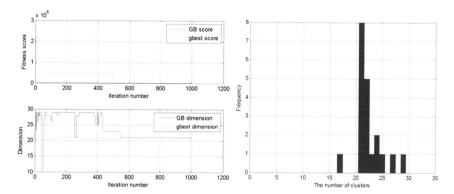

Fig. 7.13 Fitness score (*left top*) and dimension (*left bottom*) plots vs. iteration number for a typical MD PSO run. The resulting histogram plot (*right*) of cluster numbers which are determined by the MD PSO method

Table 7.7 Overall performance comparison (in percent) for Flevoland dataset

Method	Training area	Testing area
ECHO [72]	–	81.30
Wavelet-based [59]	–	88.28
RBF-BP	95.50	92.05
NN [53]	**98.62**	92.87
RBF-MDPSO	95.55	**93.36**

The computational complexity of the evolutionary RBF classifier depends on the following distinct processes: the pre-processing stage, feature extraction, post-processing, and RBF network classifier with MD PSO dynamic clustering-based training. Computation times of the first three stages are deterministic while a precise computational complexity analysis for the RBF training stage is not feasible as the dynamic clustering technique based on MD PSO is in stochastic nature. All experiments in this section are performed on a computer with P-IV 2.4 GHz CPU and 1 GB RAM. Based on the experiments, for the data of San Francisco Bay area with a dimension of 900×1024 data points ($D = 921,600$), it takes 30 min to perform feature extraction and necessary pre- and post-processing stages. Most of this time is used to extract the GLCM and four texture features calculated from it. For computational complexity of RBF classifier training using MD PSO process, there are certain attributes which directly affect the complexity such as swarm size (S), the number of iteration (*IterNo*) to terminate the MD PSO process, and the dimensions of data space (D). While the problem determines D, the computational complexity can still be controlled by S and *IterNo* settings. For the same dataset, the average (over 10 runs) processing time to perform evolutionary RBF classifier training is found to be 30 min.

Table 7.8 Summary table of pixel-by-pixel classification results (in percent) of the RBF-MDPSO classifier over the training and testing data of Flevoland

	Water	Forest	Stem Beans	Potatoes	Lucerne	Wheat	Peas	Sugar Beet	Bare Soil	Grass	Rape Seed
	Training (testing) data										
Water	99(98)	0(0)	0(0)	0(0)	0(0)	0(0)	0(0)	0(0)	0(0)	0(0)	1(2)
Forest	0(0)	95(97)	0(0)	0(0)	1(0)	0(0)	1(0)	1(0)	0(0)	2(3)	0(0)
Stem Beans	0(0)	0(0)	95(97)	0(0)	5(2)	0(1)	0(0)	0(0)	0(0)	0(0)	0(0)
Potatoes	0(0)	0(0)	0(0)	99(96)	0(0)	0(0)	0(0)	1(4)	0(0)	0(0)	0(0)
Lucerne	0(0)	0(0)	2(2)	0(0)	98(97)	0(0)	0(0)	0(0)	0(0)	0(1)	0(0)
Wheat	0(0)	0(0)	0(0)	0(0)	2(4)	91(86)	4(4)	1(3)	0(0)	2(3)	3(5)
Peas	0(0)	0(0)	0(0)	0(0)	0(0)	1(0)	94(88)	2(7)	0(0)	0(0)	1(1)
Sugar Beet	0(0)	0(0)	0(0)	0(0)	0(0)	0(2)	0(1)	95(91)	0(0)	4(5)	1(1)
Bare Soil	0(0)	0(0)	0(0)	0(0)	0(0)	0(2)	0(0)	0(0)	99(97)	0(0)	0(0)
Grass	0(0)	0(0)	0(0)	0(0)	1(0)	0(1)	0(0)	2(4)	0(0)	97(95)	0(0)
Rape Seed	2(2)	0(0)	0(0)	0(0)	0(0)	2(2)	1(2)	3(2)	3(7)	0(0)	89(85)

Flevoland (750 rows x 1024 cols)

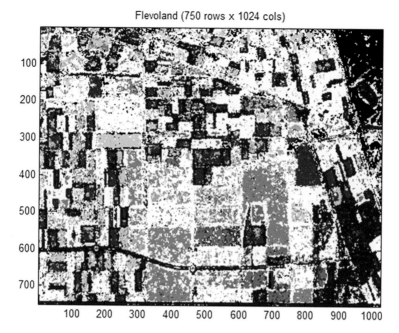

Fig. 7.14 The classification results on the L-band AIRSAR data over Flevoland

7.4 Summary and Conclusions

In this chapter, we drew the focus on evolutionary ANNs by the MD PSO with the following innovative properties:

- With the proper adaptation, MD PSO can evolve to the optimum network within an AS for a particular problem. Additionally, it provides a ranked list of all other potential configurations, indicating that any high rank configuration may be an alternative to the optimum one, yet some with low ranking, on contrary, should not be used at all.
- The evolutionary technique is generic and applicable to any type of ANNs in an AS with varying size and properties, as long as a proper hash function enumerates all configurations in the AS with respect to their complexity into proper hash indices representing the dimensions of the solution space over which MD PSO seeks for optimality.
- Due to the MD PSO's native feature of having better and faster convergence to optimum solution in low dimensions, its evolution process in any AS naturally yields to compact networks rather than large and complex ones, as long as the optimality prevails.

Experimental results over synthetic datasets and particularly over several benchmark medical diagnosis problems show that all the aforementioned properties and capabilities are achieved in an automatic way and as a result, ANNs (MLPs) evolved by MD PSO alleviates the need of human "expertise" and "knowledge" for designing a particular network; instead, such virtues may still be used in a flexible way to define only the size and perhaps some crucial properties of the AS. In this way further efficiency in terms of speed and accuracy over global search and evolution process can be achieved.

We then perform MD PSO evolution over another type of feed-forward ANNs, namely the RBF networks. For the evaluation purposes in a challenging application domain, we present a new polarimetric SAR image classification framework, which is based on an efficient formation of covariance matrix elements, H/a/A decomposition with the backscattered power (span) information, and GLCM-based texture features, and the RBF network classifier. Two different learning algorithms, the classical BP and MD PSO, were applied for the proposed classifier training/evolution. In addition to determining the correct network parameters, MD PSO also finds the best RBF network architecture (optimum number of Gaussian neurons and their centroids) within an AS and for a given input data space. The overall classification accuracies and qualitative classification maps for the San Francisco Bay and Flevoland datasets demonstrate the effectiveness of the classification framework using the evolutionary RBF network classifier. Based on the experimental results using real polarimetric SAR data, the classifier framework performs well compared to several state-of-the-art classifiers, however, more experiments using large volume of SAR data should be done for a general conclusion.

7.5 Programming Remarks and Software Packages

ClassifierTestApp is the major MD PSO test-bed application for classification and syntheses. It is a console application without a GUI over which dedicated MD PSO applications, mostly based on evolutionary classifiers and feature syntheses are implemented. In this section, its programming basics and application of the standalone evolutionary ANNs over the benchmark machine learning problems will be introduced while keeping the details of two individual applications, collective network of binary classifiers (CNBC), and evolutionary feature syntheses (EFS) to the following chapters. **ClassifierTestApp** is a single-thread application, and the source file *ClassifierTestApp.cpp* has two entry point **main()** functions separated with a compiler switch **_CNBC_TEST**. If defined, the test-bed application for CNBC will be active within the first **main()** function; otherwise, the standalone classifier implementation of evolutionary ANNs will be enabled within the second **main()** function. In this chapter, we shall only deal with the latter and the CNBC and EFS applications will be covered in Chaps. 9 and 10. Therefore, the switch **_CNBC_TEST** is not defined and the global Boolean variable **bSynthesis**, which enables EFS-based on evolutionary ANNs is set to false.

The **ClassifierTestApp** application contains four types of basic classifiers: Feed-forward ANNs (or MLPs), Radial Basis Function networks (RBFs), Support Vector Machines (SVMs), and finally, random forest (RF). Each of them is implemented in an individual static library: **MLPlibX, RBFlibX, SVMlibX,** and **RFlibX**. In this chapter the focus is drawn only on the **MLPlibX** using which a single MLP can be created, trained and used for classification. This library has been designed with three levels of hierarchy: neuron (implemented in class **CMLPneuron**), layer (implemented in class **CMLPlayer**) and finally the MLP network (implemented in class **CMLPnet**). This is a natural design since an MLP contains one or more layers and a layer contains one or more neurons. A **CMLPnet** object, therefore, can be created and initiated with certain number of layers, and neurons within, and with a particular activation function (and its corresponding derivative). As shown in Table 7.9, the constructor of the **CMLPnet** object has these as the input. The member function, **backpropagate ()** can train the MLP network and another function, **propagate ()** can propagate an input (feature) vector to obtain the actual output (class) vector.

An important property of this MLP library is the ability to encode the entire MLP parameters (the weights and biases that are stored within each **CMLPneuron** object) into an external buffer. Table 7.10 presents the function **initializeParameters()**,which assigns the externally allocated memory space (with the pointer

Table 7.9 Some members and functions of **CMLPnet** class

```
class CMLPnet
{
        CMLPnet (unsigned short* desc_layers, int layer_no,
                const ActivationFunction *fun = tanh,      const ActivationFunction *dfun = dtanh);
        void initializeParameters(float *pW, int size);
        void randomizeParameters ();
        void setInput (float* in);
        void setTargetOutput (float* optimal);
...
        void propagate ();
        void backpropagate ();
...
protected:
        float*    m_input; // input vect..
        int       m_inSize; // size of the input vect..
        float*    m_output; // output vect..
        int       m_outSize;
        float*    m_targetOutput; // target vect..
        int       m_tarSize;

        CMLPlayer**        layers; // Left-to-right list of layers within the network..
        int                layer_no;
        float              epsilon;
        float              opt_tolerance;
        float              weight_decay;
        float              momentum_term;
};
```

*pW) for all MLP parameters. In this way, any MLP network can be represented (for storage and easy initialization purposes) with a single buffer and the buffer alone will be sufficient to revive it. This is the key property to store an entire MLP AS within a single buffer and/or a binary file.

The class **CDimHash** is responsible of performing the hash function for any AS from a family of MLP networks. Recall that for a particular AS, a range is defined for the minimum and maximum number of layers, $\{L_{\min}, L_{\max}\}$ and number of neurons for hidden layer l, $\{N^l_{\min}, N^l_{\max}\}$. As presented in Table 7.11, the constructor of the class **CDimHash** receives these inputs as the variables: **int min_noL, int max_noL, int *min_noN, int *max_noN** and hashes the entire AS by the pointer array, **CENNdim** m_pDim**. Each **m_pDim[]** element stores a distinct MLP configuration in a **CENNdim** object, which basically stores the number of layers (**m_noL**) and number of neurons in each layer (**m_pNoN[]**). Therefore, once the AS is hashed, the **d**th entry (MLP configuration) can be retrieved by simply calling **GetEntry(int d)** function. The **CDimHash** object can also compute the buffer size of each individual MLP configuration in order to allocate the memory space for its parameters. This is basically needed to compute the overall buffer size needed for the entire AS and then the memory space for each consecutive MLP network can be allocated within the entire buffer. Once the evolution process is over, the evolved MLPs (their parameters) in the AS can then be stored in a binary file.

The **ClassifierTestApp** application primarily uses the static library **ClassifierLib**, to perform evolution (training) and classification over a dataset, which is partitioned into training and test sets. The library, **ClassifierLib**, then uses one of the four static libraries **MLPlibX, RBFlibX, SVMlibX,** and **RFlibX** to

Table 7.10 Some members and functions of **CMLPnet** class

```
void CMLPnet::initializeParameters (float *pW, int size)
{
        /* Reserve space for MLP parameters from an external buffer..*/
        int offset = 0;
        for (int n = 0 ; n < layer_no-1 ; ++n)
        {
                for (int neuron = 0 ; neuron < layers[n]->size ; ++neuron)
                {
                        layers[n]->neurons[neuron]->weight = &pW[offset]; // weights..
                        offset += layers[n+1]->size;
                } // for..
                if(n==0) continue;
                for (neuron = 0 ; neuron < layers[n]->size ; ++neuron)
                        layers[n]->neurons[neuron]->theta = &pW[offset++]; // and its theta..
        } // for..

        for (int neuron = 0 ; neuron < layers[n]->size ; ++neuron)
                layers[n]->neurons[neuron]->theta = &pW[offset++]; // and its theta..

}
```

Table 7.11 CDimHash class members and functions

```
class CDimHash
{
public:
        CDimHash(int min_noL, int max_noL, int *min_noN, int *max_noN);
        ~CDimHash();

        inline int              GetDim(CENNdim* dim) ;
        inline int              GetBufSize(CENNdim* dim) ;
        inline int              GetBufSize(int conf) ;
        inline int              GetTotalBufSize() {return m_totalBufSize;}

        inline CENNdim*         GetEntry(int d) {return m_pDim[d]; }
        inline int              GetSize() {return m_hashSize;}
        inline int              GetDimPos(int l) {return m_pDimPos[l-1]; }

private:
        CENNdim**    m_pDim; // Dimension Hast Table..
        int          m_hashSize; // Size of the table..
        int*         m_pDimPos; // dim. pos. of a noL conf..
        int          m_maxNoL; // max. no. of layers..
        int*         m_pMinNoN; // min. no of neurons per layer..
        int*         m_pMaxNoN; // max. no of neurons per layer..
        int          m_totalBufSize; // Total buf. size of the AS..
};
```

accomplish the task. Therefore, it is the controller library which contains six primary classes: **CMLP_BP, CMLP_PSO, CRBF_BP, CRBF_PSO, CRandomForest,** and **CSVM.** The first four are the evolutionary ANNs (MLP and RBF networks) with the two evolutionary techniques: MD PSO (**CMLP_PSO** and **CRBF_PSO**) and exhaustive BP (**CMLP_BP** and **CRBF_BP**). The last two classes are used to train RF and the network topology of the SVMs. In this chapter, we shall focus on evolutionary ANNs and particular on MLPs, therefore, **CMLP_PSO** and **CRBF_PSO** will be detailed herein. Besides these six classes for individual classifiers, the **ClassifierLib** library also has classes for implementing CNBC topology: **CNBCglobal** and **COneNBC**, which can use any of the six classes (classifier/evolution) as its binary classifier implementation. The programming details of CNBC will be covered in Chap. 9.

The classes implementing six classifiers are all inherited from the abstract base class **CGenClassifier**, which defines the API member functions and variables. Table 7.12 presents some of the member functions and variables of the **CGenClassifier**. The evolutionary technique will be determined by the choice of the class, **CMLP_PSO** or **CMLP_BP.** The evolution (training) parameters for either technique are then conducted by the **Init (train_params tp)** function with the **train_params** variable. Once the evolutionary MLP is initialized as such, it then becomes ready for the underlying evolutionary process via calling the **Train(class_params cp)** function.

Table 7.12 CGenClassifier class members and functions

```
class CLASSIFIER_API CGenClassifier
{
public:
        ...
        virtual int        Create(char* dir, char* title) = 0; // Create the classifier from file..
        virtual int        Create(float* buf) = 0; // Create classifier from bin. buffer..
        void               CreateAS(char* dir, char* title) ; // Create Arch. Space from a bas file..
        virtual void       CreateAS(float *pAS) = 0; // Create Arch. Space from an AS buf..

        void               Init(train_params tp) ; // Init. classifier(s)..
        virtual int        Train(class_params cp) = 0; // Train classifier(s)..
        virtual int        Train(float *buf) = 0; // Train classifier from bin. buffer..
        virtual int        Exit() = 0; // Clean and destroy classifier(s)..
        inline void        SetActivationMode(ActivationMode act) {m_act = act;} // act. code..

        virtual float*     Propagate(float *in_vec) = 0; // Fwd. Propagate one FV..
        virtual int        Save(char* dir, char* title) = 0; // Save classifier to a file..
        ...
protected:
        // INPUT..
        float              **train_input, **test_input ; // Train: input arrays for train and test sets..
        float              **train_output, **test_output ; // Train: output arrays for train and test sets..
        int                in_size, out_size, train_size, test_size; // Train: sizes of train and test arrays..
        int                no_run, no_iter; // Train: No. of runs and iterations per run..
        float              lp1, lp2; // Train: Learning param. 1 & 2..

        int                *minNoN, *maxNoN; // Conf: Min. and Max. no. of Neurons..
        int                max_noL, no_conf; // Conf: Max. no. of layers and conf. in ANN classifier..
        ActivationMode m_act; // Activation mode..
        // MD PSO params..
        PSOparam           _psoDef;
        SPSAparam          _saDef;
        // Output..
        ...
```

Recall that the standalone classifier implementation of evolutionary ANNs is enabled within the second **main()** function, that is given in Table 7.13. The initial dataset, which is partitioned into train and test is loaded within the function calls:

TrainProben1 ("D:\\MSVC6.0\\Source\\proben1\\diabetes\\", "diabetes1.dt");// Arrange Training I/O..
//TrainFunc ();//Function approximation..
//TrainParity (9);//n-bit parity problem..

TrainFunc() is for function approximation where any function can be used. **TrainProben1()** loads a data file (*.dt), which is nothing but a text file with a simple header. In the data file, the train and test datasets's feature vectors (FVs) along with their class vectors (CVs) with 1-to-n encoding, are separately stored. Recall that there are three *Proben1* datasets: Breast Cancer, Diabetes, and Heart Disease each of which reside in a particular folder, which is the first variable of the function call **TrainProben1()**. The second variable is the name of the dataset and there are three alternatives (e.g., "diabetes1.dt", "diabetes2.dt", or

Table 7.13 The main entry function for evolutionary ANNs and classification

```
      int main(int argc, char* argv[])
      {
      ...
      //        TrainProben1 ("D:\\MSVC6.0\\Source\\proben1\\diabetes\\", "diabetes1.dt"); // Arrange Training I/O..
                TrainFunc (); // Arrange Training I/O..
      //        TrainParity (9); // Arrange Training I/O..

                if(bSynthesis)
                { // EFS based on evol. ANNs..
                          ...
                } // if..

                minNoN[0] = in_size; minNoN[MAX_NOL-1] = out_size;
                maxNoN[0] = in_size; maxNoN[MAX_NOL-1] = out_size;

                // Put Training params..
                one_tp.bClassFile = 1; // classifier's txt file..
                ...
                memcpy(&one_tp._psoDef, &_psoDef, sizeof(PSOparam)); // PSO params..
                memcpy(&one_tp._saDef, &_saDef, sizeof(SPSAparam)); // SPSA params..

                // Put Classifier params..
                one_cp.minNoN = minNoN; one_cp.maxNoN = maxNoN; one_cp.max_noL = MAX_NOL;
                ...
      ///       pOneClassifier = new CMLP_BP(); // Create a MLP classifier with ex. BP..
                pOneClassifier = new CMLP_PSO(); // Create a MLP classifier..

                pOneClassifier->Init(one_tp) ; // this for training..
                pOneClassifier->SetActivationMode(aMode); // act. mode for MLPs..
                pOneClassifier->Train(one_cp) ; // Start Evolution.....
                ...
                pOneClassifier->Exit(); delete pOneClassifier;
                return 0;
      }
```

"diabetes3.dt") each with a random shuffling of train and test partitions. Any of these functions call basically loads them into following global variables:

float **train_input, **test_input;
float **train_output, **test_output;
int in_size, out_size, train_size, test_size = 0;

The first row of double arrays store the FVs whereas the second row of double arrays store the target CVs for the train and test datasets, respectively. The variables in the third row are the size of the FVs and CVs (**in_size, out_size**) of each individual data entry, and the size of the train, and test datasets (**train_size, test_size**). Once they are loaded from the data file (or from the function), they are stored into a **train_params** object, **one_cp,** along with the rest of the training parameters. Recall that this object will then be passed to the **CMLP_PSO** or **CMLP_BP** object by calling the **Init (train_params tp)** function, where the ANN evolution will be performed by the MD PSO process accordingly. Recall that the AS is defined by the minimum and maximum number of layers, $\{L_{\min}, L_{\max}\}$ and the two range arrays, $R_{\min} = \left\{N_1, N_{\min}^1, \ldots, N_{\min}^{L_{\max}-1}, N_O\right\}$ and

$R_{\max} = \{N_1, N_{\max}^1, \ldots, N_{\max}^{L_{\max}-1}, N_O\}$, one for minimum and the other for maximum number of neurons allowed for each layer of a MLP. Letting $L_{\min} = 2$, these parameters (called as classifier parameters as they define the AS for ANN classifiers), are then assigned into a **class_params** object, **one_cp**, by calling:

//Put Classifier params..
one_cp.minNoN = minNoN;   **one_cp.maxNoN = maxNoN;**
one_cp.max_noL = MAX_NOL;

After both **one_tp** and **one_cp** objects are fully prepared, the classifier object can be created with the underlying evolution technique, initialized (with the training parameters), and evolution process can be initiated. Once it is completed, the object can then be cleaned and destroyed. As shown in Table 7.13, all of these are accomplished by calling the following code:

pOneClassifier = new CMLP_PSO(); // Create a MLP classifier..
pOneClassifier->Init(one_tp); // this for training..
pOneClassifier->SetActivationMode(aMode); // act. mode for MLPs..
pOneClassifier->Train(one_cp); // Start Evolution.....
...
pOneClassifier->Exit(); delete pOneClassifier; // Clean up..

Note that there is a piece of test code that is discarded from the compilation. The first part of this code is to test the I/O functions of the classifier, i.e., first to retrieve the AS buffer (**pOneClassifier->GetClassBuffer(size);**) and then to save it into a file (with the name: ***_buf.mlp**). The second part involves creating a new classifier object with the AS buffer obtained from the current evolution process. Recall that the AS buffer is the best solution found by the underlying evolutionary search process, in this case MD PSO since the classifier object was **CMLP_PSO**. Note that this is not a re-evolution process from scratch, rather a new MD PSO process where the previous best solution (stored within the AS buffer) is injected into the process (by calling: **pTest->Train(buf);**). Thus, the new MD PSO process will take the previous best solution into account to search for an "even-better" one. Finally, the last part of the code loads the AS from the file, which was saved earlier with the **pTest->Save(dir, tit);** function call and as in the previous process, is injected into the new MD PSO process. Note that all three test code examples cleans up the internal memory by calling the **pTest->Exit();**

As mentioned earlier, there are two types of evolutionary processes; one is the ANN evolution as initiated with the **pOneClassifier->Train(one_cp)** call and another with the injection of the last best solution as initiated with **pTest->Train(buf);** call. We shall now cover the former implementation over the object **CMLP_PSO** (MD PSO evolutionary process over an AS of MLPs). The Reader will find it straightforward to see the similar kind of approach on the other implementation and for other objects.

Table 7.14 presents the function **CMLP_PSO::Train(class_params cp)** where the MD PSO evolutionary process over an AS of MLPs is implemented. Recall that the evolution (training) parameters are already fed into the **CMLP_PSO** object and several function calls can be made each with a different AS (**class_params**), which is immediately stored within the object and used to create the AS object with the call:

CSolSpace<float>::s_pDimHash = new   CDimHash(min_noL,   max_noL, minNoN, maxNoN);

Note that the AS object is a static pointer of the **CSolSpace** class since it is primarily needed by the MD PSO process to find the true solution space dimension (the memory space needed for weights and biases for that particular MLP configuration) from the hash function. There can be several MD PSO runs (i.e., **no_run**) and within the loop of each run, a new MD PSO object is first created, initialized with the MD PSO parameters (and SPSA parameters if needed) and the **PropagateENN()** is set as the fitness function. Note that the dimensional range, $[D_{min}, D_{max}]$, is set to **[0, no_conf]**, where **no_conf** is the number of configurations within the AS. The dimension 0 corresponds to an SLP and the highest dimension,

Table 7.14 The function evolving MLPs by MD PSO

```
int      CMLP_PSO::Train(class_params cp)
{ // Evolve MLPs and find/save the "best" one..
        char fname[511];
        minNoN = cp.minNoN; max_noL = cp.max_noL;
        maxNoN = (cp.maxNoN) ? cp.maxNoN : cp.minNoN;
        int min_noL = (cp.maxNoN) ? 2 : max_noL;

        train_sizeX = this->train_size; // For PropagateENN fn..
        train_inputX = train_input; train_outputX = train_output; // For PropagateENN fn..
        actF == m_act; // For PropagateENN fn..

        CSolSpace<float>::s_pDimHash = new CDimHash(min_noL, max_noL, minNoN, maxNoN);
        no_conf = CSolSpace<float>::s_pDimHash->GetSize(); // No. of possible MLP conf.
        train_err = (_ERROR **) malloc(no_conf*sizeof(_ERROR *));
        test_err = (_ERROR **) malloc(no_conf*sizeof(_ERROR *));
        ...
        bool bRandParams = CreateArSp (); // Create/Check AS..

        for(int run=0; run<no_run; run++)
        { // For all runs..
                CPSO_MD<CSolSpace<float>,float>   *pPSO   =   new   CPSO_MD<CSolSpace<float>,float>
(_psoDef._noAgent, _psoDef._maxNoIter, _psoDef._eCutOff, _psoDef._mode);
                pPSO->Init(0,  no_conf,  _psoDef._vdMin,  _psoDef._vdMax,  -2,  2,  _psoDef._xvMin,
_psoDef._xvMax);
                pPSO->SetSAparam(_saDef); // Set SA params..
                pPSO->SetFitnessFn( PropagateENN ); /*** PUT Fn here ***/

                if(!bRandParams) // if there is a Ar. Space present..
                        pPSO->Inject(m_pArSpBuf+m_hdrAS); // inject Ar. space to one of the PSO particles..

                pPSO->Perform(); // Evol. MLP training..

                // 1. Fill out all conf. error statistics and save the best MLP conf.s into AS..
                ...
```

no_conf-1, corresponds to the MLP with maximum number of hidden layers and neurons within.

The member Boolean parameter, **bRandParams**, determines if there is an AS buffer already present within the object, from an earlier run. If so, then the AS buffer, which is nothing but the best solution found in the previous run is then injected into the current MD PSO object via **pPSO->Inject(m_pArSp-Buf+m_hdrAS)** call.

Table 7.15 presents the fitness function **PropagateENN()** in the **CMLP_PSO** class. Recall that the position, **pPos,** of each MD PSO particle in its current dimension, **pPos->GetDim()**, represents the potential solution, i.e., the MLP with a configuration corresponding to that dimension. The configuration of the MLP can be retrieved from the AS object by calling:

CENNdim* pConf = CSolSpace<float>::s_pDimHash->GetEntry(pPos->Get-Dim());//get conf..

A new MLP is then created with the configuration stored in **pConf** and the parameters of the MLP are assigned from the MD PSO particle position by calling:

CMLPnet *net = new CMLPnet (pConf->m_pNoN, pConf->m_noL);
net->initializeParameters(pPos->GetPos(), pPos->GetSize());

The rest of the function simply performs the forward propagation of the features of the training dataset and computes the average training MSE, which is the fitness

Table 7.15 The fitness function of MD PSO process in the **CMLP_PSO** class

```
double CMLP_PSO::PropagateENN(CSolSpace<float> *pPos, int iter)
{
        CENNdim* pConf = CSolSpace<float>::s_pDimHash->GetEntry(pPos->GetDim()); // get conf..

    // Create MLP of pConf with weights and theta from ..
    CMLPnet *net = new CMLPnet (pConf->m_pNoN, pConf->m_noL);
    net->initializeParameters(pPos->GetPos(), pPos->GetSize());
    if(actF == _sigm) net->setActivationFunction(All, sigm, dsigm); // Set Act. Fun. as sigm..
    else net->setActivationFunction(All, tanh, dtanh); // Set Act. Fun. as tanh..

    float error_sum = 0.0;
    for (int n = 0 ; n < train_sizeX ; ++n)
    { // For each sample in the training input..

            net->setInput (train_inputX[n]);
            net->propagate ();
            float* output = net->getOutput ();
            net->setTargetOutput (train_outputX[n]);
            error_sum += net->errorTerm ();
    } // for..

    error_sum /= (train_sizeX * pConf->m_pNoN[pConf->m_noL-1]); // Unit norm.

    delete net; // Clean up..
    return error_sum;
}
```

Table 7.16 The constructors of the **CSolSpace** class

```
    template <class X> CSolSpace<X>::CSolSpace(int nDim)
    {
           m_hDim = nDim; // hash index..
           CENNdim* pConf = CSolSpace<float>::s_pDimHash->GetEntry(m_hDim); // get conf..
           m_nDim = 0;
           for(int l=1; l<pConf->m_noL; ++l)
               m_nDim += pConf->m_pNoN[l] * (1+pConf->m_pNoN[l-1]) ; // Weights + Theta per layer..

           m_pPos=new X[m_nDim]; // All weights in m_nDim dim. array..
           m_pInGB = (bool*) calloc(m_nDim, sizeof(bool));
    }

    template <class X> CSolSpace<X>::CSolSpace(int nDim, X min, X max)
    {
           m_min = min; m_max = max;
           m_hDim = nDim; // hash index..
           CENNdim* pConf = CSolSpace<float>::s_pDimHash->GetEntry(m_hDim); // get conf..
           m_nDim = 0;
           for(int l=1; l<pConf->m_noL; ++l)
               m_nDim += pConf->m_pNoN[l] * (1+pConf->m_pNoN[l-1]) ; // Weights + Theta per layer..

           m_pPos=new X[m_nDim]; // All weights in m_nDim dim. array..
           m_pInGB = (bool*) calloc(m_nDim, sizeof(bool));
           for(int i=0; i<m_nDim; i++)
           {
               m_pPos[i] = RAND(min, max); // randomize weights & theta in betw. {min, max}
           } // for..
    }
```

to be minimized by the MD PSO process by seeking the best (optimal) MLP configuration.

The MD PSO library is almost identical to the one in *PSOTestApp* application except the **CSolSpace** class implementation. Table 7.16 shows the two constructors of the **CSolSpace** class. It is evident that the dimension of an MD PSO particle within the dimensional range is used as the hash index (**m_hDim**) and the true solution space dimension, **m_nDim**, is equivalent to the total memory space for all MLP parameters (weights and biases). This is where the AS object (**CSolSpace<float>::s_pDimHash**) is used to retrieve the number of neurons in each layer and use them to compute **m_nDim** as in the for loop. As a common operation with the earlier **CSolSpace** class implementation, once the space is allocated for the positional component of the **CSolSpace** object, then it is randomized within the positional range passed as variable: **[min, max]**.

References

1. W. McCulloch, W. Pitts, A logical calculus of the ideas immanent in nervous activity. Bull. Math. Biophys. **7**, 115–133 (1943)
2. Y. Chauvin, D.E. Rumelhart, *Back Propagation: Theory, Architectures, and Applications* (Lawrence Erlbaum Associates Publishers, Muhwah, 1995)

3. S.E. Fahlman, An empirical study of learning speed in back-propagation. Technical report, CMU-CS-88-162, Carnegie-Mellon University, Pittsburgh, 1988
4. R.S. Sutton, Two problems with back-propagation and other steepest-descent learning procedures for networks, in *Proceedings of the 8th Annual Conference Cognitive Science Society*, Erlbaum, Hillsdale, 1986, pp. 823–831
5. S. Haykin, *Neural Networks: A Comprehensive Foundation* (Prentice hall, Englewood Cliffs, 1998)
6. M.N. Jadid, D.R. Fairbairn, Predicting moment-curvature parameters from experimental data. Eng. Appl. Artif. Intell. **9**(3), 309–319 (1996)
7. T. Masters, *Practical Neural Network Recipes in C++* (Academic Press, Boston, 1994)
8. R. Hecht-Nielsen, *Neurocomputing* (Addison-Wesley, Reading, 1990)
9. N. Burgess, A constructive algorithm that converges for real-valued input patterns. Int. J. Neural Sys. **5**(1), 59–66 (1994)
10. M. Frean, The upstart algorithm: A method for constructing and training feed-forward neural networks. Neural Comput. **2**(2), 198–209 (1990)
11. Y. LeCun, J.S. Denker, S.A. Solla, Optimal brain damage. Adv. Neural Inf. Process. Syst. **2**, 598–605 (1990)
12. R. Reed, Pruning algorithms—a survey. IEEE Trans. Neural Netw. **4**(5), 740–747 (1993)
13. P.J. Angeline, G.M. Sauders, J.B. Pollack, An evolutionary algorithm that constructs recurrent neural networks. IEEE Trans. Neural Netw. **5**, 54–65 (1994)
14. G.F. Miller, P.M. Todd, S.U. Hegde, Designing neural networks using genetic algorithms, in *Proceedings of the 3rd International Conference Genetic Algorithms Their Applications*, 1989, pp. 379–384
15. A. Antoniou, W.-S. Lu, *Practical Optimization, Algorithms and Engineering Applications* (Springer, New York, 2007)
16. D. Goldberg, *Genetic Algorithms in Search, Optimization and Machine Learning* (Addison-Wesley, Reading, 1989), pp. 1–25
17. J. Koza, *Genetic Programming: On the Programming of Computers by Means of Natural Selection* (MIT Press, Cambridge, MA, 1992)
18. T. Back, F. Kursawe, Evolutionary algorithms for fuzzy logic: a brief overview, in *Fuzzy Logic and Soft Computing, World Scientific*, Singapore, 1995, pp. 3–10
19. U.M. Fayyad, G.P. Shapire, P. Smyth, R. Uthurusamy, *Advances in Knowledge Discovery and Data Mining* (MIT Press, Cambridge, 1996)
20. P. Bartlett, T. Downs, Training a neural network with a genetic algorithm, Technical report, Department of Electrical Engineering, University Queensland, Jan 1990
21. J.V. Hansen, R.D. Meservy, Learning experiments with genetic optimization of a generalized regression neural network. Decis. Support Syst. **18**(3–4), 317–325 (1996)
22. V.W. Porto, D.B. Fogel, L.J. Fogel, Alternative neural network training methods. IEEE Expert **10**, 16–22 (1995)
23. D.L. Prados, Training multilayered neural networks by replacing the least fit hidden neurons, in *Proceedings of IEEE SOUTHEASTCON' 92*, (1992), pp. 634–637
24. X. Yao, Y. Liu, A new evolutionary system for evolving artificial neural networks. IEEE Trans. Neural Netw. **8**(3), 694–713 (1997)
25. R.K. Belew, J. McInerney, N.N. Schraudolph, Evolving networks: using genetic algorithm with connectionist learning, Technical report CS90-174 revised, Computer Science Engineering Department, University of California-San Diego, Feb 1991
26. M. Carvalho, T.B. Ludermir, Particle swarm optimization of neural network architectures and weights, in *Proceedings of the 7th International Conference on Hybrid Intelligent Systems*, Washington DC, 17–19 Sep 2007, pp. 336–339
27. M. Meissner, M. Schmuker, G. Schneider, Optimized particle swarm optimization (OPSO) and its application to artificial neural network training. BMC Bioinform. **7**, 125 (2006)
28. J. Yu, L. Xi, S. Wang, An Improved Particle Swarm Optimization for Evolving Feed-forward Artificial Neural Networks. Neural Process. Lett. **26**(3), 217–231 (2007)

29. C. Zhang, H. Shao, An ANN's evolved by a new evolutionary system and its application, in *Proceedings of the 39th IEEE Conference on Decision and Control*, vol. 4 (2000), pp. 3562–3563
30. J. Salerno, Using the particle swarm optimization technique to train a recurrent neural model, in *Proceedings of IEEE Int. Conf. on Tools with Artificial Intelligence*, (1997), pp. 45–49
31. M. Settles, B. Rodebaugh, T. Soule, Comparison of genetic algorithm and particle swarm optimizer when evolving a recurrent neural network. Lecture Notes in Computer Science (LNCS) No. 2723, in *Proceedings of the Genetic and Evolutionary Computation Conference 2003 (GECCO 2003)*, (Chicago, IL, USA, 2003), pp. 151–152
32. L. Prechelt, Proben1—a set of neural network benchmark problems and benchmark rules, Technical report 21/94, Fakultät für Informatik, Universität Karlsruhe, Germany, Sept 1994
33. S. Guan, C. Bao, R. Sun, Hierarchical incremental class learning with reduced pattern training. Neural Process. Lett. **24**(2), 163–177 (2006)
34. J. Zhang, J. Zhang, T. Lok, M.R. Lyu, A hybrid particle swarm optimization-back-propagation algorithm for feed forward neural network training. Appl. Math. Comput. **185**, 1026–1037 (2007)
35. K.J. Lang, M.J. Witbrock, Learning to tell two spirals apart, in *Proceedings of the 1988 Connectionist Models Summer School*, San Mateo, 1988
36. E. Baum, K. Lang, Constructing hidden units using examples and queries, in *Advances in Neural Information Processing Systems*, vol. 3 (San Mateo, 1991), pp. 904–910
37. M.H. Hassoun, *Fundamentals of Artificial Neural Networks* (MIT Press, Cambridge, 1995)
38. R.S. Sexton, R.E. Dorsey, Reliable classification using neural networks: a genetic algorithm and back propagation comparison. Decis. Support Syst. **30**, 11–22 (2000)
39. E. Cantu-Paz, C. Kamath, An empirical comparison of combinations of evolutionary algorithms and neural networks for classification problems. IEEE Trans. Syst. Man Cybern. Part B, **35**, 915–927 (2005)
40. G.-J. Qi, X.-S. Hua, Y. Rui, J. Tang, H.-J. Zhang, Image Classification with kernelized spatial-context. IEEE Trans. Multimedia, **12**(4), 278–287 (2010). doi:10.1109/TMM.2010.2046270
41. E. Pottier, J.S. Lee, Unsupervised classification scheme of PolSAR images based on the complex Wishart distribution and the H/A/α. Polarimetric decomposition theorem, in *Proceedings of the 3rd European Conference on Synthetic Aperture Radar (EUSAR 2000)*, (Munich, Germany, 2000), pp. 265–268
42. J.J. van Zyl, Unsupervised classification of scattering mechanisms using radar polarimetry data. IEEE Trans. Geosci. Remote Sens. **27**, 36–45 (1989)
43. J.S. Lee, M.R. Grunes, T. Ainsworth, L.-J. Du, D. Schuler, S.R. Cloude, Unsupervised classification using polarimetric decomposition and the complex Wishart classifier. IEEE Trans. Geosci. Remote Sens. **37**(5), 2249–2257 (1999)
44. Y. Wu, K. Ji, W. Yu, Y. Su, Region-based classification of polarimetric SAR images using Wishart MRF. IEEE Geosci. Rem. Sens. Lett. **5**(4), 668–672 (2008)
45. Z. Ye, C.-C. Lu, Wavelet-Based Unsupervised SAR Image Segmentation Using Hidden Markov Tree Models, in *Proceedings of the 16th International Conference on Pattern Recognition (ICPR'02)*, (2002), pp. 729–732
46. C.P. Tan, K.S. Lim, H.T. Ewe, Image processing in polarimetric SAR images using a hybrid entropy decomposition and maximum likelihood (EDML), in *Proceedings of the International Symposium on Image and Signal Processing and Analysis (ISPA)*, Sep 2007, pp. 418–422
47. T. Ince, Unsupervised classification of polarimetric SAR image with dynamic clustering: an image processing approach. Adv. Eng. Softw. **41**(4), 636–646 (2010)
48. J.A. Kong, A.A. Swartz, H.A. Yueh, L.M. Novak, R.T. Shin, Identification of terrain cover using the optimum polarimetric classifier. J. Electromagn. Waves Applicat. **2**(2), 171–194 (1988)
49. J.S. Lee, M.R. Grunes, R. Kwok, Classification of multi-look polarimetric SAR imagery based on complex Wishart distribution. Int. J. Rem. Sens. **15**(11), 2299–2311 (1994)

50. S.R. Cloude, E. Pottier, An entropy based classification scheme for land applications of polarimetric SAR. IEEE Trans. Geosci. Remote Sens. **35**, 68–78 (1997)
51. K.U. Khan, J. Yang, W. Zhang, Unsupervised classification of polarimetric SAR images by EM algorithm. IEICE Trans. Commun. **90**(12), 3632–3642 (2007)
52. T.N. Tran, R. Wehrens, D.H. Hoekman, L.M.C. Buydens, Initialization of Markov random field clustering of large remote sensing images. IEEE Trans. Geosci. Remote Sens. **43**(8), 1912–1919 (2005)
53. Y.D. Zhang, L.-N. Wu, G. Wei, A new classifier for polarimetric SAR images, Progress in Electromagnetics Research, PIER 94 (2009), pp. 83–104
54. L. Zhang, B. Zou, J. Zhang, Y. Zhang, Classification of polarimetric SAR image based on support vector machine using multiple-component scattering model and texture features, Eurasip J. Adv. Sig. Process. (2010). doi:10.1155/2010/960831
55. T. Ince, Unsupervised classification of polarimetric SAR image with dynamic clustering: an image processing approach. Adv. Eng. Softw. **41**(4), 636–646 (2010)
56. D.A. Clausi, An Analysis of Co-occurrance Texture Statistics as a Function of Grey Level Quantization. Canadian J. Remote Sens. **28**(1), 45–62 (2002)
57. K. Ersahin, B. Scheuchl, I. Cumming, Incorporating texture information into polarimetric radar classification using neural networks, in *Proceedings of the IEEE International Geoscience and Remote Sensing Symposium*, Anchorage, USA, Sept 2004, pp. 560–563
58. L. Ferro-Famil, E. Pottier, J.S. Lee, Unsupervised classification of multifrequency and fully polarimetric SAR images based on the H/A/Alpha-Wishart classifier. IEEE Trans. Geosci. Remote Sens. **39**(11), 2332–2342 (2001)
59. S. Fukuda, H. Hirosawa, A wavelet-based texture feature set applied to classification of multifrequency polarimetric SAR images. IEEE Trans. Geosci. Remote Sens. **37**(5), 2282–2286 (1999)
60. E. Krogager, J. Dall, S. Madsen, Properties of sphere, diplane and helix decomposition, in *Proceedings of the 3rd International Workshop on Radar Polarimetry*, 1995
61. J.-S. Lee, E. Pottier, *Polarimetric Radar Imaging: From Basics to Applications*, in Optical Science and Engineering, vol. 142 (CRC Press, Boca Raton, 2009)
62. S.R. Cloude, E. Pottier, A review of target decomposition theorems in radar polarimetry. IEEE Trans. Geosci. Remote Sens. **34**(2), 498–518 (1996)
63. C. Fang, H. Wen, W. Yirong, An improved Cloude–Pottier decomposition using h/α/span and complex Wishart classifier for polarimetric SAR classification, in *Proceedings of CIE*, Oct 2006, pp. 1–4
64. S. Pittner, S.V. Kamarthi, Feature extraction from wavelet coefficients for pattern recognition tasks. IEEE Trans. Pattern Anal. Machine Intell. **21**, 83–88 (1999)
65. Y. Chauvin, D.E. Rumelhart, *Back propagation: Theory, architectures, and applications*, (Lawrence Erlbaum Associates Publishers, UK, 1995)
66. M. Riedmiller, H. Braun, A direct adaptive method for faster back propagation learning: the RPROP algorithm, in *Proceedings of the IEEE International Conference on Neural Networks*, 1993, pp. 586–591
67. U.S. Geological Survey Images [Online], http://terraserver-usa.com/
68. J.S. Lee, M.R. Grunes, G. de Grandi, Polarimetric SAR speckle filtering and its implications for classification. IEEE Trans. Geosci. Remote Sens. **37**(5), 2363–2373 (1999)
69. Y. Shi, R.C. Eberhart, A modified particle swarm optimizer, in *Proceedings of the IEEE Congress on Evolutionary Computation*, 1998, pp. 69–73
70. S. Kiranyaz, T. Ince, A. Yildirim, M. Gabbouj, Evolutionary artificial neural networks by multi-dimensional particle swarm optimization. Neural Netw. **22**, 1448–1462. doi:10.1016/j.neunet.2009.05.013, Dec. 2009
71. T.L. Ainsworth, J.P. Kelly, J.-S. Lee, Classification comparisons between dual-pol, compact polarimetric and quad-pol SAR imagery. ISPRS J. Photogram. Remote Sens. **64**, 464–471 (2009)
72. E. Chen, Z. Li, Y. Pang, X. Tian, Quantitative evaluation of polarimetric classification for agricultural crop mapping. Photogram. Eng. Remote Sens. **73**(3), 279–284 (2007)

Chapter 8
Personalized ECG Classification

Science is the systematic classification of experience.

George Henry Lewes

Each individual heartbeat in the cardiac cycle of the recorded electrocardiogram (ECG) waveform shows the time evolution of the heart's electrical activity, which is made of distinct electrical depolarization–repolarization patterns of the heart. Any disorder of heart rate or rhythm, or change in the morphological pattern is an indication of an arrhythmia, which could be detected by analysis of the recorded ECG waveform. Real-time automated ECG analysis in clinical settings is of great assistance to clinicians in detecting cardiac arrhythmias, which often arise as a consequence of a cardiac disease and may be life-threatening and require immediate therapy.

In this chapter, first a generic and patient-specific classification system designed for robust and accurate detection of ECG heartbeat patterns is presented. An overview of this system is shown in Fig. 8.1. An extensive feature extraction will be presented which utilizes morphological wavelet transform features, which are projected onto a lower dimensional feature space using principal component analysis (PCA), and temporal features from the ECG data. Due to its time–frequency localization properties, the wavelet transform is an efficient tool for analyzing nonstationary ECG signals which can be used to decompose such an ECG signal according to scale, thus allowing separation of the relevant ECG waveform morphology descriptors from the noise, interference, baseline drift, and amplitude variation of the original signal. For the pattern recognition unit, as detailed in the previous chapter, feedforward and fully connected ANNs, which are optimally designed for each patient by MD PSO evolutionary search technique, are employed. This is indeed a great advantage in terms of classification accuracy, although many promising ANN-based techniques have been applied to ECG signal classification, these classifier systems have not performed well in practice and their results have generally been limited to relatively small datasets mainly because such systems have in general static (fixed) network structures for the classifiers. On the other hand, the approach discussed in this chapter which is based on patient-specific architecture by means of an evolutionary classifier design will show a significant performance improvement over such conventional global classifier

S. Kiranyaz et al., *Multidimensional Particle Swarm Optimization for Machine Learning and Pattern Recognition*, Adaptation, Learning, and Optimization 15, DOI: 10.1007/978-3-642-37846-1_8, © Springer-Verlag Berlin Heidelberg 2014

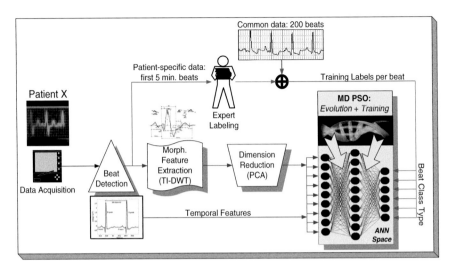

Fig. 8.1 Patient-specific ECG classification system

systems. By using relatively limited common and patient-specific training data, this evolutionary classification system can adapt to significant inter-patient variations in ECG patterns by training the optimal network structure, and thus can achieve a high accuracy over large datasets. The classification experiments, which will be discussed in Sect. 8.1.3.2, over a benchmark database will demonstrate that this system achieves such average accuracies and sensitivities better than most of the current state of the art algorithms for detection of ventricular ectopic beats (VEBs) and supra-VEBs (SVEBs). Finally, due to its parameter–invariant nature, this system is highly generic, and thus applicable to any ECG dataset.

The second half of the chapter presents a personalized long-term ECG classification framework, which addresses the problem within a long-term ECG signal, known as *Holter* register, recorded from an individual patient. Due to the massive amount of ECG beats in a *Holter* register, visual inspection is quite difficult and cumbersome, if not impossible. Therefore, the presented system helps professionals to quickly and accurately diagnose any latent heart disease by examining only the representative beats (the so-called master key-beats) each of which is automatically extracted from a time frame of homogeneous (similar) beats. In order to accomplish this, dynamic clustering in high dimensional data/feature spaces is conducted as an optimization problem and we naturally use MD PSO with FGBF as detailed in Chap. 6 for finding the optimal (number of) clusters, with respect to a given validity index function. As a result, this unsupervised classification system helps professionals to quickly and accurately diagnose any latent heart disease by examining only the representative beats.

8.1 ECG Classification by Evolutionary Artificial Neural Networks

8.1.1 Introduction and Motivation

Automated classification of ECG beats is a challenging problem as the morphological and temporal characteristics of ECG signals show significant variations for different patients and under different temporal and physical conditions [1]. Many algorithms for automatic detection and classification of ECG heartbeat patterns have been presented in the literature including signal processing techniques such as frequency analysis [2], wavelet transform [3, 4], and filter banks [5], statistical [6] and heuristic approaches [7], hidden Markov models [8], support vector machines [9], artificial neural networks (ANNs) [1], and mixture of experts method [10]. In general, ECG classifier systems based on past approaches have not performed well in practice because of their important common drawback of having an inconsistent performance when classifying a new patient's ECG waveform. This makes them unreliable to be widely used clinically, and causes severe degradation in their accuracy and efficiency for larger datasets, [11, 12]. Moreover, the Association for the Advancement of Medical Instrumentation (AAMI) provides standards and recommended practices for reporting performance results of automated arrhythmia detection algorithms [13]. However, despite quite many ECG classification methods proposed in the literature, only few [10, 14, 15] have in fact used the AAMI standards as well as the complete data from the benchmark MIT-BIH arrhythmia database.

The performance of ECG pattern classification strongly depends on the characterization power of the features extracted from the ECG data and on the design of the classifier (classification model or network structure and parameters). Due to its time–frequency localization properties, the wavelet transform is an efficient tool for analyzing nonstationary ECG signals [16]. The wavelet transform can be used to decompose an ECG signal according to scale, allowing separation of the relevant ECG waveform morphology descriptors from the noise, interference, baseline drift, and amplitude variation in the original signal. Several researchers have previously used wavelet transform coefficients at appropriate scales as morphological feature vectors rather than the original signal time series and achieved good classification performance [4, 17]. Accordingly, in the current work the utilized feature extraction technique employs the translation-invariant dyadic wavelet transform (TI-DWT) in order to extract effectively the morphological information from ECG data. Furthermore, the dimension of the input morphological feature vector is reduced by projecting it onto a lower dimensional feature space using PCA in order to significantly reduce redundancies in such a high dimensional data space. The lower dimensional morphological feature vector is then combined with two critical temporal features related to inter-beat time interval to improve accuracy and robustness of the classification as suggested by the results of previous studies [14].

As discussed in the previous chapters, ANNs are powerful tools for pattern recognition as they have the capability to learn complex, nonlinear surfaces among different classes, and such ability can therefore be the key for ECG beat recognition and classification [18]. Although many promising ANN-based techniques have been applied to ECG signal classification, [17–20], the global classifiers based on a static (fixed) ANN have not performed well in practice. On the other hand, algorithms based on patient–adaptive architecture have demonstrated significant performance improvement over conventional global classifiers [10, 12, 15]. Among all, one particular approach, a personalized ECG heartbeat pattern classifier based on evolvable block-based neural networks (BbNN) using *Hermite* transform coefficients [15], achieved such a performance that is significantly higher than the others. Although this recent work clearly demonstrates the advantage of using evolutionary ANNs, which can be automatically designed according to the problem (patient's ECG data), serious drawbacks and limitations still remain. For instance, there are around 10–15 parameters/thresholds that need to be set empirically with respect to the dataset used and this obviously brings about the issue of robustness when it is used for a different dataset. Another drawback can occur due to the specific ANN structure proposed, i.e., the BbNN, which requires equal sizes for input and output layers. Even more critical is the back propagation (BP) method, used for training, and genetic algorithm (GA), for evolving the network structure, both have certain deficiencies [21]. Recall in particular that, BP is likely to get trapped into a local minimum, making it entirely dependent on the initial (weight) settings.

As demonstrated in the previous chapter, in order to address such deficiencies and drawbacks, MD PSO technique can be used to search for the optimal network configuration specifically for each patient and according to the patient's ECG data. On the contrary to the specific BbNN structure used in [15] with the aforementioned problems, MD PSO is used to evolve traditional ANNs and so the focus is particularly drawn on automatic design of the MLPs. Such an evolutionary approach makes this system *generic*, that is no assumption is made about the number of (hidden) layers and in fact none of the network properties (e.g., feedforward or not, differentiable activation function or not, etc.) is an inherent constraint. Recall that as long as the potential network configurations are transformed into a hash (dimension) table with a proper hash function where indices represent the solution space dimensions of the particles, MD PSO can then seek both positional and dimensional optima in an interleaved PSO process. This approach aims to achieve a high level of *robustness* with respect to the variations of the dataset, since the system is designed with a minimum set of parameters, and in such a way that their significant variations should not show a major impact on the overall performance. Above all, using standard ANNs such as traditional MLPs, instead of specific architectures (e.g., BbNN in [15]) further contributes to the generic nature of this system and in short, all these objectives are meant to make it applicable to any ECG dataset without any modifications (such as tuning the parameters or changing the feature vectors, ANN types, etc.).

8.1.2 ECG Data Processing

8.1.2.1 ECG Data

In this section, the MIT-BIH arrhythmia database [22] is used for training and performance evaluation of the patient-specific ECG classifier. The database contains 48 records, each containing two-channel ECG signals for 30-min duration selected from 24-h recordings of 47 individuals. Continuous ECG signals are band-pass filtered at 0.1–100 Hz and then digitized at 360 Hz. The database contains annotation for both timing information and beat class information verified by independent experts. In the current work, so as to comply with the AAMI ECAR-1987 recommended practice [13], we used 44 records from the MIT-BIH arrhythmia database, excluding 4 records which contain paced heartbeats. The first 20 records (numbered in the range of 100–124), which include representative samples of routine clinical recordings, are used to select representative beats to be included in the common training data. The remaining 24 records (numbered in the range of 200–234), contain ventricular, junctional, and supraventricular arrhythmias. A total of 83,648 beats from all 44 records are used as test patterns for performance evaluation. AAMI recommends that each ECG beat be classified into the following five heartbeat types: N (beats originating in the sinus mode), S (supraventricular ectopic beats), V (ventricular ectopic beats), F (fusion beats), and Q (unclassifiable beats). For all records, we used the modified-lead II signals and the labels to locate beats in ECG data. The beat detection process is beyond the scope of this chapter, as many highly accurate (>99 %) beat detection algorithms have been reported in the literature, [16, 23].

8.1.2.2 Feature Extraction Methodology

As suggested by the results from numerous previous works [10, 14, 23], both morphological and temporal features are extracted and combined into a single feature vector for each heartbeat to improve accuracy and robustness of the classifier. The wavelet transform is used to extract morphological information from the ECG data. The time-domain ECG signatures were first normalized by subtracting the mean voltage before transforming into time-scale domain using the dyadic wavelet transform (DWT). According to the wavelet transform theory, the multiresolution representation of the ECG signal is achieved by convolving the signal with scaled and translated versions of a mother wavelet. For practical applications, such as processing of sampled and quantized raw ECG signals, the discrete wavelet transform can be computed by scaling the wavelet at the dyadic sequence $(2^j)_{j \in Z}$ and translating it on a dyadic grid whose interval is proportional to 2^{-j}. The discrete WT is not only complete but also nonredundant unlike the continuous WT. Moreover, the wavelet transform of a discrete signal can be efficiently calculated using the decomposition by a two-channel multirate filter bank (the pyramid decomposition). However, due to the rate-change operators in the filter bank, the discrete WT is not time invariant but actually very sensitive to the alignment of the signal in time [25].

To address the time-varying problem of wavelet transforms, Mallat proposed a new algorithm for wavelet representation of a signal, which is invariant to time shifts [26]. According to this algorithm, which is called a TI-DWT, only the scale parameter is sampled along the dyadic sequence $(2^j)_{j \in Z}$ and the wavelet transform is calculated for each point in time. TI-DWTs pioneered by Mallat have been successfully applied to pattern recognition [26]. The fast TI-DWT algorithm, whose computational complexity is $O(N \log N)$, can be implemented using a recursive filter tree architecture [26]. In this study, we selected a quadratic spline wavelet with compact support and one vanishing moment, as defined in [26]. The same wavelet function has already been successfully applied to QRS detection in [16], achieving a 99.8 % QRS detection rate for the MIT-BIH arrhythmia database. In the presented ECG classification system, using a wavelet-based beat detector such as in [16] allows the same wavelet transform block to operate directly on the raw input ECG signal for beat detection and then morphological feature extraction, thus making the system more efficient and robust.

Figure 8.2 shows sample beat waveforms, including Normal (N), PVC (V), and APC (S) AAMI heartbeat classes, selected from record 201, modified-lead II from the MIT-BIH arrhythmia database and their corresponding TI-DWT decompositions computed for the first five scales. While wavelet-based morphological features provide effective discrimination capability between normal and some

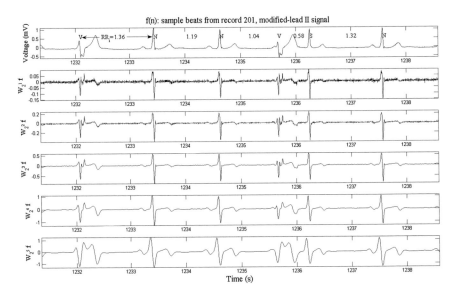

Fig. 8.2 Sample beat waveforms, including normal (*N*), PVC (*V*), and APC (*S*) AAMI heartbeat classes, selected from record 201 modified-lead II from the MIT/BIH arrhythmia database and corresponding TI-DWT decompositions for the first five scales

abnormal heartbeats (i.e., PVC beats), two temporal features (i.e., the R–R time interval and R–R time interval ratio) contribute to the discriminating power of wavelet-based features, especially in discriminating morphologically similar heartbeat patterns (i.e., Normal and APC beats).

In Fig. 8.3 (top), the estimated power spectrum of windowed ECG signal (a 500 ms long *Hanning* window is applied before FFT to suppress high-frequency components due to discontinuities in the end-points) from record 201 for N, V, and S beats is plotted, while equivalent frequency responses of FIR filters, $Q_j(w)$, for the first five scales at the native 360 Hz sampling frequency of the MIT-BIH data are illustrated at the bottom part of the figure. After analyzing the DWT decompositions of different ECG waveforms in the database, and according to the power spectra of ECG signal (the QRS complex, the P- and T-waves), noise, and artifact in [28], we selected $W_{2^4}f$ (at scale 2^4) signal as morphological features of each heartbeat waveform. Based on the -3 dB bandwidth of the equivalent $Q_4(w)$ filter (3.9–22.5 Hz) in Fig. 8.3 (bottom), $W_{2^4}f$ signal is expected to contain most of QRS complex energy and the least amount of high-frequency noise and low-frequency baseline wander. The fourth scale decomposition together with RR-interval timing information was previously shown to be the best performing feature set for DWT-based PVC beat classification in [4]. Therefore, a 180-sample morphological feature vector is extracted per heartbeat from DWT of the ECG signal at scale 2^4

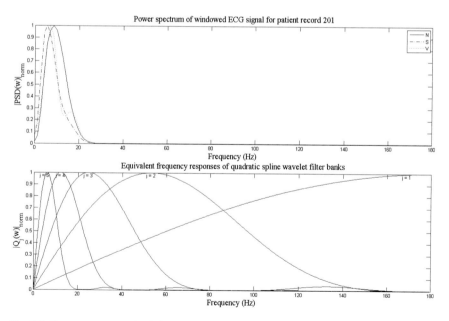

Fig. 8.3 Power spectrum of windowed ECG signal from record 201 for normal (*N*), PVC (*V*), and APC (*S*) AAMI heartbeat classes, and equivalent frequency responses of FIR digital filters for a quadratic spline wavelet at 360 Hz sampling rate

by selecting a 500 ms window centered at the R-peak (found by using the beat annotation file). Each feature vector is then normalized to have a zero mean and a unit variance to eliminate the effect of dc offset and amplitude biases.

8.1.2.3 Preprocessing by Principal Component Analysis

The wavelet-based morphological features in the training set are post-processed using PCA to reduce dimensionality (and redundancy) of input feature vectors. PCA, also known as the Karhunen–Loéve transform (KLT), is a well-known statistical method that has been used for data analysis, data compression, redundancy and dimensionality reduction, and feature extraction. PCA is the optimal linear transformation, which finds a projection of the input pattern vectors onto a lower dimensional feature space that retains the maximum amount of energy among all possible linear transformations of the pattern space. To describe the basic procedure of PCA, let F be a feature matrix of size $K \times N$, whose rows are wavelet features of size $1 \times N$ each belonging to one of K heartbeats in the training data. First, the covariance matrix C_F of this feature matrix is computed as,

$$C_F = E\{(F - m)(F - m)^t\}, \tag{8.1}$$

where m is the mean pattern vector. From the eigen-decomposition of C_F, which is a $K \times K$ symmetric and positive-definite matrix, the principal components taken as the eigenvectors corresponding to the largest eigenvalues are selected, and the morphological feature vectors are then projected onto these principal components (KL basis functions). In this work, 9 principal components which contain about 95 % of the overall energy in the original feature matrix are selected to form a resultant compact morphological feature vector for each heartbeat signal. In this case, the PCA reduced the dimensionality of morphological features by a factor of 20. Figure 8.4 shows a scatter plot of Normal, PVC, and APC beats from record 201 in terms of the first and third principal components and inter-beat time interval. It is worth noting that dimensionality reduction of the input information improves efficiency of the learning for a NN classifier due to a smaller number of input nodes [29].

The data used for training the individual patient classifier consist of two parts: Global (common to each patient) and local (patient-specific) training patterns. While patient-specific data contain the first 5 min segment of each patient's ECG record and is used as part of the training data to perform patient adaptation, the global dataset contains a relatively small number of representative beats from each class in the training files and helps the classifier learn other arrhythmia patterns that are not included in the patient-specific data. This practice conforms to the AAMI recommended procedure allowing the usage of at most a 5 min section from the beginning of each patient's recording for training [13].

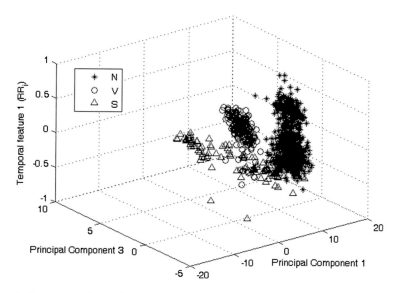

Fig. 8.4 Scatter plot of normal (*N*), PVC (*V*), and APC (*S*) beats from record 201 in terms of the first and third principal components and RR$_i$ time interval

8.1.3 Experimental Results

First, we shall demonstrate the *optimality* of the networks (with respect to the training MSE), which are automatically evolved by the MD PSO method according to the training set of an individual patient record in the benchmark database. We shall then present the overall results obtained from the ECG classification experiments and perform comparative evaluations against several state-of-the-art techniques in this field. Finally, the robustness of this system against variations of major parameters will be evaluated.

8.1.3.1 MD PSO Optimality Evaluation

In order to determine which network architectures are optimal (whether it is global or local) for a particular problem, exhaustive BP training is applied over every network configuration in the architecture space defined. As mentioned earlier, BP is a gradient descent algorithm and thus for a single run, it is susceptible to get trapped in the nearest local minimum. However, performing it a large number of times (e.g., $K = 500$) with randomized initial parameters eventually increases the chance of converging to (a close vicinity of) the global minimum of the fitness function. Note that even though K is kept quite high, there is still no guarantee of converging to the global optimum with BP; however, the idea is to obtain the "trend" of best performances achievable with every configuration under equal training conditions. In this way, the optimality of the networks evolved by MD PSO can be justified under the assumed criterion.

Due to the reasoning given earlier, the architecture space is defined over MLPs (possibly including one SLP) with the following activation function: *hyperbolic tangent* $\left(\tanh(x) = \frac{e^x - e^{-x}}{e^x + e^{-x}} \right)$. The input and output layer sizes are determined by the problem. A learning parameter for BP $\lambda = 0.001$ is used and the number of iterations is 10,000. The default PSO parameters for MD PSO are used here, specifically, the swarm size, $S = 100$, and velocity ranges $V_{max} = -V_{min} = X_{max}/2$, and $VD_{max} = -VD_{min} = D_{max}/2$. The dimension range is determined by the architecture space defined and the position range is set as $X_{max} = -X_{min} = 2$. Unless stated otherwise, these parameters are used in all experiments presented in this section.

In order to show the optimality of the network configurations evolved by MD PSO with respect to the MSE criterion, the "limited" architecture space is used first. Recall that this AS involves 41 ANNs ($R^1 : R^1_{min} = \{N_I, 1, 1, N_O\}$ and $R^1_{max} = \{N_I, 8, 4, N_O\}$) containing the simplest 1-, 2-, or 3-layer MLPs with $L^1_{min} = 1$, $L^1_{max} = 3$, and $N_I = 11$, $N_O = 5$. Then, one of the most challenging records is selected among the MIT-BIH arrhythmia database, belonging to the patient record 222. For this record, 100 MD PSO runs are performed with 100 particles, each of which terminates at the end of 1,000 epochs (iterations). At the end of each run, the best fitness score (minimum MSE) achieved, $f\left(x\hat{y}^{dbest}\right)$, by the particle with the index *gbest(dbest)* at the optimum dimension *dbest* is retained. The histogram of *dbest*, which is a hash index indicating a particular network configuration in R^1, eventually identifies the (near-) optimal configuration(s).

Figure 8.5 shows *dbest* histogram and the error statistics plot from the exhaustive BP training data of patient record 222. From the minimum mean-square error (*mMSE*) plot of the exhaustive BP training on top, it is clear that only four distinct sets of network configurations can achieve training *mMSEs* below 0.1. The corresponding indices (dimensions) of these four optimal networks are *dbest* = 9, 25, 33, and 41, where MD PSO managed to evolve either exactly to them or comes to a neighboring configuration, i.e., a near-optimal solution. MD PSO can in fact achieve the best (lowest) training *MSEs* for two sets of configurations: *dbest* = 33 and 41 (including their 3 close neighbors). These are 3-layer MLPs; *dbest* = 33 is for 11 × 8 × 3 × 5 and *dbest* = 41 is for 11 × 8 × 4 × 5. All MD PSO runs evolved either to *dbest* = 25 (corresponding to configuration 11 × 8 × 2 × 5) or to its neighbors with a slightly worse than the best configurations. MD PSO runs, which evolved to the simplest MLPs with a single hidden layer (i.e., *dbest* = 8 and 9 are for the MLPs 11 × 7 × 5 and 11 × 8 × 5) achieved the worst *mMSE, about* 15 % *higher than for dbest* = 33 and 41. The reason of MD PSO evolutions to those slightly worse configurations (for *dbest* = 25 and particularly for *dbest* = 9) is that MD PSO or PSO in general performs better in low dimensions. Furthermore, premature convergence is still a problem in PSO when the search space is in high dimensions. Therefore, MD PSO naturally favors a low-dimension solution

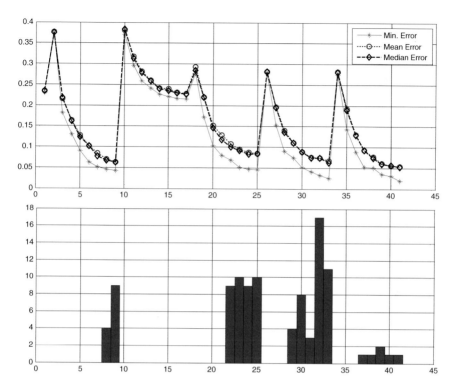

Fig. 8.5 Error (MSE) statistics from exhaustive BP training (*top*) and *dbest* histogram from 100 MD PSO evolutions (*bottom*) for patient record 222

when it exhibits a competitive performance compared to a higher dimension counterpart. Such a natural tendency eventually yields the evolution process to compact network configurations in the architecture space rather than the complex ones, as long as optimality prevails.

8.1.3.2 Classification Performance

We performed classification experiments on 44 records of the MIT-BIH arrhythmia database, which includes a total of 100,389 beats to be classified into five heartbeat types following the AAMI convention. For the classification experiments in this section, the common part of the training dataset contains a total of 245 representative beats, including 75 from each type N, -S, and -V beats, and all (13) type-F and (7) type-Q beats, *randomly* sampled from each class from the first 20 records (picked from the range 100 to 124) of the MIT-BIH database. The patient-specific training data includes the beats from the first 5 min of the corresponding patient's ECG record. Patient-specific feedforward MLP networks are trained with a total of 245 common training beats and a variable number of patient-specific

Table 8.1 Summary table of beat-by-beat classification results for all 44 records in the MIT/BIH arrhythmia database

Ground truth	Actual Classification Results				
	N	S	V	F	Q
N	73,019 (40,532)	991 (776)	513 (382)	98 (56)	29 (20)
S	686 (672)	1,568 (1,441)	205 (197)	5 (5)	6 (5)
V	462 (392)	333 (299)	4,993 (4,022)	79 (75)	32 (32)
F	168 (164)	28 (26)	48 (46)	379 (378)	2 (2)
Q	8 (6)	1 (0)	3 (1)	1 (1)	1 (0)

Classification results for the testing dataset only (24 records from the range 200 to 234) are shown in parenthesis

beats depending on the patient's heart rate, so only less than 1 % of the total beats are used for training each neural network. The remaining beats (25 min) of each record, in which 24 out of 44 records are completely new to the classifier, are used as test patterns for performance evaluation.

Table 8.1 summarizes beat-by-beat classification results of ECG heartbeat patterns for all test records. Classification performance is measured using the four standard metrics found in the literature [10]: Classification accuracy (Acc), sensitivity (Sen), specificity (Spe), and positive predictivity (Ppr). While accuracy measures the overall system performance over all classes of beats, the other metrics are specific to each class and they measure the ability of the classification algorithm to distinguish certain events (i.e., VEBs or SVEBs) from nonevents (i.e., nonVEBs or nonSVEBs). The respective definitions of these four common metrics using true positive (TP), true negative (TN), false positive (FP), and false negative (FN) are as follows: *Accuracy* is the ratio of the number of correctly classified patterns to the total number of patterns classified, $Acc = (TP + TN)/(TP + TN + FP + FN)$; *Sensitivity* is the rate of correctly classified events among all events, $Sen = TP/(TP + FN)$; *Specificity* is the rate of correctly classified nonevents among all nonevents, $Spe = TN/(TN + FP)$; and *Positive Predictivity* is the rate of correctly classified events in all detected events, $Ppr = TP/(TP + FP)$. Since there is a large variation in the number of beats from different classes in the training/testing data (i.e., 39,465/50,354 type-N, 1,277/5,716 type-V, and 190/2,571 type-S beats), sensitivity, specificity, and positive predictivity are more relevant performance criteria for medical diagnosis applications.

The system presented in this section is compared with three state-of-the-art methods, [10, 14] and [15], which comply with the AAMI standards and use *all* records from the MIT-BIH arrhythmia database. For comparing the performance results, the problem of VEB and SVEB detection is considered individually. The VEB and SVEB classification results over all 44 records are summarized in Table 8.2. The performance results for VEB detection in the first four rows of Table 8.2 are based on 11 test recordings (200, 202, 210, 213, 214, 219, 221, 228, 231, 233, and 234) that are common to all four methods. For SVEB detection, comparison results are based on 14 common recordings (with the addition of records 212, 222, and 232) between the presented system and the methods in [14] and [15].

Table 8.2 VEB and SVEB classification performance of the presented method and comparison with the three major algorithms from the literature

Methods	VEB				SVEB			
	Acc	Sen	Spe	Ppr	Acc	Sen	Spe	Ppr
Hu et al. [10][1]	94.8	78.9	96.8	75.8	N/A	N/A	N/A	N/A
Chazal et al. [14][1]	96.4	77.5	98.9	90.6	92.4	76.4	93.2	38.7
Jiang and Kong [15][1]	98.8	94.3	99.4	95.8	97.5	74.9	98.8	78.8
Presented[1]	97.9	90.3	98.8	92.2	96.1	81.8	98.5	63.4
Jiang and Kong [15][2]	98.1	86.6	99.3	93.3	96.6	50.6	98.8	67.9
Presented[2]	97.6	83.4	98.1	87.4	96.1	62.1	98.5	56.7
Presented[3]	98.3	84.6	98.7	87.4	97.4	63.5	99.0	53.7

[1] The comparison results are based on 11 common recordings for VEB detection and 14 common recordings for SVEB detection
[2] The VEB and SVEB detection results are compared for 24 common testing records only
[3] The VEB and SVEB detection results of the presented system for all training and testing records

Several interesting observations can be made from these results. First, for SVEB detection, sensitivity and positive predictivity rates are comparably lower than VEB detection, while a high specificity performance is achieved. The reason for the worse classifier performance in detecting SVEBs is that SVEB class is under-represented in the training data and hence more SVEB beats are misclassified as normal beats. Overall, the performance of the presented system in VEB and SVEB detection is significantly better than [10] and [14] for all measures and is comparable to the results obtained with evolvable BbNNs in [15]. Moreover, it is observed that this system achieves comparable performance over the training and testing set of patient records. It is worth noting that the number of training beats used for each patient's classifier was less than 2 % of all beats in the training dataset and the resulting classifiers designed by the MD PSO process have improved generalization ability, i.e., the same low number of design parameters are used for all networks.

8.1.3.3 Robustness

In order to investigate the robustness of the ECG classification system presented in this chapter against the variations of the few PSO parameters used, such as the swarm size, S, the iteration number I, and to evaluate the effect of the architecture space (and hence the characteristics of the ANNs used), we performed four classification experiments over the MIT-BIH arrhythmia database (I–IV) and their classification accuracy results per VEB and SVEB, are presented in Table 8.3. Experiments I–III are performed over the same architecture space, with 1-, 2-. and 3-layer MLP architectures defined by $R^1_{min} = \{11, 8, 4, 5\}$ $R^1_{min} = \{11, 16, 8, 5\}$. Between I and II, the swarm size, and between II and III, the iteration number is changed significantly, whereas in IV an entirely different architecture space containing 4-layer MLPs is used. From the table, it is quite evident that the effects of such major variations over the classification accuracy are insignificant. Therefore,

Table 8.3 VEB and SVEB classification accuracy of the classification system for different PSO parameters and architecture spaces

Percentage (%)	I	II	III	IV
VEB	98.3	98.2	98.3	98.0
SVEB	97.4	97.3	97.1	97.4

I: $R_{min}^1 = \{11,8,4,5\}$, $R_{min}^1 = \{11,16,8,5\}$, $S = 100$, $I = 500$
II: $R_{min}^1 - \{11,8,4,5\}$, $R_{min}^1 = \{11,16,8,5\}$, $S = 250$, $I = 200$
III: $R_{min}^1 = \{11,8,4,5\}$, $R_{min}^1 = \{11,16,8,5\}$, $S = 80$, $I = 200$
IV: $R_{min}^1 = \{11,6,6,3,5\}$, $R_{min}^1 = \{11,12,10,5,5\}$, $S = 400$, $I = 500$

any set of common PSO parameters within a reasonable range can be conveniently used. Furthermore, for this ECG database, the choice of the architecture space does not affect the overall performance, yet any other ECG dataset containing more challenging ECG data might require the architecture spaces such as in IV, in order to obtain a better generalization capability.

8.2 Classification of Holter Registers

Holter registers [27] are ambulatory ECG recordings with a typical duration of 24–48 h and they are particularly useful for detecting some heart diseases such as *cardiac arrhythmias, silent myocardial ischemia, transient ischemic episodes, and for arrhythmic risk assessment of patients,* all of which may not be detected by a short-time ECG [31]. Yet any process that requires humans or even an expert cardiologist to examine more than small amount of data can be highly error prone. A single record of a *Holter* register is usually more than 100,000 beats, which make the visual inspection almost infeasible, if not impossible. Therefore, the need for automatic techniques for analyzing such a massive data is imminent and in that, it is crucial not to leave out significant beats since the diagnosis may depend on just a few of them. However, the dynamic nature and intra-signal variation in a typical *Holter* register is quite low and the abnormal beats, which may indicate the presence of a potential disease, can be scattered along the signal. So utilizing a dynamic clustering technique based on MD PSO with FGBF, a systematic approach is developed, which can summarize a long-term ECG record by discovering the so-called master key-beats that are the representative or the prototype beats from different clusters. With a great reduction in effort, the cardiologist can then perform a quick and accurate diagnosis by examining and labeling only the master key-beats, which in duration are no longer than few minutes of ECG record. The expert labels over the master key-beats are then back propagated over the entire ECG record to obtain a patient-specific, long-term ECG classification. As the main application of the current work, this systematic approach is then applied over a real (benchmark) dataset, which contains seven long-term ECG recordings [32].

8.2.1 The Related Work

ECG analysis has proven to be an important method routinely used in clinical practice for continuous monitoring of cardiac activities. ECG analysis can be used to detect cardiac arrhythmias, which often arise as a consequence of a cardiac disease and may be life-threatening and require immediate therapy. According to Kushner and Yin [30], with an estimated 300 million of ECGs performed every year, there is clearly a need for accurate and reliable ECG interpretation. Computer analysis of ECG data can be of great assistance to the experts in detecting cardiac abnormalities both for real time clinical monitoring and long-term (24–48 h) monitoring in intensive care unit (ICU) and ambulatory settings. Many computer-based methods have been proposed for automated analysis and interpretation of ECGs. However, automated classification of ECG beats is a challenging problem as the morphological and temporal characteristics of ECG signals show significant variations for different patients and under different temporal and physical conditions [1]. This is the reason in practice for underperformance of many fully automated ECG processing systems, which hence make them unreliable to be widely used clinically [12]. Additionally, it is known that both accuracy and efficiency of these systems degrade significantly for larger datasets [11].

Long-term continuous ECG monitoring and recording, also known as Holter ECG or Holter register [27], is needed for detection of some diseases, such as *cardiac arrhythmias, transient ischemic episodes* and *silent myocardial ischemia*, and for arrhythmic risk assessment of patients [31]. Since visual analysis of long-term recordings of the heart activity, with more than 100,000 ECG beats in a single recording, is difficult to diagnose and can be highly error prone, automated computer analysis is of major importance. In the past, a number of methods have been proposed for feature extraction from ECG signals including heartbeat temporal intervals [14], time-domain morphological features [14], frequency domain features [2], wavelet transform features [4], and Hermite transform coefficients [33]. Accordingly, several techniques have been developed by researchers for long-term ECG data analysis. In [33], a method for unsupervised characterization of ECG signals is presented. Their approach involves Hermite function representation of ECG beats (specifically QRS complex) and self-organized neural networks (SOMs) for beat clustering. Application to all (48) 30-min records from the MIT-BIH arrhythmia database results in 25 clusters and by classifying each cluster according to an expert's annotation of one typical beat a total misclassification error of 1.5 % is achieved. The method proposed in [34] consists of nonlinear temporal alignment, trace segmentation as feature extraction and *k-medians* as clustering algorithm. Its primary goal is to extract accurately significant beats, which can be examined by a physician for the diagnosis. From the results of experimental studies using 27 registers (of total 27,412 beats) from the MIT-BIH database, *k-medians* performs better than the Max–Min clustering algorithm achieving a clustering error of ~ 7 % in the best case. The work in [7] describes a new approach for analyzing large amounts of cardiovascular data, for example multiple days of continuous high-resolution ECG data, based on symbolic representations of cardiovascular signals and morphology-based Max–Min clustering. It was tested over cardiologist-annotated ECG data (30 min. recordings) from 48 patients from the MIT-BIH arrhythmia database achieving 98.6 % overall correct classification. This approach has the advantage of using no a priori knowledge about disease states allowing for discovery of unexpected events (patterns). The goal of the work in [36] is to achieve better clustering analysis of ECG complexes using an ant colony optimization (ACO) based clustering algorithm. In this study, time-dependent morphological parameters extracted from two consecutive periods of an ECG signal are used as specific features. The method is tested using a total of 8,771 ECG periods taken from the MIT-BIH database resulting in a total sensitivity of 94.4 % to all six arrhythmia types.

Most of these techniques mainly suffer from the usage of suboptimal clustering algorithms, such as Max–Min in [7], *k-medians* in [34] and SOMs in [33], some of which require a priori setting of some thresholds or parameters, such as $\theta = 50$ in [7]. Particularly, the performance of the approach in [33] is limited by the ability of small number of Hermite expansion coefficients used for the approximation of the heartbeats. It is worth noting that although all these techniques claim to address the problem of long-term (Holter) ECG classification, none has really been applied to a real Holter register, probably due to such limitations.

8.2.2 Personalized Long-Term ECG Classification: A Systematic Approach

The section presents the systematic approach for personalized classification of long-term ECG data. As the overview shown in Fig. 8.6, such an approach addresses the problem within the entire lifetime of a long-term ECG signal recorded from an individual patient, i.e., starting with acquisition and pre-processing, to the temporal segmentation, followed with a master key-beat extraction with 2-pass dynamic clustering, and finally classification of the entire ECG data by back propagating the expert cardiologist labels over the master key-beats. As a *personalized* approach, the objective is to minimize the amount of data from each individual patient by selecting the most relevant data, which will be subject to manual classification, as much as possible so that the cardiologist can quickly and accurately diagnose any latent disease by examining only the representative beats (the master key-beats) each from a cluster of homogeneous (similar) beats. This justifies the application of the dynamic clustering technique based on MD PSO with FGBF, which is designed to extract the optimal (number of) clusters within a diverse dataset. Recall that *optimality* here can only be assessed according to the *validity index* function, the *feature extraction* (data representation), and the *distance* (similarity) *metric* used. Therefore, the performance of the clustering technique can be further improved by using better alternatives than the basic and simple

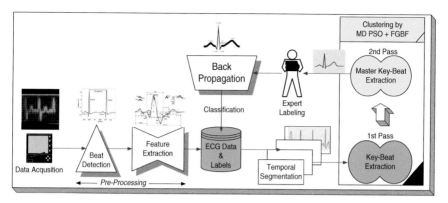

Fig. 8.6 The overview of the systematic approach for long-term ECG classification

ones used in the current work with the purpose of demonstrating the basic performance level of this systematic approach. For both passes, the dynamic clustering operation is performed using the same validity index given in Eq. (6.1) with $\alpha = 1$. Recall that this is the simplest form, which is entirely parameter free and in addition, L_2 *Minkowski* norm (*Euclidean*) is used as the distance metric in the feature space.

As shown in Fig. 8.6, after the data acquisition is completed, the pre-processing stage basically contains beat detection and feature extraction of the sampled and quantized ECG signal. Before beat detection, all ECG signals are filtered to remove baseline wander, unwanted power-line interference and high-frequency noise from the original signal. This filtering unit can be utilized as part of heartbeat detection process (for example, the detectors based on wavelet transforms [16]). For all records, we used the modified-lead II signals and utilized the annotation information (provided with the MIT-BIH database [22]) to locate beats in ECG signals. Beat detection process is beyond the scope of this chapter, as many beat detection algorithms achieving over 99 % accuracy have been reported in the literature, e.g., [16] and [24]. Before feature extraction, the ECG signal is normalized to have a zero mean and unit variance to eliminate the effect of dc offset and amplitude biases. After the beat detection over *quasiperiodic* ECG signals by using RR-intervals, morphological and temporal features are extracted for each beat as suggested in [14] and combined into a single characteristic feature vector for each heartbeat. As shown in Fig. 8.7, temporal features relating to heartbeat *fiducial* point intervals and morphology of the ECG signals are extracted by sampling the signals. They are calculated separately for the first lead signals for each heartbeat. Since the detection of some arrhythmia (such as *Bradycardia, Tachycardia, and premature ventricular contraction*) depends on the timing sequence of two or more ECG signal periods [35], four temporal features are considered in our study. They are extracted from heartbeat *fiducial* point intervals (RR-intervals), as follows:

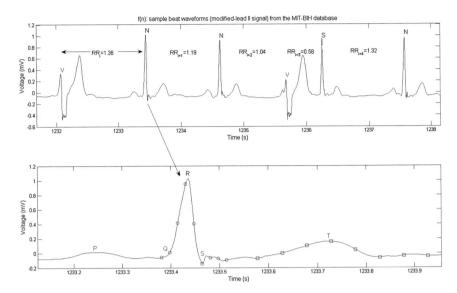

Fig. 8.7 Sample beat waveforms, including normal (*N*), PVC (*V*), and APC (*S*) AAMI [13] heartbeat classes from the MIT-BIH database. Heartbeat fiducial point intervals (RR-intervals) and ECG morphology features (samples of QRS complex and T-wave) are extracted

1. *pre-RR-interval* the RR-interval between a given heartbeat and the previous heartbeat,
2. *post-RR-interval* the RR-interval between a given heartbeat and the following heartbeat,
3. *local average RR-interval* the average of the ten RR-intervals surrounding a heartbeat,
4. *average-RR interval* the mean of the RR-intervals for an ECG recording.

In addition to temporal features, ECG morphology features are extracted from two sampling windows in each heartbeat formation. The sampling windows are formed based on the heartbeat *fiducial* points (maximum of R-wave or minimum of S-wave in Fig. 8.7). Specifically, the morphology of the QRS complex is extracted using a 150-ms window and 60-Hz sampling rate, resulting in nine ECG samples as features. The eight ECG samples representing the low-frequency T-wave morphology are extracted using a 350-ms window and 20-Hz sampling rate. The final feature vector for each heartbeat is then formed by combining 17 morphological and 4 temporal interval features.

Once the 21 dimensional (21-D) feature vectors composed from the temporal and morphological characteristics of ECG beats are extracted, the entire ECG data is temporally segmented into fixed size frames (segments) for achieving mainly two objectives. On the one hand, the massive size of ECG data makes it almost infeasible to perform an efficient clustering and on the other hand, *outliers*, which are significantly different from the typical (normal) beats and thus may indicate the presence of an abnormal heart activity, may get lost due to their low frequency of occurrences. Therefore, we adopt a typical approach, which is frequently performed in audio processing, that is, temporally segmenting data into homogeneous frames.

Due to the dynamic characteristics of an audio signal, the frame duration is typically chosen between 20 and 50 ms in order to get as a homogeneous signal as possible, i.e., [6]. Accordingly, for a *Holter* register with 24–48 h long, we choose ~5 min long (300 beats) duration for time segments since the intra-segment variation along the time axis is often quite low. So performing a clustering operation within such homogeneous segments will yield only one or few clusters except perhaps the transition segments where a *change*, morphological or temporal, occurs on the normal form of the ECG signal. No matter how minor or insignificant duration this abnormal *change* might take, in such a limited time segment, the MD PSO-based dynamic clustering technique can separate those "different" beats from the normal ones and group them into a distinct cluster. One key-beat, which is the *closest* to the cluster centroid with respect to the distance metric used in 21-D feature space, is then chosen as the "prototype" to represent all beats in that cluster. Since the optimal number of clusters is extracted within each time segment, only necessary and sufficient number of key-beats is thus used to represent all 300 beats in a time segment. Note that the possibility of missing *outliers* is thus reduced significantly with this approach since one key-beat is equally selected either from an *outlier* or a typical cluster without considering their size. Yet redundancy among the key-beats of consecutive segments still exists, since it is highly probable that similar key-beats shall occur among different segments. This is the main reason for having the second pass, which performs dynamic clustering over key-beats to obtain finally the master key-beats. They are basically the "elite" prototypes representing all possible physiological heart activities occurring during a long-term ECG record.

Since this is a personalized approach, each patient has, in general, normal beats with possibly one or few abnormal periods, indicating a potential heart disease or disorder. Therefore, ideally speaking only a few master key-beats would be expected at the end, each representing a cluster of similar beats from each type. For instance, one cluster may contain *ventricular* beats arising from ventricular cavities in the heart and another may contain only *junctional* beats arising from atrioventricular junction of the heart. Yet, due to the lack of discrimination power of the morphological or temporal features or the similarity (distance) metric used, the dynamic clustering operation may create more than one cluster for each anomaly. Furthermore, the normal beats have a broad range of morphological characteristics [7] and within a long time span of 24 h or longer, it is obvious that the temporal characteristics of the normal beats may significantly vary too. Therefore, it is reasonable to represent normal beats with multiple clusters rather than only one. In short, several master key-beats may represent the same physiological type of heart activity. The presentation of the master key-beats to the expert cardiologist can be performed with any appropriate way as this is a *visualization* detail, and hence beyond the scope of this work. Finally, the overall classification of the entire ECG data can be automatically accomplished by back propagating the master key-beats' labels in such a way that a beat closest to a particular master key-beat (using the same distance metric in 21-D feature space) is assigned to its label.

8.2.3 Experimental Results

The systematic approach presented in this section is applied to long-term ECG data in the Physionet MIT-BIH long-term database [22], which contains six two-channel ECG signals sampled at 128 Hz per channel with 12-bit resolution, and one three-channel ECG sampled at 128 Hz per channel with 10-bit resolution. The duration of the 7 recordings changes from 14 to 24 h each and a total of 668,486 heartbeats in the whole database are used in this study. The database contains annotation for both timing and beat class informations manually reviewed by independent experts. The WFDB (Waveform Database) software package with library functions (from PhysioToolkit [35]) is used for reading digitized signals with annotations. In this study, for all records, we used the first lead signals and utilized the annotation to locate beats in ECG signals. The CVI, the feature extraction, and the distance metric are already presented in Sect. 8.2.2 and the typical PSO parameters used are as follows. Due to the massive size of data, to speed up the process we set $iterNo = 500$ and $S = 100$, respectively. Due to the same reason mentioned earlier the use of cutoff error as a termination criterion is avoided. The positional range, $[X_{min}, X_{max}]$ is automatically set to $[-1, 1]$, as the range of the feature normalization, ± 1, and the rest of the range limits are same as in the previous section.

Following the preprocessing that consists of formation of heartbeats using the RR-intervals and the feature extraction thereafter, the patient's long-term ECG data is temporally segmented into homogenous frames of 300 beats (~ 5 min duration) as described in Sect. 8.2.2. The dynamic clustering is then performed in 21-D feature space to extract optimal number of clusters within each time frame. The number of clusters, that is identical to the number of key-beats found automatically for each frame depends on distinct physiological heartbeat types in each patient's ECG record. Figures 8.8 and 8.9 present excerpts from patients 14,046 and 14,172 showing a short sequence of ECG and the extracted key-beats. Note that in each case, the key-beats selected by the clustering algorithm show distinct morphological and temporal heartbeat interval characteristics. In addition, significant morphological (and possibly temporal interval) differences between the same type of beats from one patient's ECG to another are also visible. As a result, this systematic approach by temporal segmentation and the dynamic clustering technique produces such key-beats that represent all possible physiological heart activities in patient's ECG data. Therefore, finding the true number of clusters by the systematic approach is the key factor that makes a major difference from some

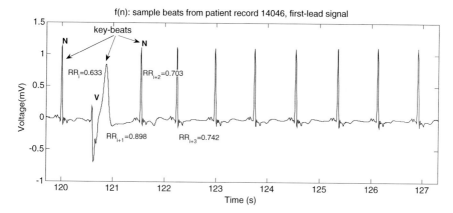

Fig. 8.8 Excerpt of raw ECG data from patient record 14,046 in the MIT-BIH long-term database. The three key-beats, having morphological and RR-interval differences, are chosen by the systematic approach presented

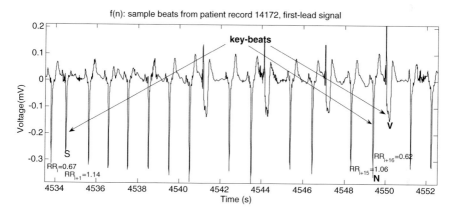

Fig. 8.9 Excerpt of raw ECG data from patient record 14,172 in the MIT-BIH long-term database. The key-beats extracted by the systematic approach are indicated

earlier works such as [34] and [7], both of which iteratively determines this number by an empirical threshold parameter.

Table 8.4 shows the overall results of the systematic approach over all patients from the MIT-BIH ong-term ECG database. Labels manually annotated by the experts are used only for the master key-beats selected by the system. The classification of the entire ECG data, or in other words, the labeling of all beats contained therein is then automatically accomplished by the BP of the master key-beat labels, as explained in Sect. 8.2.2. The performance results tabulated in Table 8.4 are calculated based on the differences between the labels generated by the systematic approach presented and the expert supplied labels provided with the database. The AAMI provides standards and recommended practices for reporting

Table 8.4 Overall results for each patient in the MIT-BIH long-term database using the systematic approach presented

Patient	N	S	V	F	Q	Accuracy (%)
14,046	105,308/105,405	0/1	9,675/9,765	34/95	0/0	99.79
14,134	38,614/38,766	0/29	9,769/9,835	641/994	0/0	98.80
14,149	144,508/144,534	0/0	235/264	0/0	0/0	99.96
14,157	83,340/83,412	6/244	4,352/4,368	53/63	0/0	99.62
14,172	58,126/58,315	77/1003	6,517/6,527	0/1	0/0	98.41
14,184	77,946/78,096	0/39	23,094/23,383	2/11	0/0	99.53
15,814	91,532/91,617	6/34	9,680/9,941	1,427/1,744	0/0	99.32
Total	599,374/600,145	89/1350	63,322/64,083	2,157/2,908	0/0	
Weighted average						99.48

For each class, the number of correctly detected beats is shown relative to the total beats originally present

performance results of automated arrhythmia detection algorithms [13]. In this study, according to the AAMI recommended practice, each ECG beat is classified into the following five heartbeat types: N (beats originating in the sinus mode), S (supraventricular ectopic beats), V (ventricular ectopic beats), F (fusion beats), and Q (unclassifiable beats). For this study, the systematic approach labeled heartbeats consistent with the cardiologist supplied annotations over ∼99.5 % of the time for a total of 668,486 beats.

From the results in Table 8.4, the systematic approach performed with a high accuracy for detection of normal (N) and ventricular (V) groups of beats. Specifically, accurate detection of premature ventricular contractions (PVCs) from the ventricular group (V) in long-term ECG data is essential for patients with heart disease, since it may lead to possible life-threatening cardiac conditions [35]. On the other hand, for supraventricular ectopic (S) beats and some cases of fusion of ventricular and normal (F) beats, the systematic approach presented did not form a separate cluster corresponding to each type of beat due to the fact that their morphological and temporal features are indeed quite similar to normal (N) beats. Therefore, we can conclude that a more accurate separation of both supraventricular and fusion beats from the normal beats requires a more effective feature extraction technique than the one used in the current work. The (average) classification error, ∼0.5 %, can further be divided into *critical* and *noncritical* errors, which can be defined as follows: All normal (N) beats that are misclassified as one of the anomaly classes (S, V, or F) contribute to *noncritical* error because the expert cardiologist, who is false alerted by the presence of such beats that indicate a potential cardiac disease, can review them and correct the classification. This is also true for such misclassification among the anomaly classes since those beats are anyway classified as *not normal*, but with a wrong class type and thus they shall be all subject to expert's attention following with a manual correction. The *critical* errors occur when a beat in one of the anomaly classes is misclassified as normal since this is the case where the expert is not alerted and kept unnoticed by the presence of a potential heart disease. So the consequences of such *critical*

Table 8.5 The overall confusion matrix

Truth	Classification results				
	N	S	V	F	Q
N	**599,374**	254	280	237	0
S	1,256	**89**	5	0	0
V	456	3	**63,322**	302	0
F	221	0	530	**2,157**	0
Q	0	0	0	0	**0**

errors might be fatal. According to the overall confusion matrix given in Table 8.5, the (average) *critical* error level is below 0.3 % where major part of critical errors occurred for the beats within class S due to the above-mentioned reasons specific to the feature extraction method used in this study.

Since MD PSO is in stochastic nature, to test the *repeatability* and *robustness* of the systematic approach presented, we performed 10 independent runs on patient record 14,172, from which we obtained the lowest performance with the average classification accuracy 98.41 %. All runs led to similar accuracy levels with only a slight deviation of 0.2 %. It is also worth mentioning that with an ordinary computer, the extraction of key-beats in a ~5 min. time frame typically takes less than 1.5 min. Therefore, this systematic approach is quite suitable for a real-time application, that is, the key-beats can be extracted in real time with a proper hardware implementation during the recording of a *Holter* ECG.

8.3 Summary and Conclusions

In this chapter, we presented an automated patient-specific ECG heartbeat classifier, which is based on an efficient formation of morphological and temporal features from the ECG data and evolutionary neural network processing of the input patterns individually for each patient. The TI-DWT and the PCA are the principal signal processing tools employed in the feature extraction scheme. The wavelet-based morphology features are extracted from the ECG data and are further reduced to a lower dimensional feature vector using PCA technique. Then, by combining compact morphological features with the two critical temporal features, the resultant feature vector to represent each ECG heartbeat is finally obtained with a proper normalization. In the core of the patient-specific approach lies the evolutionary ANNs which are introduced in the previous chapter, that is, for each individual patient, an optimal feedforward ANN will be evolved using the MD PSO technique.

The results of the classification experiments, which are performed over the benchmark MIT-BIH arrhythmia database show that the such a patient-specific approach can achieve average accuracies and sensitivities better than most of the existing algorithms for classification of ECG heartbeat patterns according to the

AAMI standards. An overall average accuracy of 98.3 % and an average sensitivity of 84.6 % for VEB detection, and an average accuracy of 97.4 % and an average sensitivity of 63.5 % for SVEB detection were achieved over all 44 patient records from the MIT-BIH database. The overall results promise a significant improvement over other major techniques in the literature with the exception of the BbNN-based personalized ECG classifier in [15], which gave comparable results over the test set of 24 records. However, recall that it used many critical parameters, BP method to train the classifier, which has certain drawbacks, and a specific ANN architecture, which may not suit feature vectors in higher dimensions. The approach presented in this chapter is based only on well-known, standard techniques such as DWT and PCA, while using the most typical ANN structure, the MLPs. Experimental results approve that its performance is not affected significantly by variations of the few parameters used. Therefore, the resulting classifier successfully achieves the main design objectives, i.e., maintaining a robust and generic architecture with superior classification performance. As a result it can be conveniently applied to any ECG database "as is", alleviating the need of human "expertise" and "knowledge" for designing a particular ANN.

In the second part of the chapter, the dynamic clustering scheme based on MD PSO with FGBF is used in the core of a long term, personalized ECG classification system. Such a systematic approach basically addresses the inspection and infeasibility problems within the entire lifetime of a long-term ECG signal recorded from an individual patient and it is tested over a real (benchmark) database containing a total of 668,486 (manually) labeled heartbeats. To our knowledge, this is the first work ever applied to a real *Holter* database, since most of the earlier works tested only over 30 min regular ECG records from MIT-BIH arrhythmia database. Again, as a *personalized* approach with manual labeling of only a few minutes long ECG data from each patient, we achieved an average of 99.5 % classification accuracy. In a typical *Holter* register of 24 h long, selection of the right prototype beats, which can yield such a high accuracy level and a great reduction in effort, is mainly due to two key operations. The first one, the so-called temporal segmentation, partitions the entire data into homogenous time segments that can be represented by *minimal* amount of key-beats. The following two-pass dynamic clustering operation first extracts the *true* number of key-beats and then the master key-beats among them. In both operations, such delicate classification accuracy indicates that, the utilized dynamic clustering technique is able to extract the optimal (number of) clusters in a 21-D feature (data) space. Although the underlying optimization technique (MD PSO) is stochastic in nature, repeating the classification process several times over the benchmark dataset yields almost identical accuracy levels with insignificant variations (~ 0.2 %), thus indicating a high level of robustness for finding the optimal solutions.

Moreover, such a systematic approach apparently promises a high level of *insensitivity* to the length (duration) of the data since the duration of the time segments is fixed and the number of clusters (master key-beats) found in the second pass is not related at all with the number of key-beats in the first pass. Although this system is intended and purposefully developed for long-term data to help professionals focus on the most relevant data, it can also provide efficient and robust solutions for much shorter ECG datasets too. Besides classification, with some proper annotation, master key-beats can also be used for the *summarization* of any long-term ECG data for a fast and efficient visual inspection, and they can further be useful for indexing *Holter* databases, for a fast and accurate information retrieval. On the other hand, 0.5 % error rate, although may seem quite insignificant for a short ECG dataset, can still be practically high for *Holter* registers because it corresponds to several hundreds of misclassified beats, some of which might be important for a critical diagnosis. Yet, recall that the optimality of the clustering algorithm depends on the CVI, the feature extraction method and the distance metric, in that, we use simple and typical ones so as to obtain a basic or unbiased performance level. Therefore, by using for instance a more efficient CVI and better alternatives for distance metric may further improve the performance.

8.4 Programming Remarks and Software Packages

As discussed in Sect. 7.5, **ClassifierTestApp** is the major MD PSO test-bed application for evolutionary ANNs. The patient-specific ECG classification can thus be performed using this application with the same data format, i.e., the "*.dt" files for each patient. In overall, there are 44 data files (patients) with the naming convention, "ECGfeatures2_p*ID*.dt" where *ID* corresponds to the patient index.

On the other hand, the classification of Holter registers is a typical case of feature clustering in N-D feature space. As discussed in Sect. 6.4, the MD PSO test-bed application, **PSOTestApp**, is used to perform feature clustering for Holter register classification. The **CPSOcluster** class has already been described in detail in Sects. 4.4.2 and 6.4 (with the FGBF application). Recall that this class was for 2-D clustering over simple 2-D synthetic data spaces. For the N-D clustering operations over data files, the option: "6. N-D Clustering over *.dt (data) files...", should be selected from the main GUI of the application, and the main dialog object from **CPSOtestAppDlg** class will then use the object from **CPSOclusterND** class in **void CPSOtestAppDlg::OnDopso()** function to perform N-D clustering. Both classes, **CPSOcluster** and **CPSOclusterND**, have almost an identical class structure with the same API functions besides the few differences such as data classes (**CPixel2D** and **CVectorNDopt**) and temporal data segmentation performed for **CPSOclusterND** class where clustering is individually (and independently) performed for each time segment. Since the data classes (the template class represented by **class X**) are different, this makes the main difference over the template-based MD PSO object, **CPSO_MD<T,X>** and the solution space object, **CSolSpace<X>**. The MD PSO object declaration is as follows:

CPSO_MD<CSolSpace<CVectorNDopt>, CVectorNDopt>*m_pPSO; // The current MDPSO obj...

Table 8.6 The function **CPSOclusterND::PSOThread()**

```
void CPSOclusterND::PSOThread()
{
        m_pFL->Reset();
        for(pFL = NULL; (pFL = m_pFL->Next()); )
        { // For all dt files, apply PSO clustering..
                // Read the data file header..
                ...
                vec_size = in_size + out_size; // we use all the entries = input + classification,  for clustering..
                int no_segm = (training_examples + validation_examples) / SEGM_SIZE; // no of segments to be
        clustered..
                ...
                for(m_segm=0; m_segm<no_segm; ++m_segm)
                {         // for each time segment, perform clustering..
                        int beats_left = training_examples + validation_examples - m_segm*SEGM_SIZE ; //
        no. of beats left for clustering..
                        s_vaSize = (beats_left < 2*SEGM_SIZE) ? beats_left : SEGM_SIZE ; // size of beats for
        key-beat clustering..
                        s_pVectorArr = new CVectorNDopt*[s_vaSize]; // Create the FV array..
                        // Read Training beat vector data..
                        ...
                        for(int rep = 0; rep < m_psoParam._repNo; ++rep)
                        { // Repeat MD PSO clustering..
                                ...
                                ...
                        } // for..
                } // for s..
        } // for..
```

Besides these differences, the clustering operation is pretty much same as explained in Sects. 4.4.2 and 6.4. There are again two CVI functions (identical to their 2-D counterparts) implemented: **ValidityIndex()** and **ValidityIndex2()** where the latter is again in use. For each time segment, the N-D feature vectors are read from the (next) data file in the file queue, and stored in the static pointer array, **s_pVectorArr**, and once the clustering operation has been performed, the cluster centroids, the so-called key beats' FVs, are stored in the ***_KEY[*segmID*].dt** files. Table 8.6 presents the **CPSOclusterND::PSOThread()** function with its basic blocks. Note that there are a total of **no_segm** segments each of which contains a fixed number (**SEGM_SIZE** or less) of beat FVs. The *segmID* corresponds to the member variable **m_segm**. The key beats, once extracted by the MD PSO object, are then retrieved and stored in the local member function, **GetResults()** in that there is a sample code for *K-means* clustering method, which is repeated by numerous times (defined by **KMEANS_REP**) among which the best achieved fitness score is then compared with the clustering result of the MD PSO operation. For an unbiased comparison, the parameter K is obviously set to the optimum number of clusters found in the MD PSO operation.

References

1. R. Hoekema, G.J.H. Uijen, Oosterom van A, Geometrical aspects of the interindividual variability of multilead ECG recordings. IEEE Trans Biomed Eng **48**(5), 551–559 (2001)
2. K. Minami, H. Nakajima, T. Toyoshima, Real-Time discrimination of ventricular tachyarrhythmia with Fourier-transform neural network. IEEE Trans Biomed Eng **46**(2), 179–185 (1999)
3. L.Y. Shyu, Y.H. Wu, W.C. Hu, Using wavelet transform and fuzzy neural network for VPC detection from the holter ECG. IEEE Trans Biomed Eng **51**(7), 1269–1273 (2004)
4. O.T. Inan, L. Giovangrandi, G.T. Kovacs, Robust neural-network based classification of PVCs using wavelet transform and timing interval features. IEEE Trans Biomed Eng **53**(12), 2507–2515 (2006)
5. X. Alfonso, T.Q. Nguyen, ECG beat detection using filter banks. IEEE Trans Biomed Eng **46**(2), 192–202 (1999)
6. J.L. Willems, E. Lesaffre, Comparison of multigroup logisitic and linear discriminant ECG and VCG classification. J. Electrocardiol. **20**, 83–92 (1987)
7. Z. Syed, J. Guttag, C. Stultz, Clustering and symbolic analysis of cardiovascular signals: discovery and visualization of medically relevant patterns in long-term data using limited prior knowledge. EURASIP J. Appl. Sign. Proces. pp. 1–16 (2007). Article ID 67938. doi:10.1155/2007/67938
8. D.A. Coast, R.M. Stern, G.G. Cano, S.A. Briller, An approach to cardiac arrhythmia analysis using hidden Markov models. IEEE Trans Biomed Eng **37**(9), 826–836 (1990)
9. S. Osowski, L.T. Hoai, T. Markiewicz, Support vector machine based expert system for reliable heartbeat recognition. IEEE Trans Biomed Eng **51**(4), 582–589 (2004)
10. Y. Hu, S. Palreddy, W.J. Tompkins, A patient-adaptable ECG beat classifier using a mixture of experts approach. IEEE Trans Biomed Eng **44**(9), 891–900 (1997)
11. S.C. Lee, Using a Translation-Invariant Neural Network to Diagnose Heart Arrhythmia, in *IEEE Proceedings of Conference Neural Information Processing Systems*, Nov 1989
12. P. de Chazal, R.B. Reilly, A patient-adapting heartbeat classifier using ECG morphology and heartbeat interval features. IEEE Trans Biomed Eng **53**(12), 2535–2543 (2006)
13. Recommended practice for testing and reporting performance results of ventricular arrhythmia detection algorithms, Association for the Advancement of Medical Instrumentation. (Arlington, VA, 1987)
14. P. de Chazal, M. O'Dwyer, R.B. Reilly, Automatic classification of heartbeats using ECG morphology and heartbeat interval features. IEEE Trans Biomed Eng **51**(7), 1196–1206 (2004)
15. W. Jiang, S.G. Kong, Block-based neural networks for personalized ECG signal classification. IEEE Trans. Neural Networks, **18**(6), 1750–1761 (2007)
16. C. Li, C.X. Zheng, C.F. Tai, Detection of ECG characteristic points using wavelet transforms. IEEE Trans. Biomed. Eng. **42**(1), 21–28 (1995)
17. R. Silipo et al., ST-T segment change recognition using artificial neural networks and principal component analysis, in *Computers in Cardiology*, 1995, pp. 213–216
18. R. Silipo, C. Marchesi, Artificial neural networks for automatic ECG analysis. IEEE Trans Sign Proces **46**(5), 1417–1425 (1998)
19. L.Y. Shyu, Y.H. Wu, W.C. Hu, Using wavelet transform and fuzzy neural network for VPC detection from the holter ECG. IEEE Trans. Biomed. Eng. **51**(7), 1269–1273 (2004)
20. X. Yao, Y. Liu, A new evolutionary system for evolving artificial neural networks. IEEE Trans Neural Networks **8**(3), 694–713 (1997)
21. R. Mark, G. Moody, MIT-BIH arrhythmia database directory [Online], http://ecg.mit.edu/dbinfo.html
22. J. Pan, W.J. Tompkins, A real-time QRS detection algorithm. IEEE Trans Biomed Eng **32**(3), 230–236 (1985)
23. Y.H. Hu, W.J. Tompkins, J.L. Urrusti, V.X. Afonso, Applications of artificial neural networks for ECG signal detection and classification. J. Electrocardiol. pp. 66–73 (1994)

24. S.G. Mallat, *A Wavelet Tour of Signal Processing*, 2nd edn. (Academic Press, San Diego, 1999)
25. S.G. Mallat, S. Zhong, Characterization of signals from multiscale edges. IEEE Trans Pattern Anal Machine Intell **14**, 710–732 (1992)
26. N.V. Thakor, J.G. Webster, W.J. Tompkins, Estimation of QRS complex power spectra for design of a QRS filter. IEEE Trans Biomed Eng **BME-31**, 702–705 (1984)
27. M. Paoletti, C. Marchesi, Discovering dangerous patterns in long-term ambulatory ECG recordings using a fast QRS detection algorithm and explorative data analysis. Comput. Methods Progr. Biomedicine **82**, 20–30 (2006)
28. S. Pittner, S.V. Kamarthi, Feature extraction from wavelet coefficients for pattern recognition tasks. IEEE Trans Pattern Anal Machine Intell **21**, 83–88 (1999)
29. N.J. Holter, New methods for heart studies. Science **134**(3486), 1214–1220 (1961)
30. M. Lagerholm, C. Peterson, G. Braccini, L. Edenbrandt, L. Sörnmo, Clustering ECG complexes using Hermite functions and self-organizing maps. IEEE Trans. Biomed. Eng. **47**(7), 838–848 (2000)
31. PhysioBank, MIT-BIH long-term database directory [Online], http://www.physionet.org/physiobank/database/ltdb/
32. P. Mele, Improving electrocardiogram interpretation in the clinical setting, J. Electrocardiol. (2008). doi:10.1016/j.jelectrocard.2008.04.003
33. D. Cuesta-Frau, J.C. Pe'rez-Corte's, G. Andreu-Garci'a, Clustering of electrocardiograph signals in computer-aided Holter analysis. Comput Methods Programs Biomed **72**(3), 179–196 (2003)
34. M. Korurek, A. Nizam, A new arrhythmia clustering technique based on ant colony optimization, J. Biomed. Inform. (2008). doi:10.1016/j.jbi.2008.01.014
35. PhysioToolkit, The WFDB software package [Online], http://www.physionet.org/physiotools/wfdb.shtml
36. W.J. Tompkins, J.G. Webster, *Design of microcomputer-based medical instrumentation* (Prentice Hall Inc, Englewood Cliffs, 1981), pp. 398–3999. ISBN 0-13-201244-8

Chapter 9
Image Classification and Retrieval by Collective Network of Binary Classifiers

> *It is not the strongest of the species that survives, nor the most intelligent that survives. It is the one that is the most adaptable to change.*
>
> Charles Darwin

Multimedia collections are growing in a tremendous pace as the *modus operandi* for information creation, exchange, and storage in our modern era. This creates an urgent need for means and ways to manage them efficiently. Earlier attempts such as text-based indexing and information retrieval systems show drastic limitations and require infeasible laborious work. The efforts are thus focused on a content-based management approach; yet, we are still at the early stages of the development of techniques to guarantee efficiency and effectiveness in content-based multimedia systems. The peculiar nature of multimedia information, such as difficulty of semantic indexing, complex multimedia identification, and difficulty of adaptation to different applications are the main factors hindering breakthroughs in this area.

This chapter focuses primarily on content-based image retrieval (CBIR) and classification. For classifying and indexing a large image content data reserve, the key questions are: (1) how to select certain features so as to achieve the highest discrimination over certain classes, (2) how to combine them in the most effective way, (3) which distance metric to apply, (4) how to find the optimal classifier configuration for the classification problem at hand, and (5) how to scale/adapt the classifier if a large number of classes/features are present and finally, (6) how to train the classifier efficiently to maximize the classification accuracy. These questions still remain unanswered. The current state-of-the-art classifiers such as SVMs, RF, ANNs, etc., cannot cope with such requirements since a single classifier, no matter how powerful and well-trained it can be, cannot discriminate efficiently a vast amount of classes, over an indefinitely large set of features, where both classes and features are not static, but rather dynamic such as the case of media content repositories. Therefore, in order to address these problems and to maximize the classification accuracy which will in turn boost the retrieval performance, we shall focus on a global framework design that embodies a collective network of evolutionary classifiers. In this way, we shall achieve as compact classifiers as possible, which can be evolved in a much more efficient way than a

S. Kiranyaz et al., *Multidimensional Particle Swarm Optimization for Machine Learning and Pattern Recognition*, Adaptation, Learning, and Optimization 15, DOI: 10.1007/978-3-642-37846-1_9, © Springer-Verlag Berlin Heidelberg 2014

single but complex classifier, and the optimum classifier for the classification problem at hand can be searched with the underlying evolutionary technique. At a given time, this allows the creation and designation of a dedicated classifier for discriminating a certain class type from the others based on a single feature. Each incremental evolution session will "learn" from the current best classifier and can improve it further. Moreover, with each incremental evolution, new classes/features can also be introduced which signals the collective classifier network to create new corresponding networks and classifiers within to adapt dynamically to the change. In this way the collective classifier network will be able to dynamically scale itself to the indexing requirements of the media content data reserve while striving for maximizing the classification and retrieval accuracies for a better user experience.

9.1 The Era of CBIR

The CBIR has been an active research field for which several feature extraction, classification, and retrieval techniques have been proposed up to date. However, when the database size varies in time and usually grows larger, it is a common fact that the overall classification and retrieval performance significantly deteriorates. Due to the reasoning given earlier, the current state-of-the-art classifiers such as support vector machines (SVMs) [1, 2], Bayesian Classifiers, random forests (RFs) [3], and artificial neural networks (ANNs) cannot provide optimal or even feasible solutions to this problem. This fact drew the attention toward classifier networks (or ensembles of classifiers). An earlier ensemble of classifier type of approach, Learn++ [4], incorporates an ensemble of weak classifiers, which can perform incremental learning of new classes; however, albeit at a steep cost, i.e., learning new classes requires an increasingly large number of classifiers for each new class to be learned. The resource allocating network with long-term memory (RAN-LTM) [5] can avoid this problem by using a single RBF network, which can be incrementally trained by "memory items" stored in a long-term memory. However, RAN-LTM has a fixed output structure and thus, is not able to accommodate a varying number of classes. For the incremental learning problem when new classes are dynamically introduced, some hierarchical techniques such as [6] and [7] have been proposed. They basically separate a single class from the previous classes within each hierarchical step, which builds up on its previous step. One major drawback of this approach is parallelization since the addition of N new classes will result in N steps of adding one class at a time. Furthermore, the possible removal of an existing class is not supported and hence requires retraining of the entire classifier structure from scratch. None of the ensemble of classier methods proposed so far can support feature scalability and thus a new feature extraction will eventually make the current classifier ensemble obsolete and require a new design (and re-training) from scratch.

Another major question that still remains in CBIR is how to narrow the "Semantic Gap" between the low-level visual features that are automatically extracted from images and the high-level semantics and content-description by humans. Among a wide variety of features proposed in the literature, none can really address this problem alone. So the focus has been drawn on fusing several features in the most effective way since whatever type of classifiers is used, the increased feature space may eventually cause the "Curse of Dimensionality" phenomenon that significantly degrades the classification accuracy. In [8], three MPEG-7 visual features, color layout descriptor (CLD), scalable color descriptor (SCD), and edge histogram descriptor (EHD) are fused to train several classifiers (SVMs, KNN, and Falcon-ART) for a small database with only two-classes and 767 images. This work has clearly demonstrated that the highest classification accuracy has been obtained with the proper feature fusion. This is indeed an expected outcome since each feature may have a certain level of discrimination for a particular class. In another recent work [9], this fact has been, once again, confirmed where the authors fused three MPEG-7 features: SCD, Homogenous Texture Descriptor and EHD and trained SVM over the same database. Basic color (12 dominant colors) and texture (DWT using quadrature mirror filters) features were used in [10] to annotate image databases using ensemble of classifiers (SVMs and Bayes Point Machines). Although the classification is performed over a large image database with 25 K images and 116 classes from Corel repository, the authors used above 80 % of the database for training and another database to evaluate and manually optimize various kernel and classifier parameters. In [11] SVMs together with 2-D Hidden Markov Model (HMM) are used to discriminate image classes in an integrated model. Two features, 50-D SIFT (with a dimension reduction by PCA) and 9-D color moments (CM) are used individually in two datasets using 80 % of the images for training, and the classification accuracies are compared. In all these image classification works and many alike, besides the aforementioned key problems there are other drawbacks and limitations, e.g., they can work with only a limited feature set to avoid the "Curse of Dimensionality" and they used the major part of the database, some as high as 80 % or even higher, for training to sustain a certain level of classification accuracy. They are all image classification methods for static databases assuming a fixed GTD and fixed set of features where feature and class scalability and dynamic adaptability have not been considered.

In order to address these problems and hence to maximize the classification accuracy which will in turn boost the CBIR performance, in this chapter, we shall focus on a global framework design that embodies a collective networks of (evolutionary) binary classifiers (CNBC). In this way, we shall demonstrate that one can achieve as compact classifiers as possible, which can be evolved and trained in a much more efficient way than a single but complex classifier, and the optimum classifier for the classification problem in hand can be *searched* with an underlying evolutionary technique, e.g., as described in Chap. 7. At a given time, this allows the creation and designation of a dedicated classifier for discriminating a certain class type from the others based on a single feature. The CNBC can support varying and large set of visual features among which it optimally selects, weights, and fuses the most discriminative ones for a particular class. Each NBC is devoted to a unique class and further encapsulates a set of *evolutionary binary classifiers* (BCs), each of which is optimally chosen within the architecture space (AS), discriminating the class of the NBC with a unique feature. For instance for an NBC evolved for the *sunset* class, it will most likely select and use mainly color features (rather than texture and edge features) and the most descriptive color feature elements (i.e., color histogram bins) for discriminating this particular class (i.e., red, yellow, and black), are weighted higher than the others.

The CNBC framework is primarily designed to increase the retrieval accuracy in CBIR on "dynamic databases" where variations do occur at any time in terms of (new) images, classes, features, and users' relevance feedback. Whenever a "change" occurs, the CNBC dynamically and "optimally" adapts itself to the change by means of its topology and the underlying evolutionary method. Therefore, it is not a "just another classifier" as no static (or fixed) classifier including the state-of-the-art competitors such as ANNs, SVMs, and RF can really do that, e.g., if (a) new feature(s) or class(es) are introduced, one needs to reform and retrain a new (static) classifier from scratch—which is a waste of time and information so far accumulated. This topology shall prove useful for dynamic, content/data adaptive classification in CBIR. In that sense, note that comparative evaluations against SVMs and RFs are performed only in "static databases" to show that the CNBC has a comparable or better performance –despite of the fact that neither CBIR nor classification in *static* databases is not the primary objective. Furthermore, the CNBC is not an "ensemble of classifiers" simply because: At any time t, (1) it contains an individual network of classifier (NBC) for each class, (2) each NBC contains several evolving classifiers, each is optimally selected among a family of classifiers (the so-called architecture space, AS) and their number is determined by the number of features present at that time, (3) the number of NBCs is also determined by the number of classes at time t. So the entire CNBC body is subject to change whenever there is a need for it. The ensemble of classifiers usually contains a certain number of "static" classifiers and their input/output (features and classes) must be fixed in advance; therefore, similar to standalone static classifiers, they cannot be used for dynamic databases.

9.2 Content-Based Image Classification and Retrieval Framework

This section describes in detail the image classification framework; the collective network of (evolutionary) binary classifiers (CNBC), which uses user-defined ground truth data (GTD) as the train dataset[1] to configure its internal structure and

---

[1] As a common term, we shall still use "train dataset" or "train partition" to refer to the dataset over which the CNBC is evolved.

to evolve its binary classifiers (BCs) individually. Before going into details of CNBC, a general overview for this novel classification topology will be introduced in Sect. 9.2.1.

9.2.1 Overview of the Framework

The image classification framework is designed to dynamically adapt (or scale) to any change and update in an image database. As shown in Fig. 9.1, new image(s) can be inserted into the database for which the user may introduce new class(es) via relevance feedbacks, and/or new feature(s) can be extracted, at any convenient time. As long as the user provides new ground-truth data (GTD) for the new classes, the existing classifier body, CNBC, can incrementally be evolved if the

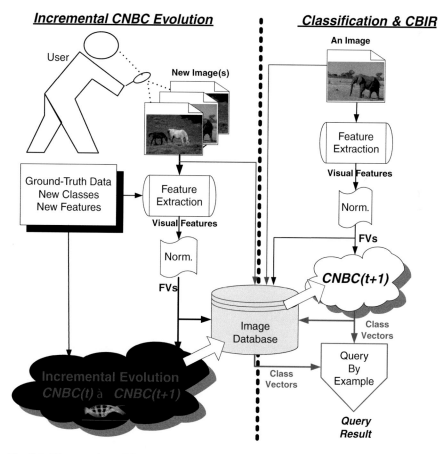

Fig. 9.1 The overview of the CBIR framework

need arises, i.e., if the existing CNBC (or some NBCs in it) fails to classify these new classes accurately enough.

MUVIS system [12] is used to extract a large set of low-level visual features that are properly indexed in the database along with the images. Unit normalized feature vectors (FVs) formed from those features are fed into the input layer of the CNBC where the user provided GTD is converted to target class vectors (CVs) to perform an incremental evolution operation. The user can also set the number of evolutionary runs or the desired level of classification. Any CNBC instance can directly be used to classify a new image and/or to perform content-based image queries, the result of which can be evaluated by the user who may introduce new class(es), yielding another incremental evolution operation, and so on. This is an ongoing cycle of human-classifier interaction, which gradually *adapts* CNBC to the user's class definitions. New low-level features can also be extracted to improve the discrimination among classes, which signals CNBC to adapt to the change simultaneously. In short, dynamic class and feature scalability are the key-objectives aimed within the CNBC design. Before going into the details of the CNBC framework, Sect. 9.2.2 will first introduce the evolutionary update mechanism that keeps *the best* classifier networks in the AS during numerous incremental evolutionary runs.

9.2.2 Evolutionary Update in the Architecture Space

Since the evolutionary technique, MD PSO, is a stochastic optimization method, in order to improve the probability of convergence to the global optimum, several evolutionary runs can be performed. Let N_R be the number of runs and N_C be the number of configurations in the architecture space (AS). For each run the objective is to find the optimal classifier within AS with respect to a pre-defined criterion. Note that along with the best classifier, all other configurations in AS are also subject to evolution and therefore, they are continuously re-trained with each run. So during this ongoing process, between any two consecutive runs, any network configuration can replace the current best one in AS if it improves the fitness function. This is also true for the alternative *exhaustive search*, i.e., each network configuration in AS is trained by N_R back-propagation (BP) runs and the same evolutionary update rule applies.

Figure 9.2 demonstrates an evolutionary update operation over a sample AS containing 5 MLP configurations. The tables in the figure show the training MSE which is the criterion used to select the optimal configuration at each run. The best runs for each configuration are highlighted and the best configuration in each run is tagged with '*'. Note that at the end of the three runs, the overall best network with MSE = 0.10 has the configuration: $15 \times 2 \times 2$ and thus used as the classifier until another configuration beats it in a future run during an evolutionary update. In this way, each binary classifier configuration in AS can only *evolve* to a better state, which is the main purpose of this evolutionary update mechanism.

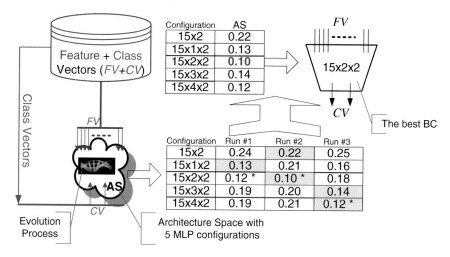

Fig. 9.2 Evolutionary update in a sample AS for MLP configuration arrays $R_{min} = \{15, 1, 2\}$ and $R_{max} = \{15, 4, 2\}$ where $N_R = 3$ and $N_C = 5$. The best runs for each configurations are highlighted and the best configuration in each run is tagged with '*'

9.2.3 The Classifier Framework: Collective Network of Binary Classifiers

9.2.3.1 The Topology

To achieve scalability with respect to a varying number of classes and features, a dedicated framework encapsulating NBCs is developed, where NBCs can *evolve* continuously with the ongoing evolution sessions, i.e., cumulating the user supplied GTD for images and thus forming the train dataset. Each NBC corresponds to a *unique* image class while striving to discriminate only that class from the rest of the classes in the database. Moreover, each NBC shall contain a varying number of evolutionary BCs in the input layer where each BC performs binary classification using a single feature. Each FV of a particular feature is only fed to its corresponding binary classifier in each NBC. Therefore, whenever a new feature is extracted, its corresponding binary classifier will be created, evolved (using the available GTD), and inserted into each NBC, yet keeping each of the other binary classifiers "as is". On the other hand, whenever an existing feature is removed, the corresponding binary classifier is simply removed from each NBC in the CNBC. In this way *scalability* with respect to a varying number of features is achieved and the overall system can adapt to the change without requiring re-forming and re-evolving from scratch.

Each NBC has a "fuser" binary classifier in the output layer, which collects and fuses the binary outputs of all binary classifiers in the input layer and generates a single binary output, indicating the relevancy of each FV of an image belonging to

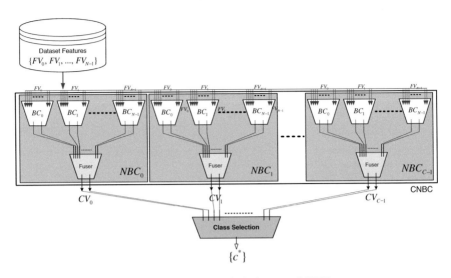

Fig. 9.3 Topology of the CNBC framework with C classes and N FVs

the NBC's corresponding class. Furthermore, CNBC is also *scalable* to any number of classes since whenever a new class is defined by the user, a new corresponding NBC can simply be created (and evolved) only for this class without requiring any need for change in the other NBCs as long as they can accurately discriminate the new class from their corresponding classes. This allows the overall system to adapt dynamically to the varying number of classes. As shown in Fig. 9.3, the main idea in this approach is to use as many classifiers as necessary, so as to divide a large-scale learning problem into many NBC units along with the binary classifiers within, and thus prevent the need of using complex classifiers as the performance of both training and evolution processes degrade significantly as the complexity rises due to the well-known *curse of dimensionality* phenomenon. A major benefit of this approach with respect to efficient evolution process is that the configurations in the AS can be kept as *compact* as possible avoiding unfeasibly large storage and training time requirements. This is a significant advantage especially for the training methods performing local searches, such as BP since the amount of deceiving local minima is significantly lower in the error space for such simple and compact ANNs. Especially, when BP is applied exhaustively, the probability of finding the optimum solution is significantly increased. Also note that evolving CNBC may reduce the computation time significantly since it contains simple and compact classifier networks, each of which can be individually evolved by a separate CPU and hence the overall computation time can be reduced as much as desired. This in practice leads to a significantly less complex classifier compared to training a single but more complex ANN classifier.

In order to increase the classification accuracy, a dedicated class selection technique is applied in CNBC. A *1-of-n* encoding scheme is used in all BCs, therefore, the output layer size of all binary classifiers is always two. Let $CV_{c,1}$ and $CV_{c,2}$ be the first and second output of the cth BC's class vector (CV). The class selection in 1-of-n encoding scheme can simply be performed by comparing the individual outputs, e.g., say a positive output if $CV_{c,2} > CV_{c,1}$, and vice versa for negative output. This is also true for the fuser BC, whose output is the output of its NBC. The FVs of each dataset item are fed into each NBC in the CNBC. Each FV drives through (via forward propagation) its corresponding binary classifier in the input layer of the NBC. The outputs of these binary classifiers are then fed into the fuser binary classifier of each NBC to produce class vectors (CVs). The class selection block shown in Fig. 9.3 collects these outputs and selects the positive class(es) of the CNBC as the final outcome. This selection scheme, first of all, differs with respect to the dataset class type, i.e., the dataset can be called "uni-class", if an item in the dataset can belong to $only$ one class, otherwise it is called "multi-class". Therefore, in a uni-class dataset there must be $only$ one class, c^*, selected as the positive outcome whereas in a multi-class dataset, there can be one or more NBCs, $\{c^*\}$, with a positive outcome. In the class selection scheme the $winner$-$takes$-all strategy is utilized. Assume without loss of generality that a CV of $\{0, 1\}$ or $\{-1, 1\}$ corresponds to a positive outcome where $CV_{c,2} - CV_{c,1}$ is maximum. Therefore, for uni-class datasets, the positive class index, c^*, ("the winner") is determined as follows:

$$c^* = \arg\max_{c \in [0, C-1]} (CV_{c,2} - CV_{c,1}) \tag{9.1}$$

In this way the erroneous cases (false negative and false positives) where no or more than one NBC exists with a positive outcome can be properly handled. However, for multi-class datasets the $winner$-$takes$-all strategy can only be applied when no NBC yields a positive outcome, i.e., $CV_{c,2} \le CV_{c,1} \; \forall c \in [0, C-1]$, otherwise for an input set of FVs belonging to a dataset item, multiple NBCs with positive outcome may indicate multiple true-positives and hence cannot be further pruned. As a result, for a multi-class dataset the (set of) positive class indices, $\{c^*\}$, is selected as follows:

$$\{c^*\} = \begin{pmatrix} \arg\max_{c \in [0, C-1]} (CV_{c,2} - CV_{c,1}) & if \; CV_{c,2} \le CV_{c,1} \; \forall c \in [0, C-1] \\ \{\arg_{c \in [0, C-1]} (CV_{c,2} > CV_{c,1})\} & else \end{pmatrix}$$

$$\tag{9.2}$$

9.2.3.2 Evolution of the CNBC

The evolution of a subset of the NBCs or the entire CNBC is performed for each NBC individually with a two-phase operation, as illustrated in Fig. 9.4. As explained earlier, using the FVs and the target class vectors (CVs) of the training dataset, the evolution process of each binary classifier in a NBC is performed within the current AS in order to find the optimal binary classifier configuration with respect to a given criterion (e.g., training/validation MSE or classification error, CE). During the evolution, only NBCs associated with those classes represented in the training dataset are evolved. If the training dataset contains new classes, which does not yet have a corresponding NBC, a new NBC is created for each, and evolved using the training dataset.

In Phase 1, see top of Fig. 9.4, the binary classifiers of each NBC are first evolved given an input set of FVs and a target CV. Recall that each CV is associated with a unique NBC and the fuser binary classifiers are not used in this phase. Once an evolution session is over, the AS of each binary classifier is then recorded so as to be used for potential (incremental) evolution sessions in the future.

Recall further that each evolution process may contain several runs and according to the aforementioned evolutionary update rule, the best configuration achieved will be used as the classifier. Hence once the evolution process is completed for all binary classifiers in the input layer (Phase 1), the best binary classifier configurations are used to forward propagate all FVs of the items in the training dataset to compose the FV for the fuser binary classifier from their output CVs, so as to evolve the fuser binary classifier in the second phase. Apart from the difference in the generation of the FVs, the evolutionary method (and update) of the fuser binary classifier is same as any other binary classifier. In this phase, the

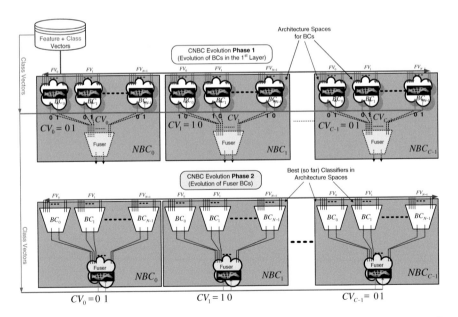

Fig. 9.4 Illustration of the two-phase evolution session over BCs' architecture spaces in each NBC

fuser binary classifier learns the *significance* of each individual binary classifier (and its feature) for the discrimination of that particular class. This can be viewed as the adaptation of the entire feature space to discriminate a specific class in a large dataset. Alternatively, this can be viewed as an efficient *feature selection* scheme over the set of FVs, by selecting the most discriminative FVs for a given class. The fuser BC, if properly evolved and trained, can then "weight" each binary classifier (with its FV), accordingly. In this way the feature (and its BC) shall optimally be "fused" according to its discrimination power with respect to each class. In short the CNBC, if properly evolved, shall learn the significance (or the discrimination power) of each FV and its individual components.

In this chapter, the image databases are considered as *uni-class* where one sample can belong to only one class, and during the evolution process, each positive sample of one class can be used as a negative sample for all others. However, if there is a large number of classes, an *uneven distribution* of positive and negative samples per class, may bias the evolution process. In order to prevent this, a negative sample selection is performed in such a way that for each positive sample, the number of negative samples (per positive sample) will be limited according to a predetermined positive-to-negative ratio (PNR). The selection of negative samples is performed with respect to the closest proximity to the positive sample so that the classifier can be evolved by discriminating those negative samples (from the positive one) which have the highest potential for producing a false-positive. Therefore, if properly trained, the classifier can draw the "best possible" boundary between positive and negative samples, which shall in turn improve the classification accuracy. The features of those selected items and the classes will form the FVs of the training dataset, over which the CNBC body can be created and evolved.

9.2.3.3 Incremental Evolution of the CNBC

Incremental evolution is the ability to evolve each classifier in the network dynamically so as to cope with changes that may occur at any time in a dynamic database. To achieve this, the CNBC framework is designed to carry out continuous "incremental" evolution sessions where each session may further improve the classification performance of each binary classifier using the advantage of the "evolutionary updates". The main difference between the initial and the subsequent (incremental) evolution sessions is the initialization of the evolution process: the former uses *random* initialization whereas the latter starts from the last AS parameters of each classifier in each BC. Note that the training dataset used for the incremental evolution sessions may be different from the previous ones, and each session may contain several runs. Thus, the evolutionary update rule compares the performance of the last recorded and the *current* (after the run) network over the *current* training dataset.

During each incremental evolution phase, existing NBCs are *incrementally* evolved *only if* they cannot accurately classify the training samples of the new classes. In that, an empirical threshold level (e.g., 95 %) is used to determine the level of classification accuracy required for the *new* GTD encountered. The NBCs for the new classes are directly evolved *without* such verification and they use the available (or *log*) GTD (the positive samples) of the existing NBCs during their evolution process as negative samples. Therefore, for each evolution session new and log GTD are individually fused to evolve both new (initially) and existing NBCs (incrementally).

Consequently, the MD PSO evolutionary technique used for evolving MLPs is initialized with the current AS parameters of the network. That is the swarm particles are randomly initialized (as in the initial evolutionary step) except that one of the particles (without loss of generality, we assume the first particle with $a = 0$) has its personal best set to the optimal solution found in the previous evolutionary session. For MD PSO evolution over MLPs, this can be expressed as,

$$xy_0^d(0) \leftarrow \Psi^d \left\{ \{w_{jk}^0\}, \{w_{jk}^1\}, \{\theta_k^1\}, \{w_{jk}^2\}, \{\theta_k^2\}, \ldots, \{w_{jk}^{o-1}\}, \{\theta_k^{o-1}\}, \{\theta_k^o\} \right\} \quad \forall d$$
$$\in [1, N_C]$$

$$(3)$$

where $\left\{w_{jk}^l\right\}$ and $\{\theta_k^l\}$ represent the sets of weights and biases of layer l of the MLP network, Ψ^d, which is the dth (MLP) configuration retrieved from the last AS record. It is expected that especially at the early stages of the MD PSO run, the first particle is likely to be the *gbest* particle in every dimension, guiding the swarm toward the last solution otherwise keeping the process independent and unconstrained. Particularly, if the training dataset is considerably different in the incremental evolution sessions, it is quite probable that MD PSO can converge to a new solution while taking the past solution (experience) into account.

For the alternative evolutionary technique, the exhaustive search via repetitive BP training of each network in the AS, the first step of an incremental training will simply be the initialization of the weights $\left\{w_{jk}^l\right\}$ and biases $\{\theta_k^l\}$ with the parameters retrieved from the last record of the AS of that BC. Starting from this as the initial point, and using the *current* training dataset with the target CVs, the BP algorithm can then perform its gradient descent in the error space.

9.3 Results and Discussions

During the classification and CBIR experiments in this section, three major properties of the CNBC will be demonstrated: (1) the (incremental) evolutionary and dynamic adaptability virtues. For instance, when new classes and features are introduced, whether or not it can adapt itself to the change with a "minimal" effort, i.e., the (incremental) evolution is applied only if there is a need for it and if so, it uses the advantage of the previous (accumulated) knowledge, (2) the competitive performance level with the state-of-the-art classifiers (SVMs and RFs) on *static* databases and to demonstrate its parameter independence as the default parameters are used for CNBC evolution while the best possible internal parameters are searched and used for the competitors for training, and finally, (3) the application of the classifier to improve CBIR performance on dynamically varying image databases (the main objective). On the other hand, since CBIR is the main application domain, designing a classifier network for massive databases with millions of images is *not* our objective due to the well-known "semantic gap" phenomenon. Note that this cannot be achieved only with the classifier design since the low-level features have extremely limited or in practice no discrimination power on such magnitudes. Obviously, designing such highly discriminative features that can provide the description power needed is beyond the scope of this chapter; however, in some cases, CNBC can also achieve a superior classification performance thanks to its "Divide and Conquer" approach. For instance, in a synthetic aperture radar (SAR) classification [13] where the SAR features have indeed a high degree of discrimination, CNBC was evolved using the GTD only from 0.02 % of SAR image pixels with a total number exceeding a million and achieved higher than 99.6 % classification accuracy.

9.3.1 Database Creation and Feature Extraction

MUVIS framework [12], is used to create and index the following two image databases by extracting 14 features for each image.

(1) *Corel_10* Image Database: There are 1,000 medium resolution (384 × 256 pixels) images obtained from Corel repository [14] covering 10 diverse classes: 1—*Natives*, 2—*Beach*, 3—*Architecture*, 4—*Bus*, 5—*Dino Art*, 6—*Elephant*, 7—*Flower*, 8—*Horse*, 9—*Mountain*, and 10—*Food*.

(2) *Corel_Caltech_30* Image Database: There are 4,245 images from 30 diverse classes that are obtained from both Corel and Caltech [15] image repositories.

As detailed in Table 9.1, some of the basic color (e.g., MPEG-7 Dominant Color Descriptor, HSV color histogram and Color Structure Descriptor [16]), texture (e.g. Gabor [17], Local Binary Pattern [18], and Ordinal Co-occurrence Matrix [19]), and edge (e.g., Edge Histogram Direction [16]) features, are extracted. Some of them are created with different parameters to extract several features and the total feature dimension is obtained as 2,335. Such a high feature space dimension can thus give us an opportunity to test the performance of the CNBC framework against the "curse-of-dimensionality" and the *scalability* with respect to the varying number of features.

Table 9.1 14 Features extracted per MUVIS database

FV	Feature	Parameters	Dim.
1	HSV color histogram	$H = 6, S = 2, V = 2$	24
		$H = 8, S = 4, V = 4$	
		$N_{DC}^{max} = 6, T_A = 2\%, T_S = 15$	
2		$N_{DC}^{max} = 8, T_A = 2\%, T_S = 15$	128
3	Dominant color descriptor		27
4			35
5	Color structure descriptor	32 bins	32
6		64 bins	64
7		128 bins	128
8		256 bins	256
9		512 bins	512
10		1024 bins	1,024
11	Local binary pattern		16
12	Gabor	Scale = 4, orient. = 6	48
13	Ordinal co-occurence	$d = 3, o = 4$	36
14	Edge histogram dir.		5

9.3.2 Classification Results

Both databases are partitioned in such a way that the majority (55 %) of the items is spared for testing and the rest was used for evolving the CNBC. The evolution (and training) parameters are as follows: For MD PSO, we use the termination criteria as the combination of the maximum number of iterations allowed ($iterNo = 100$) and the cut-off error ($\varepsilon_C = 10^{-4}$). Other parameters were empirically set as: the swarm size, $S = 50$, $V_{max} = x_{max}/5 = 0.2$ and $VD_{max} = 5$. For exhaustive BP, the learning parameter is set as $\lambda = 0.01$ and the iteration number is 20. We use the typical activation function: *hyperbolic tangent* ($\tanh(x) = \frac{e^x - e^{-x}}{e^x + e^{-x}}$). For the AS, we used simple configurations with the following range arrays: $R_{min} = \{N_i, 8, 2\}$ and $R_{max} = \{N_i, 16, 2\}$, which indicate that besides the single layer perceptron (SLP), all MLPs have only a single hidden layer, i.e., $L_{max} = 2$, with no more than 16 hidden neurons. Besides the SLP, the hash function enumerates all MLP configurations in the AS, as shown in Table 9.2. Finally, for both evolution methods, the exhaustive BP and MD PSO, $N_R = 10$ independent runs are performed. Note that for exhaustive BP, this corresponds to 10 runs for each configuration in the AS.

9.3.2.1 Feature Scalability and Comparative Evaluations

In order to demonstrate the feature scalability property of the CNBC, we evolved three CNBCs individually using 7 FVs (FVs 1, 3, 5, 11, 12, 13, and 14 in Table 9.1 with a total dimension of 188), 10 FVs (those 7 plus FVs 6, 7, and 8 with total

Table 9.2 The architecture space used for MLPs

Dim.	Conf.	Dim.	Conf.	Dim.	Conf.
0	$N_i \times 2$	6	$N_i \times 6 \times 2$	12	$N_i \times 12 \times 2$
1	$N_i \times 1 \times 2$	7	$N_i \times 7 \times 2$	13	$N_i \times 13 \times 2$
2	$N_i \times 2 \times 2$	8	$N_i \times 8 \times 2$	14	$N_i \times 14 \times 2$
3	$N_i \times 3 \times 2$	9	$N_i \times 9 \times 2$	15	$N_i \times 15 \times 2$
4	$N_i \times 4 \times 2$	10	$N_i \times 10 \times 2$	16	$N_i \times 16 \times 2$
5	$N_i \times 5 \times 2$	11	$N_i \times 11 \times 2$		

dimension of 636) and 14 FVs (all FVs with a total dimension of 2,335) features. Therefore, the first CNBC has $7 + 1 = 8$, the second one has $10 + 1 = 11$, and finally the third one has $14 + 1 = 15$ binary classifiers in each NBC. As for the competing methods, we selected the two most powerful classifier networks, namely SVMs and RF despite the fact that they are static classifiers that cannot adapt dynamically to the changes in features, classes, and any update in training dataset. Therefore, as the features are populated from 7 to 14, new SVM networks and RFs are trained whereas CNBC dynamically adapts itself as mentioned earlier. For SVM networks, we employ the libSVM library [2] using the *one-against-one* topology [1]. Since in the CNBC framework, the optimal binary classifier configuration within each NBC is determined by the underlying evolutionary method, in order to provide a fair comparison, all possible classifier kernels and parameters are also determined for the SVMs and RFs. For SVMs, all standard kernel types such as *linear, polynomial*, radial basis function (*RBF*), and *sigmoid,* are individually used while searching for the best internal SVM parameters, e.g., the respectable penalty parameter, $C = 2^n$; for $n = 0,...,3$ and parameter $\gamma = 2^{-n}$; for $n = 0,...,3$, whenever applicable to the kernel type. For the RF, the best number of trees within the forest is also searched from 10 to 50 in steps of 10.

Table 9.3 presents the average classification performances achieved by 10-fold random train/test set partitions in *Corel_10* database by the competing techniques against the CNBC that is evolved by both evolutionary methods (exhaustive BP and MD PSO) over the sample AS given in Table 9.2. It is evident that all SVM networks with different kernels and RF suffered from the increased feature space dimension, as a natural consequence of "Curse of Dimensionality". Particularly, SVMs with *RBF* and *sigmoid* kernels cannot be trained properly and thus exhibits severe classification degradation in the test set. Henceforth with 14 features, the best classification accuracy has been achieved by the CNBC evolved with the exhaustive BP. We can thus foretell that the performance gap may even get widened if more features are involved. Between two evolutionary techniques, the results indicate that MD PSO achieves the lowest MSE and classification error levels (and hence the best results) within the training set whereas the opposite is true for the exhaustive BP within the test set. CNBC in general demonstrates a solid improvement against the major feature dimension increase [i.e., from 7 (188-D) to 14 subfeatures (2335-D)] since the classification performance does not show any deterioration, on the contrary, with both evolutionary techniques a better

Table 9.3 Average classification performance of each evolution method per feature set by 10-fold random train/test set partitions in *Corel_10* database

Feature set	Classifier	Train MSE	Train CE	Test MSE	Test CE
7 sub-features	SVM (Linear)	0	0	3.56	**13.76**
	SVM (Polynom.)	0.28	0	3.53	13.8
	SVM (RBF)	0	0	4.35	16.87
	SVM (SIGMOID)	3.51	12.78	4.96	18.07
	Random Forest	0.04	0.2	4.96	17.58
	CNBC (MD PSO)	0.52	2.42	1.33	16.49
	CNBC (BP)	0.47	5.56	**1.21**	16.44
10 sub-features	SVM (Linear)	0	0	3.48	**13.84**
	SVM (Polynom.)	0.16	0.58	3.87	14.65
	SVM (RBF)	0	0	6.92	29.87
	SVM (SIGMOID)	16.15	83.22	16.99	86.86
	Random Forest	0.06	0.3	4.82	16.33
	CNBC (MD PSO)	0.88	5.56	5.43	15.91
	CNBC (BP)	0.36	4.22	**1.19**	14.43
14 sub-features	SVM (Linear)	0	0	3.59	14.8
	SVM (Polynom.)	0	0	3.59	14.56
	SVM (RBF)	0	0	10.55	40.45
	SVM (SIGMOID)	19.25	88.7	20.43	91.07
	Random Forest	0.09	0.47	4.76	17.09
	CNBC (MD PSO)	0.44	5.52	6.33	14.41
	CNBC (BP)	0.37	4.56	**1.21**	**13.43**

The best classification performances in the test set are highlighted

performance is achieved, thus demonstrating an enhanced generalization ability. This is an expected outcome since CNBC can benefit from the additional discrimination capability of each new feature.

Table 9.4 presents the confusion matrix of the best classification result over the test set, i.e., achieved by the exhaustive BP method using 14 features. It is worth noting that the major source of error results from the confusion between the 2nd (*Beach*) and 9th (*Mountain*) classes where low-level features cannot really discriminate the content due to excessive color and texture similarities among those classes. This is also true for the 6th class (*Elephant*) from which the background of some images share a high similarity with both classes (2nd and 9th).

9.3.2.2 Incremental CNBC Evolutions

The CNBC evolutions so far performed are much alike to the (batch) training of traditional classifiers where the training data (the features) and (number of) classes are all fixed and the entire GTD is used during the training (evolution). As detailed earlier, incremental evolutions can be performed whenever new features/classes

Table 9.4 Confusion matrix of the evolution method, which produced the best (lowest) test classification error in Table 9.3

Actual		1	2	3	4	5	6	7	8	9	10
Truth	1	**42**	2	1	1	0	5	0	0	1	3
	2	2	**37**	4	1	0	0	0	1	9	1
	3	2	3	**46**	1	0	1	0	0	1	1
	4	2	0	0	**53**	0	0	0	0	0	0
	5	0	0	0	0	**55**	0	0	0	0	0
	6	2	4	2	0	0	**37**	0	1	8	1
	7	1	0	0	0	0	0	**53**	1	0	0
	8	0	0	0	0	0	0	0	**55**	0	0
	9	0	8	1	0	0	0	0	0	**46**	0
	10	1	1	0	1	1	2	1	0	2	**46**

can be introduced and the CNBC can dynamically create new binary classifiers and/or NBCs as the need arises. In order to evaluate the incremental evolution performance, a fixed set of 10 features (FVs with indices 1, 2, 3, 4, 5, 6, 11, 12, 13, and 14 in Table 9.1 with a total dimension of 415) are used, and the GTD is divided into three distinct partitions, each of which contains 5 (classes 1–5), 3 (classes 6–8), and 2 (classes 9 and 10) classes, respectively. Therefore, three stages of incremental evolutions have been performed where at each stage except the first one, the CNBC is further evolved using the *new* and the *log* GTD. During the second phase, three out of five existing NBCs failed the verification test (performed below 95 % classification accuracy with the new GTD of classes 6–8) and thus they were *incrementally* evolved. Finally at the third phase, 4 out of 8 existing NBCs did not undergo for incremental evolution since they passed the verification test over the training dataset of those new classes (9 and 10) while the others failed and had to be incrementally evolved.

Table 9.5 presents the confusion matrices achieved at each incremental evolution stage over the test sets of the GTD partitions involved. It is worth noting that the major source of error results from the confusion between the 2nd (*Beach*), 3rd (*Architecture*), and 9th (*Mountain*) classes where low-level features cannot really discriminate the classes due to excessive color and texture similarities among them. This is the reason class 2 and class 3 have undergone incremental evolution at each stage; however, a significant lack of discrimination still prevailed. On the other hand, at the end of stage 3, high classification accuracies are achieved for classes 1, 4, and particularly 5 (and also for 7, 8, and 10).

Table 9.6 presents the final classification performance of each evolution method per feature set. The results indicate only slight losses on both training and test classification accuracies, which can be expected since the incremental evolution was purposefully skipped for some NBCs whenever they surpass 95 % classification accuracy over the training dataset of the new classes. This means, for instance, some NBCs (e.g., the one corresponds to class 4, the *Bus*) evolved with only over a fraction of the entire training dataset.

Table 9.5 Test dataset confusion matrices for evolution stages 1 (top), 2 (middle) and 3 (bottom)

Actual		1	2	3	4	5
Truth	1	**47**	3	2	1	2
	2	1	**49**	5	0	0
	3	1	14	**36**	4	0
	4	2	3	1	**49**	0
	5	0	0	0	0	**55**

Actual		1	2	3	4	5	6	7	8
Truth	1	**44**	2	3	0	1	1	3	1
	2	1	**43**	8	0	1	1	0	1
	3	3	4	**41**	2	0	2	3	0
	4	1	3	2	**48**	0	0	1	0
	5	0	0	0	0	**55**	0	0	0
	6	2	11	3	0	2	**32**	5	0
	7	1	0	1	1	0	0	**52**	0
	8	0	0	0	0	1	0	0	**54**

Actual		1	2	3	4	5	6	7	8	9	10
Truth	1	**38**	0	0	1	2	3	0	1	3	7
	2	3	**23**	2	0	0	1	0	1	24	1
	3	8	0	**22**	1	0	2	1	0	19	2
	4	1	2	0	**39**	0	1	0	0	10	2
	5	0	0	0	0	**55**	0	0	0	0	0
	6	4	2	0	0	3	**38**	0	0	8	0
	7	0	1	0	0	1	1	**50**	0	2	0
	8	0	0	0	0	1	0	0	**54**	0	0
	9	1	5	2	1	0	3	0	0	**43**	0
	10	0	1	0	0	3	0	0	0	0	**51**

Table 9.6 Final classification performance of the 3-stage incremental evolution for each evolution method and feature set for *Corel_10* database

Feature set	Evol. method	Train MSE	Train CE	Test MSE	Test CE
7 sub-features	MD PSO	1.36	6.89	2.61	28.63
	Exhaustive BP	**0.82**	**4.22**	**1.85**	**21.63**
14 sub-feature	MD PSO	1.23	**6.66**	2.39	26.36
	Exhaustive BP	**0.91**	7.55	**1.83**	**20.81**

9.3.2.3 CNBC Class Scalability

Finally, the CNBC evolution for *Corel_Caltech_30* database allows testing and evaluation of its classification performance when the number of classes along with the database size is significantly increased. For both evolution techniques, we used the same parameters as presented earlier except that the number of epochs

Table 9.7 Classification performance of each evolution method per feature set for *Corel_Caltech_30* database

Feature set	Evol. method	Train MSE	Train CE	Test MSE	Test CE
7 subfeatures	MD PSO	0.54	8.1	2.3	**33.40**
	Exhaustive BP	**0.24**	**2.95**	**2.16**	34.67
14 subfeature	MD PSO	0.33	5.47	**2.52**	36.33
	Exhaustive BP	**0.074**	**1.31**	2.69	**33.86**

(iterations) for BP and MD PSO were increased to 200 and 500. Table 9.7 presents the classification performances of each evolution method per feature set. As compared with the results from *Corel_10* database in Table 9.3, it is evident that both evolution methods achieved a similar classification performance over the training set (i.e., similar train classification errors) while certain degradation occurs in the classification performance in the test set (i.e., 10–15 % increase in the test classification errors). This is an expected outcome since the lack of discrimination within those low-level features can eventually yield a poorer generalization especially when the number of classes is tripled. This eventually brought the fact that higher discrimination power with the addition of new powerful features is needed so as to achieve a similar test classification performance in large image databases.

9.3.3   CBIR Results

The traditional retrieval process in MUVIS is based on the query by example (QBE) paradigm. The (sub-) features of the query item are used for (dis-) similarity measurement among all the features of the visual items in the database. Ranking the database items according to their similarity distances yields the retrieval result. The traditional (dis-) similarity measurement in MUVIS is accomplished by applying a distance metric such as L2 (*Euclidean*) between the FVs of the query and each database item. When CNBC is used for the purpose of retrieval, the same (L2) distance metric is now applied to the class vectors at the output layer of the CNBC ($10 \times 2 = 20$-D for *Corel_10* and $30 \times 2 = 60$-D for *Corel_Caltech_30* databases). In order to evaluate the retrieval performances with and without CNBC, we use average precision (AP) and an unbiased and a limited formulation of the *Normalized Modified Retrieval Rank* (NMRR(q)), which is defined in MPEG-7 as the retrieval performance criteria per query (q). It combines both of the traditional hit-miss counters; *Precision—Recall,* and further takes the ranking information into account as given in the following expression:

$$\text{AVR}(q) = \frac{\sum\limits_{k=1}^{N(q)} R(k)}{N(q)} \quad \text{and} \quad W = 2N(q)$$

$$\text{NMRR}(q) = \frac{2AVR(q) - N(q) - 1}{2W - N(q) + 1} \leq 1 \tag{4}$$

$$\text{ANMRR} = \frac{\sum\limits_{q=1}^{Q} \text{NMRR}(q)}{Q} \leq 1$$

where $N(q)$ is the minimum number of relevant (via *ground-truth*) images in a set of Q retrieval experiments, $R(k)$ is the rank of the kth relevant retrieval within a window of W retrievals, which are taken into consideration for each query, q. If there are less than $N(q)$ relevant retrievals among W then a rank of $W + 1$ is assigned for the remaining (missing) ones. $AVR(q)$ is the average rank obtained from the query, q. Since each query item is selected within the database, the first retrieval will always be the item queried and this obviously yields a biased $\text{NMRR}(q)$ calculation and it is, therefore, excluded from ranking. Hence the first relevant retrieval ($R(1)$) is ranked by counting the number of irrelevant images beforehand and note that if all $N(q)$ retrievals are relevant, then $\text{NMRR}(q) = 0$, achieving the best retrieval performance. On the other hand, if none of the relevant items can be retrieved among W then $\text{NMRR}(q) = 1$, indicating the worst case. Therefore, the lower $\text{NMRR}(q)$ is the better (more relevant) the retrieval is, for the query, q. Both performance criteria are computed by querying *all* images in the database (i.e., batch query) and within a retrieval window equal to the number of ground truth images, $N(q)$ for each query q. This henceforth makes the *average precision AP* identical to the average recall.

Over each database, six batch queries are performed to compute the average retrieval performances, four with and two without using CNBC. Whenever used, CNBC is evolved with the MD PSO and the exhaustive BP, the former with 7 and the latter with 14 subfeatures, respectively. As listed in Table 9.8, it is evident that the CNBC can significantly enhance the retrieval performance regardless of the evolution method, the feature set, and the database size. The results (without CNBC) in the table also confirm the enhanced discrimination obtained from the

Table 9.8 Retrieval performances (%) of the four batch queries in each MUVIS databases

Feature set	Retrieval method	Corel_10		Corel_Caltech_30	
		ANMRR	AP	ANMRR	AP
7 subfeatures	CNBC (MD PSO)	31.09	65.01	**43.04**	**54.47**
	CNBC (BP)	**23.86**	**74.26**	46.44	52.21
	Traditional	55.81	42.15	60.21	37.80
14 subfeature	CNBC (MD PSO)	29.84	67.93	33.29	61.12
	CNBC (BP)	**22.21**	**76.20**	**32.00**	**65.37**
	Traditional	47.19	50.38	62.94	34.92

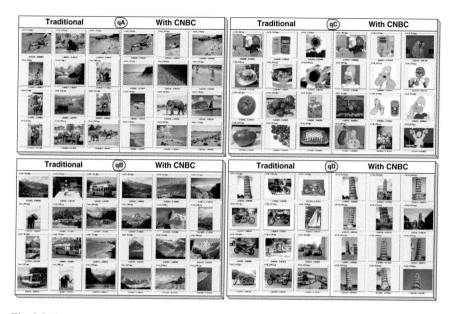

Fig. 9.5 8 sample queries in *Corel_10* (qA and qB), and *Corel_Caltech_30* (qC and qD) databases with and without CNBC. The *top-left* image is the query image

larger feature set, which led to better classification performance and in turn, leads to a better retrieval performance.

For visual evaluation, Fig. 9.5 presents four typical retrieval results with and without CNBC. All query images are selected among the test set and the query is processed within the entire database. Table 9.9 presents the retrieval performances obtained from each batch query operation of the CNBCs that are incrementally evolved where the corresponding confusion matrices at each stage are presented in Table 9.5. It is evident that the final CNBC (at the end of stage 3) can significantly enhance the retrieval performance compared to traditional query method. It is also interesting to note that this is also true for the immature CNBC at stage 2, which demonstrates a superior performance even though it is not yet fully evolved with all the classes in the database.

Figure 9.6 presents two sample retrievals of two queries from classes 2 (Beach) and 6 (Elephants) where each query operation is performed at the end of each incremental evolution stage. In the first query (qA), at stage 1 CNBC failed to retrieve relevant images since it is not yet evolved with the GTD of this class

Table 9.9 Retrieval performances per incremental evolution stage and traditional (without CNBC) method

	Stage—1	Stage—2	Stage—3	Traditional
ANMRR (%)	54.5	33.36	**24.13**	46.32
AP (%)	44.07	65.42	**73.91**	51.23

Fig. 9.6 Two sample retrievals of sample queries qA and qB, performed at each stage from classes 2 and 6. The *top-left* is the query image

(class 6). At stage 2, a precision level of, $P = 59$ %, is achieved where there are still several irrelevant retrievals as shown in the figure. Only after the last incremental evolution at stage 3, a high precision level of 85 % is achieved without any irrelevant image in the first 12 retrievals. In the second query (qB), the retrieval performance improves smoothly with each incremental evolutionary stage. It is evident that despite the NBC corresponding to class 2 has been initially evolved in stage 1, it can only provide a limited retrieval performance ($P = 42$ %) for such queries like the one shown in the figure; while it takes 2 more incremental evolution sessions to gather the maturity level for a reasonable retrieval ($P = 80$ %).

9.4 Summary and Conclusions

In this chapter, a collective network of binary classifiers (CNBC) framework is introduced to address the problem of adaptive content-based image classification within large and dynamic image databases. CNBC adopts a *Divide and Conquer* approach, which reduces both feature and class vector dimensions for individual classifiers significantly, thus enabling the use of compact classifiers. The optimum classifier configuration for each classification problem at hand and for each feature is searched individually and at a given time, this allows to create a dedicated classifier for discriminating a certain class type from the others with the use of a single feature. Each (incremental) evolution session "learns" from the current best classifier and can improve it further using the new ground-truth data (GTD), possibly finding another configuration in the architecture space (AS) as the new "optimal" classifier. Such an evolutionary update mechanism ensures that the AS containing the best configurations, is always kept intact and that only the best configuration at any given time is used for classification and retrieval. Experimental results demonstrated that this classifier framework provides an efficient solution for the problems of *dynamic adaptability* by allowing both feature space dimensions and the number of classes in a database to vary. Whenever CNBC is evolved in a "batch" mode, it can compete and even surpass other classifiers such as support vector machine (SVM) networks with different kernels (such as RBFs or polynomials), and RF especially when the feature space dimension is quite large. This is an expected outcome since the CNBC framework can take advantage of any feature as long as it has some discrimination power for one or few classes.

9.5 Programming Remarks and Software Packages

As presented in Sect. 7.5, recall that the **ClassifierTestApp** is a single-thread application, and the source file *ClassifierTestApp.cpp* has two entry point **main()** functions separated with a compiler switch **_CNBC_TEST**. If defined, the test-bed application for CNBC will be active within the first **main()** function. Besides the CNBC, other network types and topologies such as SVM (one-against-one or one against all) can also be tested by simply assigning the global function pointer variable **TrainFn** to one of the appropriate functions, i.e.,

inline void TrainCNBC(int);
inline void TrainSVM(int);
inline void TrainRF(int);
typedef void (*TrainFunction) (int);
...
TrainFunction TrainFn = TrainCNBC;//The classifier body..
ClassifierType CNBCclasstype = _MLP_BP;//Classifier type for CNBC..

When **TrainFn = TrainCNBC** then the BC type can be set by assigning **CNBCclasstype = _MLP_BP**. Otherwise, this option is irrelevant for other types of classifiers. The other BC types are enumerated in **GenClassifier.h** file, as follows:

enum ClassifierType
{

  _MLP_BP = 0,
  _MLP_PSO,

```
_RBF_BP,
_RBF_PSO,
_SVM, //Support Vector Machines..
_RF,
_BC, //Bayesian Class..
_HMM, //Hidden Markov Models..
_RANDOM = 1000
};
```

There are other global variables that are used to assign the training and classifier parameters, as follows:

```
CNBCtrain_params cnbc_tp;
CNBCclass_params cnbc_cp;
one_class_data *train_new = NULL, *test_new = NULL;
float **train_input, **test_input;
float **train_output, **test_output;
int in_size, out_size, train_size, test_size = 0;
int *one_index;
int *clip_index = NULL, *kfs = NULL, *table = NULL;
int train_clip_no, test_clip_no;
#define TRAIN_RATE .65
#define TRAIN_RUNS 500//No of training epochs..
#define LEARNING_PARAM 0.01
#define REPEAT_NO 2//Max. no. of REPEATITIONS per MLP conf. training..
PSOparam _psoDef = {80, 201, 0.001, -10, 10, -.2, .2, FAST};//default parameters MD PSO..
SPSAparam _saDef = {1, 1, .602, .2, .101};//default SPSA parameters, a, A, alpha, c, gamma, for SAD..
#define MAX_NOL 3//Max. no. of layers in ENN
#define RF_MAX_NOL 4//Max. no. of parameters for RF
#define SVM_MAX_NOL 5//Max. no. of parameters for SVM
int minNoN[MAX_NOL] = {-1, 8, 2};//Min. No. of Layers for ANNs..
int maxNoN[MAX_NOL] = {-1, 16, 2};//Max. No. of Layers ANNs..
int svmMaxNoN[SVM_MAX_NOL] = {-1, 4, 3, 1, 2};//Max. index for SVM Kernel..
int svmMinNoN[SVM_MAX_NOL] = {-1, 1, 0, 0, 2};//Min. index for SVM Kernel..
int svmMaxNoN[SVM_MAX_NOL] = {-1, 1, 0, 1, 2};//Max. index for SVM Kernel..
int rfMinNoN[RF_MAX_NOL] = {-1, 1, 1, 2};//Min. index arr for RF conf..
int rfMaxNoN[RF_MAX_NOL] = {-1, 2, 3, 2};//Min. index arr for RF conf..
//int newClassNo[] = {5, 3, 2};//SET the no. of classes per training stage here..
int newClassNo[] = {1000};//SET the no. of classes per training stage here..
bool bUniClass = 0, bEvaluateOnly = 0, bEvaluateAll = 0;//true if the database is uni-class and true if only evaluation performed..
```

Note that several of the training and classifier parameters shown above are described earlier, such as **PSOparam, SPSAparam,** and AS parameters such as **int min_noL, int max_noL, int *min_noN, int *max_noN.** Moreover, the training constants such as **TRAIN_RATE** (database partition for training), **TRAIN_RUNS** (number of epochs for BP), **LEARNING_PARAM** (learning parameter for BP), and **REPEAT_NO** (number of repetitions for each BC training/evolution).These are all related to each individual BC training and BC configuration. There are three Boolean parameters. The first one should be set according to the dataset type: *true* for uni-class (or uni-label) or *false* for multi-class (or multi-label). The second Boolean, **bEvaluateOnly**, is set to *true* only to test (evaluate the performance) of an existing CNBC without (incremental) training/evolution. If there is no CNBC yet evolved, then this parameter should be set to *false*. An important data structure where the information such as positive and negative item lists, class index, and some operators are stored in **one_class_data**. When the GTD of a dataset is loaded, the individual class information for train and test datasets is stored using this data structure. Finally, there are two CNBC-related data structures and their global variables: **CNBCtrain_params,** and **CNBC-class_params cnbc**. As the other variables, they are also all declared in the header file, **GenClassifier.h** file. Table 9.10 presents these three important data structures.

The entry function, **main()** is quite straightforward:

```
int main(int argc, char* argv[])
{

    memset(&cnbc_cp, 0, sizeof(CNBCclass_params));//init..
    //RunDTfiles();
    RunFeXfiles(-1);
    return 0;
}
```

In this application, a CNBC can be evolved either data (*.dt) by calling the function: **RunDTfiles()** or a MUVIS feature (or the so-called FeX) file by calling the function: **RunFeXfiles(-1)**. In this section we shall describe the latter process; however, the former is also quite similar and simpler than the latter since data files contain both features and GTD in a single file. In a MUVIS image database, there is a folder called "\Images\" where all the raw data, the images, reside. There is also an image database descriptor file, "*.idbs" where database properties (such as date and time of creation, version, number of images, features, and their descriptions) are stored. For instance consider the sample image database file, "dbsMC.idbs" (under the folder "\dbs60mc\"), as follows:

```
v 1.8
20:41:02, Thursday, January 12, 2012
IS = NONE
noIm = 60 visF = 1 nSEG = 0
CNBC 1 19
```

Table 9.10 The CBNC specific data structures

```
class CLASSIFIER_API one_class_data
{
public:
        one_class_data();
        ~one_class_data();
        one_class_data(int, int, int ci = -1);
        one_class_data(one_class_data&);
        void operator= (one_class_data&);
        void operator>> (one_class_data&);
        void operator+= (one_class_data&);
        void NegAlloc();
        void Clean();
        void Verify();

        int      class_ind; // class index id..
        int*     pos_item_list; // list of items that belong to this class..
        int      pos_item_no; // no. of pos. items..
        int*     neg_item_list; // list of items that belong to this class..
        int      neg_item_no; // no. of neg. items..
};

struct CNBCtrain_params
{ // Common training params for CNBC..
        float    *inputFV; // ptr. of all FVs..
        int      sizeFV, item_no; // size of one FV and total no. of items..
        int      *ptrSF; // ptr of each SF in one FV..
        int      noSF; // no. of sub-features = no. of BCs in each NBC..
        bool     bUniClass; // true if this is uniclass dataset, false if multi-class..
        ActivationMode actMode; // Activation Mode for MLPs = Range for output layer..
        train_params     tp; // training parameters of each BC in NBCs except train/test data..
};

struct CNBCclass_params
{ // Spesific classifier (AS) params related with each CNBC training session..
        class_params     cp; // classifier AS parameters for each BC in NBCs..
        char             dir[255], title[255]; // CNBC file dir. and title..
        CNBCfileType     eCNBCfile; // true if CNBC file is generated..

        int      next_class_id; // Class ID assigned for the next NBC created..
        one_class_data *train_log, *train_new; // old (from log file) and new training classes for training..
        int      no_train_class, no_trainLog_class; // no. of classes to be used for the training session.
        one_class_data *test_new; // test classes to be used for testing the training session..
        int      no_test_class; // no. of test classes to be used for testing the training session.
};
```

```
7.000000  6.000000  2.000000  2.000000  6.000000   10.000000  0.020000
0.010000  0.010000  1.000000  5.000000  32.000000  4.000000   11.000000
6.000000  4.000000  3.000000  4.000000  2.000000
```

This database has an internal MUVIS version 1.8, with the creation date below. Number of images (**noIM** = 60) is 60, number of visual features (**visF** = 1) is 1 and finally no segmentation method is applied (**nSEG** = 0). The fifth line (**CNBC 1 19**) indicates that the feature extracted for this database is CNBC, which encapsulates either 7, 10, or 14 distinct low level features. For this database, 7 features have been extracted (the first number of the 19 parameters given below). Note that under the same folder, there is a file called "dbsMC_FeX.CNBC", which contains the FVs of these 7 features within a single file. Moreover, the GTD for this database reside in the so-called Query-Batch-File (*.qbf), with the name "dbsMC CNBC.qbf". This is a simple text file with the following entries:

```
DATABASE dbsMC
QUERY_TYPE visual
CLASS_NO 5
DBGOUT 1
# Put any nonzero number to enable DbgOut signals..
QUERY 0 - 59
# this basically means to query ALL items in the database..
% this is a 5-class database with some Corel images.
CLASS 0 0 - 9
CLASS 1 10 - 19
CLASS 2 20 - 29
class 3 30 - 39
CLASS 4 40 - 59
% From now on put the multi-class entries..
CLASS 0 20 - 29, 42, 44
class 4 0, 1, 3, 8, 9, 16
```

The self-explanatory text tokens (e.g. "**DATABASE**", "**CLASS**", etc.) indicate that the database (**dbsMC**) has 5 classes, where the image indices per class are given below. For instance the first class (**CLASS 0**) contains images **0, 1, 2, ..., 9** and **20, 21, 22, ..., 29, 42** and **44** as well.

Table 9.11 presents the function **RunFeXfiles()**. The database filename and folder are indicated by variables **dir[]** and **tit_base[]**. Then all training parameters are stored within the **cnbc_tp** whereas all classifier parameters are stored in **cnbc_cp** variables, respectively. Since this is a multiclass database, the function call **CreateCNBCDataFeXgen(fname)** will load the qbf file and according to the **TRAIN_RATE**, it will fill the class entries in train and test datasets (into the **one_class_data** varibles of **train_new** and **test_new**). Once the GTD is loaded, then the function call **LoadCNBCFex(dir, tit_base)** will load the FVs in the FeX file of the database which will be pointed by the pointer, **cnbc_tp.inputFV** with a total size of **cnbc_tp.sizeFV**. Finally, the function call **IncrementalEvolution(run)**, will simulate a multi-stage CNBC evolution session where in each stage certain number of new classes will be introduced and the CNBC will be incrementally evolved. The number of new classes for each stage can be defined in the

Table 9.11 The function: **RunFeXfiles(int run)**

```
void RunFeXfiles(int run)
{
        char fname[512], tit_base[] = "dbsMC";
        char dir[] = "..\\..\\CH9\\dbs60mc\\"; // dir. of the dbs..
        sprintf(fname, "%s%s CNBC.qbf", dir, tit_base);

        ...
        // Put all Training params..
        cnbc_tp.actMode = _tanh; // act. mode for this dataset, change it with the dataset!!!

        ...
        // Put Classifier params..
        if(TrainFn == TrainCNBC)
        {
                cnbc_cp.cp.minNoN = minNoN; cnbc_cp.cp.maxNoN = maxNoN; cnbc_cp.cp.max_noL
= MAX_NOL;
                if(CNBCclasstype == _MLP_BP || CNBCclasstype == _MLP_PSO) cnbc_tp.tp.lp1 =
LEARNING_PARAM; // learning param..
        } // if..

        ...
        cnbc_cp.cp.ctype = CNBCclasstype; // Classifier/training type for BCs - valid only for a new training
session..
        cnbc_cp.eCNBCfile = _distributed; // Create a single / distributed CNBC file..
        strcpy(cnbc_cp.dir, dir); strcpy(cnbc_cp.title, tit_base); // dir and title for the best classifier..
        cnbc_cp.next_class_id = NEXT_CLASS_ID; // start from NEXT_CLASS_ID..

        if(cnbc_tp.bUniClass) CreateCNBCDataFeX(fname); // Uni-class selection for CNBC training..
        else CreateCNBCDataFeXgen(fname); // Multi-class data-FV selection for CNBC training..

        sprintf(fname, "%s%s_Fex.CNBC", dir, tit_base);
        if(!LoadCNBCFex(dir, tit_base)) return; // Load the Fex file and normalize them within [-1,1]
        ...
        IncrementalEvolution(run);
}
```

global array of **newClassNo[]**. For example, the following definition for a 10-class database:

int newClassNo[] = {5, 3, 2};//SET the no. of classes per training stage here..

will introduce GTD of the images, first from the 5 classes, then (in the 2nd stage) from the 3 classes and finally, from the 2 classes. This is to test if the CNBC can adapt to new class entries. If a batch training is desired, then a single entry should be given with the total number of classes in the database (or higher), i.e., **int newClassNo[] = {1000};**. Then the function call **IncrementalEvolution(run)** will directly copy the entries **cnbc_cp.train_new** and **cnbc_cp.test_new** from the global **one_class_data** varibles of **train_new** and **test_new**. This will yield a single (batch) CNBC evolution session by the function call: **TrainFn(run)**; with all the GTD available and the classification performance will be computed over both train and test datasets.

In each of the training functions, the appropriate classifier object should be created and used for training and testing. Table 9.12 presents the training function

Table 9.12 The function: **TrainCNBC(int run)**

```
void TrainCNBC(int run)
{
        ...
        // Create CNBC obj and start training..
        pCNBC = new CNBCglobal(); // Create the CNBC obj..

        pCNBC->CreateDistributed(cnbc_cp.dir, cnbc_cp.title, CNBCclasstype); // Create the CNBC obj.
from the file..
        if(!bEvaluateOnly) pCNBC->TrainDistributed(cnbc_cp, cnbc_tp, run) ; // Train..
        pCNBC->Exit(); delete pCNBC;

        pCNBC = new CNBCglobal(); // Create the CNBC obj..
        pCNBC->CreateDistributed(cnbc_cp.dir, cnbc_cp.title, CNBCclasstype, run); // Create the CNBC
obj. from the file..
        pCNBC->EvaluatePerformance(cnbc_cp, cnbc_tp);

        _ERROR trainErr = pCNBC->GetTrainError();
        _ERROR testErr = pCNBC->GetTestError();
        fprintf(fpPerformance, "%f\t%f\t\t",trainErr.MSE, trainErr.CE);
        fprintf(fpPerformance, "%f\t%f\n",testErr.MSE, testErr.CE);
        ...
        pCNBC->Exit(); delete pCNBC;
}
```

for CNBC: **TrainCNBC(int run)**. The global **CNBCglobal** object, **pCNBC**, is created and loads the existing CNBC body from the NBC files—if any—by calling:

pCNBC- > CreateDistributed(cnbc_cp.dir, cnbc_cp.title, CNBCclasstype, run);

Then the CNBC evolution can be initiated by, **pCNBC- > TrainDistributed(cnbc_cp, cnbc_tp, run)**, where the GTD, FVs and all other training and test parameters are conducted by variables **cnbc_cp,** and **cnbc_tp**. Recall that if some NBCs or the entire CNBC is available, this means that there will be an incremental evolution over them; otherwise, this will be the very first evolution session. The CNBC topology consists of a 3-layer class hierarchy:

CNBCglobal, → **COneNBC** → **CGenClassifier (CMLP_BP, CMLP_PSO, CRBF_BP, CRBF_PSO, CRandomForest** and **CSVM**). The **CNBCglobal** is the main CNBC class, which creates the CNBC (from the file), propagate a FV, perform (incremental) evolution, compute the classification performance (for both train and test datasets), and some other high-level I\O routines.

Table 9.13 presents the class **CNBCglobal**. Note that it encapsulates a list of individual NBCs, by the queue object **m_qNBC**. The two main API functions have two forms: standalone (**Create()** and **Train()**)and distributed (**CreateDistributed** () and **TrainDistributed** ()). The only difference lies on the file format of the CNBC; the former keeps a single CNBC file whereas the latter has a distributed format, which keeps each NBC in a separate file. The former format is now abandoned because the latter allows parallel computation for CNBC evolution

Table 9.13 The class: **CNBCglobal**

```
class CLASSIFIER_API CNBCglobal
{
public:
        CNBCglobal();
        ~CNBCglobal();

        int      Create(char* dir, char* title, ClassifierType ct = _RANDOM); // Create the CNBC from a *.cnbc file..
        int      CreateDistributed(char* dir, char* title, ClassifierType ct = _RANDOM, int cnbcId = -1); // Create
the CNBC from set of *.nbc files..
        int      RemoveNBC(int class_ind); // Remove one NBC and re-order the NBCs..
        int      Train(CNBCclass_params& cp, CNBCtrain_params& tp) ; // Train/create CNBC..
        int      TrainDistributed(CNBCclass_params& cp, CNBCtrain_params& tp, int run = -1) ; // Train/create a
distributed CNBC..

        float*   Propagate(float *in_vec) ; // Fwd. Propagate one FV..
        int      Save(char* dir, char* title); // Save the CNBC to a file..
        int      Exit() ; // Clean and destroy CNBC..

        int      GetInSize();
        int      GetOutSize();
        ClassIndex* GetItemClass() {return m_pItemClass;}
        int      GetClassInd(int nbc) {return m_qNBC->GetItem(nbc)->GetClassInd(); }

        _ERROR GetTrainError() {return trainErr;}
        _ERROR GetTestError() {return testErr;}
        inline  void EvaluatePerformance(CNBCclass_params cp, CNBCtrain_params tp); // Evaluate the CNBC
performance wrt. train/test sets..

private:
        inline  _ERROR ComputeError(float *inputFV, int sizeFV, one_class_data *pClass, int no_class); // Com-
putes train and test err. statistics if dataset is uni-class..
        inline  _ERROR ComputeErrorGEN(float *inputFV, int sizeFV, one_class_data *pClass, int no_class); //
Computes train and test err. statistics if dataset is uni-class..
        inline  int  GetNextNBCno(CNBCclass_params cp, int cold); // retrieve the next available nbc no. that is not
trained from the log file.

private:
        float                      *m_input, *m_output; // input and output arrays..
        ClassIndex                 *m_pItemClass; // Array for class per item..
        CQueueList< COneNBC >      *m_qNBC; // The list of NBCs..

        char                       *m_log, nbc_log_fname[511]; // nbc log file name..
        _ERROR                     trainErr, testErr; // current train and test CE..
};
```

where multiple **ClassifierTestApp** instances can now evolve multiple NBCs in parallel (and independent from each other) with a proper semaphore implementation.

Table 9.14 presents the code for the function, **TrainDistributed()** with the in-builts semaphore structure. Whenever called, the function first creates the NBC list—if not created earlier, e.g., by the function call: **CreateDistributed()**. Then in a for-loop, the function asks for the first "idle" NBC slot by calling:

int c = GetNextNBCno(cp, cold);//Get the next available nbc no..

Table 9.14 The member function **TrainDistributed()** of the class **CNBCglobal**

```
int     CNBCglobal::TrainDistributed(CNBCclass_params& cp, CNBCtrain_params& tp, int run)
{
        if (cp.no_train_class <= 0) return 0; // train with what?
        if(tp.tp.out_size != 2) return 0; // Only bin. classifiers with 1-to-2 out. layer..

        if(!m_qNBC) m_qNBC = new CQueueList< COneNBC >; // Training from scratch.. Create the queue..
        ...
        for(int cold = -1 ; ; )
        {
                int c = GetNextNBCno(cp, cold); // Get the next available nbc no..

                if( c == cp.no_train_class) break; // no nbc left to be trained, so get out now..

                // Step 1: Check if NBC with class id = cp.train_new[i] exists,
                // if so use it, othwise create a new NBC and append it into qNBC..
                int nbc=0; // init. the index of the nbc that is to be trained..
                m_qNBC->Reset();
                for(COneNBC* pNBC =NULL; (pNBC = m_qNBC->Next()); ++nbc)
                        if(pNBC->GetClassInd() == cp.train_new[c].class_ind) break; // found!!..

                if(!pNBC)
                {
                        pNBC = new COneNBC(cp.train_new[c].class_ind, tp.actMode); // This is a new class,
create a new NBC..
                        m_qNBC->Insert(pNBC); // and insert it into the NBC queue..
                } else if( pNBC->Verify(cp, tp, c) )
                {
                        cold = c;
                        continue; // this NBC discriminates well between pos. and neg. items.. So NOGO for train..
                } // else if..

                pNBC->Train(cp, tp, c); // Start training of this NBC..
                sprintf(tit, "%s_%d", cp.title, c); // title of the NBC file trained..
                if(cp.eCNBCfile == _distributed) pNBC->Save(dir, tit); // Save the trained NBC to file..
                cold = c;
        } // for..

        free(m_log);
        Exit(); // Clean up all..
        CreateDistributed(cp.dir, cp.title, cp.cp.ctype, run); // Create the CNBC obj. from the distr. NBC files..
        EvaluatePerformance(cp, tp); // Evaluate the CNBC performance wrt. train/test sets..

        return cp.next_class_id; // last class id. vacant..
}
```

This function call will verify and retrieve the next available NBC slot (from a log file) for the current process. If there is no log file present (for the first run of **ClassifierTestApp**), it will create the one dedicated to the current database and initially set all the NBC slots as "idle". It will then retrieve the NBC index of the first idle slot back so that the evolution/training process can begin for it. In the same time it will change its "idle" status to "started" status so that other **ClassifierTestApp** instances will not attempt any further processing on this particular NBC. Whenever the evolution process is completed, the same function will then change its "started" status to "completed" status. The **ClassifierTestApp** instances, which cannot find any more NBCs with an "idle" status, will simply sleep until all the NBCs are evolved (all NBC slots are turned to "completed" status), and the log file is deleted by the last process active. This will eventually break the for-loop and for each process, the few internal NBCs evolved will be deleted (clean-up) and instead the entire CNBC will be loaded (by the **Create-Distributed()** call) and its performance is evaluated by the **EvaluatePerformance()** call.

On the other hand, if an NBC has already been created (and evolved) by a past process and loaded in the current process, then it may undergo to an incremental evolution only if it fails to verification test. This is clear in the following code:

if(!pNBC)
{
…
} else if(pNBC- > Verify(cp, tp, c))
{

    cold = c;
    continue;//this NBC discriminates well between pos. and neg. items.. So NOGO
    for train..
}//else if..

If it does not fail the verification test (the **pNBC- > Verify(cp, tp, c)** returns *true*), then this means that there is no need for further training or incremental evolution. The process simply continues with the next NBC.

Table 9.15 presents the class **COneNBC** in the second layer hierarchy. As its name implies, each object of this class represents a single NBC, and thus encapsulates one or many BCs within. Recall that each NBC is dedicated to an individual class in the database. Therefore, besides evolving/training its BCs, performing verification tests, propagating a FV to return a CV, it is the also task of this class to separate positive and negative items for the first layer BCs (via the function: **SelectFeatures()**) and for the fuser BC (via the function: **SelectFuserFeatures()**). Moreover, it performs the selection of the negative samples with respect to the predetermined positive-to-negative ratio (PNR) by the function: **NegativeFeatureSelection()**. Once the positive and negative item lists are selected for the current NBC, for a proper training especially when BP is used, it also shuffles them by calling the function, **RandomizeFeatures()**.

Table 9.16 presents the code for the function, **Train()**, which creates and train the BC(s) of the current NBC. Recall that the first layer of each NBC has a number of BCs equivalent to the number of the database features. If there are two or more BCs, then a fuser BC is also created (and evolved) to fuse the outputs of the first layer BCs. So if this is the first time the current NBC is created, then a BC pointer array (**m_pBC**) of **CGenClassifier** is created where each pointer is spared per BC. Next, within the for-loop, first each pointer in the array will be allocated to one of the six classifiers according to the classifier-training choice within **cp.cp.ctype**. Then the function call, **SelectFeatures(cp, tp, bc, nbc)**, selects the train and test datasets' FVs for the current BC according to the class of the NBC. Recall that all

Table 9.15 The class: **COneNBC**

```
class CLASSIFIER_API COneNBC
{
public:
        COneNBC();
        COneNBC(int cid, ActivationMode act);
        virtual ~COneNBC();

        int      Create(float *buf, ClassifierType ctype=_RANDOM); // Create the NBC from buf..
        int      Create(char* dir, char* title, ClassifierType ctype); // Create the NBC from file..
        int      Train(CNBCclass_params& cp, CNBCtrain_params& tp, int nbc) ; // Train the NBC..
        int      Verify(CNBCclass_params& cp, CNBCtrain_params& tp, int nbc) ; // Verify the NBC wrt. classf.
over train_new..

        float*   Propagate(float *in_vec) ; // Fwd. Propagate one FV..
        float*   Propagate(float *in_vec, float *tmp_out) ; // Fwd. Propagate one FV..
        int      Save(char* dir, char* title); // Save the NBC to a file..
        int      Exit() ; // Clean and destroy the NBC..
        ...
private:
        inline void    SelectFeatures(CNBCclass_params& cp, CNBCtrain_params& tp, int bc, int nbc);
        inline void    NegativeFeatureSelection(train_params &tp, int pos_item_no);
        inline void    SelectFuserFeatures(CNBCclass_params& cp, CNBCtrain_params& tp, int nbc);
        inline void    RandomizeFeatures(train_params& tp);
        inline void    CleanFeatures(train_params& tp);
        inline void    CleanFuserFeatures(train_params& tp);
        inline CGenClassifier* CreateClassifier(ClassifierType ctype);

private:
        CGenClassifier      **m_pBC; // Array of BCs in the 1st layer..
        CGenClassifier      *m_pFuser; // The Fuser BC..
        float               *m_input, *m_output; // input and output arrays..
        int                 m_noBC; // No. of BCs..
        int                 m_classInd; // Class ind. of the NBC..
        ActivationMode      m_act; // Activation mode..
};
```

FVs are stored in a single chunk of memory pointed by the pointer, **cnbc_tp.inputFV**. So this function basically assigns the train and test dataset input FV pointers within **cnbc_tp.inputFV** and output FV pointers to a static outcome according to positive and negative item lists. The following code piece is taken from this function to accomplish this. Note that the first for-loop basically assigns to input and output FVs of the positive items of the training dataset whereas the second one does the same for the negative items. A similar piece of code exists for the test set too.

```
for (int n = 0 ; n < train_data->pos_item_no ; ++n)
{
int ptr = tp.sizeFV * train_data- > pos_item_list[n] + tp.ptrSF[bc];//ptr of the
```
bc.th feature vector within inputFV..
```
tp.tp.train_input[n] = tp.inputFV + ptr;//bc.th feature vector of the item poin-
```
ted by train_data- > pos_item_list[n]
```
tp.tp.train_output[n] = CGenClassifier::outV[0];//positive outcome..
}//for..
```

Table 9.16 The member function **Train()** of the class **COneNBC**

```
int      COneNBC::Train(CNBCclass_params& cp, CNBCtrain_params& tp, int nbc)
{
         // Step 0: Create the BC array if not already created..
         if(!m_noBC)
         {
                   m_pBC = (CGenClassifier**) calloc(tp.noSF, sizeof(CGenClassifier*));
                   m_noBC = tp.noSF;
         } // if..

         // Create (if not already created) and train all BCs..
         for(int bc = 0; bc < m_noBC; ++bc)
         {
                   // Create each individual classifier obj. if not already created..
                   if(!m_pBC[bc]) m_pBC[bc] = CreateClassifier(cp.cp.ctype); // Create an appropriate BC obj..

                   SelectFeatures(cp, tp, bc, nbc); // Now for this BC, arrange train and test datasets for training..
                   RandomizeFeatures(tp.tp); // Randomize training datasets..

                   m_pBC[bc]->Init(tp.tp); // Init. training with the training params..
                   m_pBC[bc]->SetActivationMode(tp.actMode); // and set the act. mode..

                   TRACE("** Training of the %d. BC started.. \n", bc);
                   m_pBC[bc]->Train(cp.cp); // Start training..

                   CleanFeatures(tp.tp); // Clean up the local FV ptrs..
         } // for..

         // Now Create and train the Fuser -if needed..
         if(tp.noSF <= 1) return 2; // no need for a fuser..

         // Create the Fuser of this NBC if not already created..
         if(!m_pFuser) m_pFuser = CreateClassifier(cp.cp.ctype); // Create an appropriate BC obj..

         SelectFuserFeatures(cp, tp, nbc); // Now for the Fuser, arrange train and test datasets for training..
         RandomizeFeatures(tp.tp); // Randomize training datasets..
         m_pFuser->Init(tp.tp); // Init. training with the training params..

         m_pFuser->Train(cp.cp); // Start training..

         CleanFuserFeatures(tp.tp); // Clean up the local FV ptrs for Fuser..
         return 1;
}
```

```
for (n = 0; n < train_data->neg_item_no; ++n)
{
int ptr = tp.sizeFV * train_data->neg_item_list[n] + tp.ptrSF[bc];//ptr of the
bc.th feature vector within inputFV..
tp.tp.train_input[n + train_data->pos_item_no] = tp.inputFV+ptr;//bc.th fea-
ture vector of the item pointed by train_data- > neg_item_list[n]
tp.tp.train_output[n + train_data- > pos_item_no] = CGenClassifi-
er::outV[1];//negative outcome...
}//for..
```

Once the training dataset FVs for both positive and negative item list are assigned, the function call, **SelectFeatures(cp, tp, bc, nbc)**, then performs the selection of the negative samples with respect to the predetermined positive-to-negative ratio (PNR) by the function call, **NegativeFeatureSelection()**.

After the train and test datasets' FVs are selected and pointed by, **tp.tp.train_input, tp.tp.train_output,** and **tp.tp.test_input, tp.tp.test_output**, the function call, **RandomizeFeatures(tp.tp);**, shuffles the entries of the train dataset for proper training. They are then fed to the BC by calling: **m_pBC[bc]->Init(tp.tp)**. Finally, with the rest of the training parameters (**tp.tp**), the training/evolution of the BC is initiated by: **m_pBC[bc]->Train(cp.cp)** with the classifier parameters (**cp.cp**). When completed, the pointers for the current BC's training FVs are cleaned and the process is then repeated in the for-loop for the next BC, until all BCs in the first layer are trained/evolved. If there are two or more BCs present in the NBC, then the fuser BC is created, its FVs are selected from the CVs of the first layer BCs for both train and test datasets. Besides this difference, the training/evolution of the fuser BC is identical to any first layer BC's.

The I\O routines of the NBCs are quite straightforward. In any time, the NBC can be saved to or read from a binary file. As can be seen in the **COne-NBC::Save()** function, with a simple preceding header covering NBC information such as number of BCs, the NBC index and the activation function used, the architecture spaces of all BCs are then stored to the file in a sequential order, with the last one for the fuser BC. To load an NBC from the file via **COne-NBC::Create(char* dir, char* title, ClassifierType ctype)** function, the binary file header is simply read to assign the NBC parameters, and each BC is initialized with their AS buffers. Recall that the entire AS information is needed for any incremental evolution session and the best configuration in the AS will always be used for classification (forward propagation of FVs).

Finally, the third hierarchical class layer is the **CGenClassifier (CMLP_BP, CMLP_PSO, CRBF_BP, CRBF_PSO, CRandomForest,** and **CSVM)** for which the programming details were covered in Sect. 7.5. Note that such a morphological structure of the base classifiers that are all inherited from the **CGenClassifier** class enables us to integrate other classifiers in the future with the common API and data structures defined within the base class.

References

1. U. Kressel, *Pairwise Classification and Support Vector Machines*, in Advances in Kernel Methods—Support Vector Learning (1999)
2. C.C. Chang, C.J. Lin, LIBSVM : a library for support vector machines (2001). Available at http://www.csie.ntu.edu.tw/~cjlin/libsvm
3. T. Zou, W. Yang, D. Dai, H. Sun, Polarimetric SAR image classification using multifeatures combination and extremely randomized clustering forests. EURASIP J. Adv. Signal Process. 2010, Article ID 465612, 9 (2010)

4. R. Polikar, L. Udpa, S. Udpa, V. Honavar, Learn ++: an incremental learning algorithm for supervised neural networks. IEEE Trans. Syst. Man Cybern. (C) **31**(4), 497–508 (2001). (Special Issue on Knowledge Management)

5. T. Ojala, M. Pietikainen, D. Harwood, A comparative study of texture measures with classification based on feature distributions. Pattern Recogn. **29**, 51–59 (1996)

6. S. Guan, C. Bao, R. Sun, Hierarchical incremental class learning with reduced pattern training. Neural Process. Lett. **24**(2), 163–177 (2006)

7. H. Jia, Y. Murphey, D. Gutchess, T. Chang, Identifying knowledge domain and incremental new class learning in SVM. IEEE Int. Joint Conf. Neural Netw. **5**, 2742–2747 (2005)

8. S. Smale, On the average number of steps of the simplex method of linear programming. Math Program **27**(3), 241–262 (1983)

9. H. Chen, Z. Gao, G. Lu, S. Li, A novel support vector machine fuzzy network for image classification using MPEG-7 visual descriptors. International conference on multimedia and information technology, MMIT '08, pp. 365–368, 30–31 Dec 2008 doi:10.1109/MMIT.2008.199

10. E. Chang, K. Goh, G. Sychay, W. Gang, CBSA: content-based soft annotation for multimodal image retrieval using Bayes point machines. IEEE Trans. Circuits Syst. Video Technol. **13**(1), 26–38 (2003) doi:10.1109/TCSVT.2002.808079

11. G.-J. Qi, X.-S. Hua, Y. Rui, J. Tang, H.-J. Zhang, image classification with kernelized spatial-context. IEEE Trans. Multimedia **12**(4), 278–287 (2010) doi:10.1109/TMM.2010.2046270

12. MUVIS http://muvis.cs.tut.fi/

13. S. Kiranyaz, T. Ince, S. Uhlmann, M. Gabbouj, Collective network of binary classifier framework for polarimetric SAR image classification: an evolutionary approach. IEEE Trans. Syst. Man Cybern.—Part B (in Press)

14. Corel Collection/Photo CD Collection (www.corel.com)

15. L. Fei–Fei, R. Fergus, P. Perona, Learning generative visual models from few training examples: an incremental Bayesian approach tested on 101 object categories. IEEE CVPR Workshop Generative-Model Based Vision **12**, 178 (2004)

16. S.G. Mallat, *A Wavelet Tour of Signal Processing*, 2nd edn. (Academic Press, San Diego, 1999)

17. B. Manjunath, P. Wu, S. Newsam, H. Shin, A texture descriptor for browsing and similarity retrieval. J. Signal Process. Image Commun. **16**, 33–43 (2000)

18. T. Ojala, M. Pietikainen, D. Harwood, A comparative study of texture measures with classification based on feature distributions. Pattern Recogn. **29**, 51–59 (1996)

19. M. Partio, B. Cramariuc, M. Gabbouj, An Ordinal co-occurrence matrix framework for texture retrieval. EURASIP J. Image Video Process. **2007**, Article ID 17358 (2007)

Chapter 10
Evolutionary Feature Synthesis

Science is spectral analysis. Art is light synthesis.

Karl Kraus

Multimedia content features (also called descriptors) play a central role in many computer vision and image processing applications. Features are various types of information extracted from the content and represent some of its characteristics or signatures. However, especially these low-level features, which can be extracted automatically usually lack the discrimination power needed for accurate content representation especially in the case of a large and varied media content data reserve. Therefore, a major objective in this chapter is to synthesize better discriminative features using an evolutionary feature synthesis (EFS) framework, which aims to enhance the discrimination power by synthesizing media content descriptors. The chapter presents an EFS framework, which applies a set of linear and nonlinear operators in an optimal way over the given features in order to synthesize highly discriminative features in an *optimal* dimensionality. The *optimality* therein is sought by the multidimensional particle swarm optimization (MD PSO) along with the fractional global best formation (FGBF) presented in Chaps. 4 and 5, respectively. We shall further demonstrate that the features synthesized by the EFS framework that is applied over only a minority of the original feature vectors exhibit a major increase in the discrimination power between different classes and a significant content-based image retrieval (CBIR) performance improvement can thus be achieved.

10.1 Introduction

In content-based image retrieval (CBIR) systems, features used to describe the image content play a key role. Low-level features automatically extracted from images must be discriminative enough to enable the highest possible distinction among images belonging to different classes in order to maximize the retrieval accuracy. A lot of work has been done to develop more descriptive features, but it is a well-known fact that there is no such feature extractor that could automatically extract features always matching the human visual perception of the image similarity, since two images belonging to the same class may be visually different and

S. Kiranyaz et al., *Multidimensional Particle Swarm Optimization for Machine Learning and Pattern Recognition*, Adaptation, Learning, and Optimization 15, DOI: 10.1007/978-3-642-37846-1_10, © Springer-Verlag Berlin Heidelberg 2014

only a higher level understanding of the image content can reveal that they should be classified into the same class. Efficient CBIR systems require a decisive solution for this well-known "Semantic Gap" problem. Most current general purpose attempts to solve this problem gather knowledge of human perception of image similarity directly from the users. For example, user labeling of the images may be exploited to select the most appropriate ones among the vast number of available feature extraction techniques or to define a discriminative set of features best matching to the human visual perception.

The efforts addressing the aforementioned problem can be categorized into two feature transformation types: feature selection and feature synthesis. The former does not change the original features; instead selects a particular sub-set of them to be used in CBIR. So no matter how efficient the feature selection method may be the final outcome is nothing but a subset of the original features and may still lack the discrimination power needed for an efficient retrieval. The latter performs a linear and/or nonlinear transformation to synthesize new features. Both transformation types require searching for an optimal set of new features among a large number of possibilities in a search space probably containing many local optima and, therefore, evolutionary algorithms (EAs) [1] such as Genetic Algorithm (GA) [2] and Genetic Programming (GP) [3] are mainly used. Recall from the earlier chapters that the common point of all EAs is that they are stochastic population-based optimization methods that can avoid being trapped in a local optimum. Thus they can find the optimal solutions; however, this is never guaranteed. In addition to the limitations of EA-based algorithms discussed in earlier chapters, another critical drawback of the existing EA-based EFS methods is that they work only in a search space with an a priori fixed dimensionality. Therefore, the optimal dimensionality for the synthesized features remains unknown.

In order to address these problems, in this chapter, we shall present an EFS technique which is entirely based on MD PSO with FGBF. The main objective is to maximize the discrimination capability of low-level features so as to achieve the best possible retrieval performance in CBIR. Recall that MD PSO can search for the optimal dimensionality of the solution space and hence voids the need of fixing the dimensionality of the synthesized features in advance. MD PSO can also work along with the FGBF to avoid the premature convergence problem. With the proper encoding scheme that encapsulates several linear and nonlinear operators (applied to a set of optimally selected features), and their scaling factors (weights), MD PSO particles can, therefore, perform an evolutionary search to determine the optimal feature synthesizer to generate new features with the optimal dimensionality. The optimality *therein* can be defined via such a fitness measure that maximizes the overall retrieval (or classification) performance.

10.2 Feature Synthesis and Selection: An Overview

EFS is still in its infancy as there are only few successful methods proposed up to date. There are some applications of PSO to feature selection. In [4] binary PSO was successfully used to select features for face recognition and in many earlier studies PSO-based feature selection has been shown to produce good classification results when combined with different classification methods (e.g., logistic regression classifier [5], K-nearest neighbor method [6], SVMs [7, 8], and back-propagation networks [9]) and applied for different classification problems (e.g., UCI Machine Learning Repository classification tasks [5, 7, 8], gene expression data classification [6], and microcalcifications in mammograms [9]).

Most existing feature synthesis systems are based on genetic programming (GP) [3]. In [10] and [11], GP is used to synthesize features for face expression recognition. The genes are composite operators represented by binary trees whose internal nodes are primitive operators and leaf nodes are primitive features. The primitive features are generated by filtering the original images using a Gabor filter bank with 4 scales and 6 orientations (i.e., 24 images per original image) and the primitive operators are selected among 37 different options. The fitness of the composite operators is evaluated in terms of the classification accuracy (in the training set) of a Bayesian classifier learned simultaneously with the composite operator. Finally, the best composite operator found is used to synthesize a feature vector for each image in the database and the corresponding Bayesian classifier is then used to classify the image into one of 7 expressions types. The expression recognition rate was slightly improved compared to similar classification methods where no feature synthesis was applied.

In [7], co-evolutionary genetic programming (CGP) is used to synthesize features for object recognition in synthetic aperture radar (SAR) images. The approach is similar to the one in [10] and [11], but separate sub-populations are utilized to produce several composite operators. However, the primitive features used in this application are only 1-dimensional properties computed from the images and thus each composite operator only produces a single 1-dimensional composite feature. The final feature vector is formed by combining the composite features evolved by different sub-populations. Although both the number of primitive features (20) and the number of classes to be recognized (≤ 5) were low, the classification accuracy obtained using the synthesized features was only occasionally better than the classification accuracy obtained directly with the primitive features.

In [12], a similar CGP approach was applied for image classification and retrieval. The original 40-D feature vectors were reduced to 10-D feature vectors. The results were compared in terms of classification accuracy against 10-D feature vectors obtained using multiple discriminant analysis (MDA) and also against a support vector machine (SVM) classifier using the original 40-D feature vectors. The databases used for testing consisted of 1,200–6,600 images from 12 to 50 classes. In all cases, the classification results obtained using the features synthesized by CGP were superior compared to features produced by MDA. Compared to the SVM classifier the results were similar or better in the case where the database classes consisted of multiple clusters in the original feature space. Also the retrieval performance was compared using CGP and MDA generated features and the CGP features were observed to yield better retrieval results than MDA features.

In [13] and [14], the CGP-based features synthesis method and the expectation–maximization (EM) algorithm were combined into co-evolutionary feature synthesized expectation–maximization (CFS-EM). The main idea is to first use a minor part of the training data to reduce the feature space dimensionality and simultaneously to learn an initial Bayesian classifier using the CGP-based feature synthesis method and then to refine the classifier by applying the EM algorithm on the whole training data (the rest of which may be unlabeled) synthesized into the lower dimensionality. The classification and retrieval results obtained by CFS-EM were both improved compared to the pure CGP approach.

In CBIR it is often not meaningful to compare the similarity of extracted features using the Euclidean distance, but a specific similarity measure is required to match the human perception of similarity. In [15], it is pointed out that combining features requiring different similarity measures (and then comparing the similarity of the synthesized features with the Euclidean distance) may not produce the desired results. Therefore, instead of synthesizing new features, they apply GP to synthesize new similarity measures. During the retrieval, the distances are first computed using the original features and similarity measures, then these distances are given as input to the synthesized similarity measure, which will determine the final distance between the images. The report a superior retrieval performance using this technique.

In a broader sense, also well-known classifiers such as Artificial Neural Networks (ANNs) and SVMs can be thought as a special kind of feature synthesizers. Commonly ANNs used as classifiers take the original feature vector as an input and, when the 1-of-C encoding is used, their output in an optimal case is a vector corresponding to the image class (e.g., for class $c = 1$, the corresponding vector is $\{1, 0, \ldots, 0\}$). Thus ANNs try to learn a feature synthesizer that transforms each feature vector in a certain class to one corner of the C-dimensional cube (where C is the number of classes). The simplest ANNs, Single-layer Perceptrons (SLPs), synthesize the original input features by forming in each neuron a weighted sum of all the input vector elements and passing it through a bounded nonlinear function (e.g., tangent hyperbolic or sigmoid) to give one of the output vector elements. Only the weights of each input feature (and a bias) are optimized via the training algorithm used (e.g., back-propagation), while the synthesis follows a fixed path, otherwise. There is no feature selection involved and the only arithmetic operations applied are the summation and the bounding with a nonlinear function. In Multi-layer Perceptrons (MLPs) such feature synthesis is repeated several times, but the principal approach remains the same. SVMs, on the other hand, with a proper choice of the kernel used, can transform the original, linearly non-separable features of a two-class problem into a higher dimensionality where linear separation is possible. There is no feature selection involved and the performance is directly dependent on the choice of the kernel function and the dimensionality.

10.3 The Evolutionary Feature Synthesis Framework

10.3.1 Motivation

As mentioned earlier, the motivation behind the EFS technique is to maximize the discrimination power of low-level features so that CBIR performance can be improved. Figure 10.1 illustrates an ideal EFS operation where 2D features of a 3-class database are successfully synthesized in such a way that significantly improved CBIR and classification performances can be achieved. Unlike in the figure, the feature vector dimensionality is also allowed to change during the feature synthesis if more discriminative features can be obtained.

The features synthesized by the existing FS methods based on GP produce only slightly improved (or in some cases even worse) results compared to original features. These methods have several limitations and the results may be significantly improved if those limitations can be properly addressed. First of all, the synthesized feature space dimensionality is kept quite low to avoid increasing the computational complexity. Also the dimensionality is fixed a priori which further decreases the probability of finding optimal synthesized features with respect to the problem at hand. In order to maximize the discrimination among classes, also the dimensionality into which the new features are synthesized should be optimized by the evolutionary search technique. Most existing systems also use only few (non)linear operators in order to avoid a high search space dimensionality due to the fact that the probability of getting trapped into a local optimum significantly increases in higher dimensionalities. Furthermore, the methods are quite dependent on a number of parameters, which must be set manually.

ANNs, which may also be seen as feature synthesizers, similarly suffer from the pre-set dimensionality of the synthesized features (e.g., C for a C-class case). Simultaneously, the limited set of operators (only summation and nonlinear

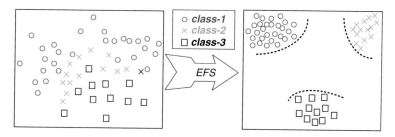

Fig. 10.1 An illustrative EFS, which is applied to 2D feature vectors of a 3-class dataset

bounding) may prevent the ANNs from finding a successful feature synthesizer for a particular set of feature vectors. The most problematic limitation is the lack of feature selection. When the dimensionality of the input feature vector rises, the number of weights to be optimized increases exponentially and the task of finding proper weights soon becomes difficult and perhaps infeasible for any training method due to the well-known "curse of dimensionality" phenomenon.

With SVM classifiers, a major drawback is the critical choice of the (non)linear kernel function along with its intrinsic parameters that may not be a proper choice for the problem at hand. Consider for instance, two sample feature synthesizers (FS-1 and FS-2) illustrated in Fig. 10.2, where for illustration purposes features are only shown in 1-D and 2-D, and only two-class problems are considered. In the case of FS-1, a SVM with a polynomial kernel in quadratic form can make the proper transformation into 3-D so that the new (synthesized) features are linearly separable. However, for FS-2, a sinusoid with a proper frequency, f, should be used instead for a better class discrimination. Therefore, searching for the right transformation (and hence for the linear and nonlinear operators within) is of paramount importance, which is not possible for static (or fixed) ANN and SVM configurations.

The primary objective of the EFS framework presented in this chapter is to address all the above-mentioned deficiencies simultaneously and to achieve the highest possible discrimination between image features belonging to different

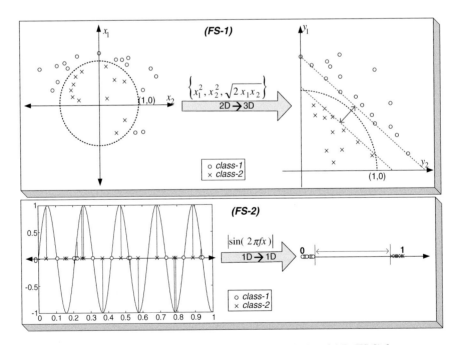

Fig. 10.2 Two sample feature synthesis performed on 2-D (FS-1) and 1-D (FS-2) feature spaces

classes. As a summary, the aim is to design an EFS framework that is able to simultaneously:

1. perform an optimal feature selection,
2. search for optimal arithmetic, linear or nonlinear, operators,
3. search for optimal weights for each feature selected,
4. search for the optimal output feature vector dimensionality,
5. use any given fitness function to measure the quality of the solution,
6. reduce the computational complexity of feature synthesis compared to ANN-based methods.

10.3.2 Evolutionary Feature Synthesis Framework

10.3.2.1 Overview

As shown in Fig. 10.3, the EFS can be performed in one or several runs where each run can further synthesize the features generated from the previous run. The number of runs, R, can be specified in advance or adaptively determined, i.e., runs are carried out until a point where the fitness improvement is no longer significant. The EFS dataset can be the entire image database or a certain subset of it where the ground truth is available. If there is more than one Feature eXtraction (FeX) module, an individual feature synthesizer can be evolved for each module and once completed, each set of features extracted by an individual FeX module can then be passed through the individual synthesizers to generate new features for CBIR.

10.3.2.2 Encoding of the MD PSO Particles

The position of each MD PSO particle in a dimension, $d \in [D_{min}, D_{max}]$, represents a potential solution, e.g., a feature synthesizer, which generates a d-dimensional feature vector using a set of applied operators over some selected features within the N-dimensional source (original) feature vector. Therefore, the jth component (dimension) of the position of particle a, in dimension d and at an iteration t, $xx_{a,j}^d(t), j \in [0, d-1]$, synthesizes the jth feature of the d-dimensional feature vector.

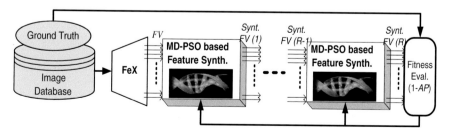

Fig. 10.3 The block diagram of Feature eXtraction (FeX) and the EFS technique with R runs

Along with the operators and the feature selection, encoding of the MD PSO particles is designed to enable a feature *scaling* mechanism with the proper weights. As illustrated in Fig. 10.4, $xx^d_{a,j}(t), j \in [0, d-1]$ is a $2K + 1$ dimensional vector encapsulating:

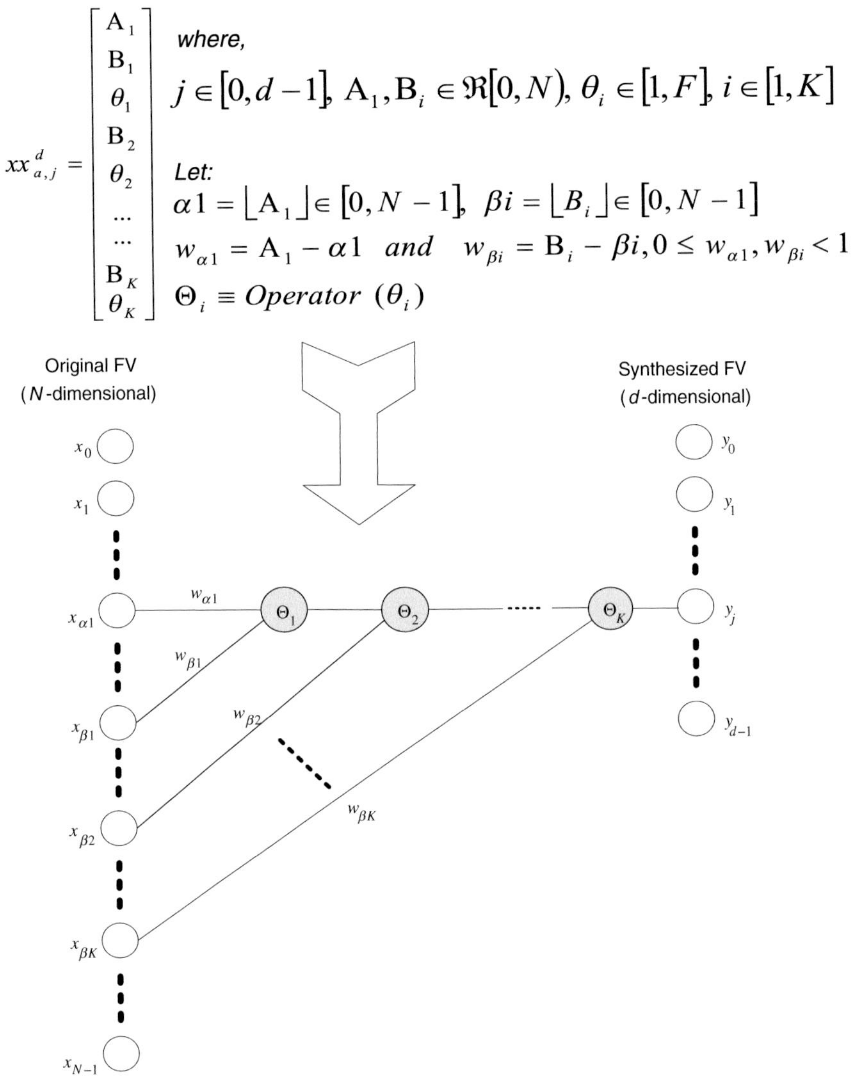

$xx^d_{a,j} = [A_1, B_1, \theta_1, B_2, \theta_2, \ldots, \ldots, B_K, \theta_K]^T$
where,
$j \in [0, d-1],\ A_1, B_i \in \Re[0, N),\ \theta_i \in [1, F],\ i \in [1, K]$
Let:
$\alpha 1 = \lfloor A_1 \rfloor \in [0, N-1],\ \beta i = \lfloor B_i \rfloor \in [0, N-1]$
$w_{\alpha 1} = A_1 - \alpha 1$ and $w_{\beta i} = B_i - \beta i, 0 \le w_{\alpha 1}, w_{\beta i} < 1$
$\Theta_i \equiv Operator(\theta_i)$
Original FV
(N-dimensional)
Synthesized FV
(d-dimensional)
x_0
x_1
$x_{\alpha 1}$
$x_{\beta 1}$
$x_{\beta 2}$
$x_{\beta K}$
x_{N-1}
$w_{\alpha 1}$
$w_{\beta 1}$
$w_{\beta 2}$
$w_{\beta K}$
Θ_1
Θ_2
Θ_K
y_0
y_1
y_j
y_{d-1}

Fig. 10.4 Encoding *j*th dimensional component of the particle *a* in dimension *d* for *K*-depth feature synthesis

Table 10.1 A sample set of $F = 18$ operators used in evolutionary synthesis

θ_i	Formula	θ_i	Formula	θ_i	Formula
0	$-A$	7	$10(A + B)$	14	$tan(100\pi\ A*B)$
1	$-B$	8	$A-B$	15	$tan(100\pi\ (A + B))$
2	$max(A,B)$	9	$A*B$	16	$0.5*exp(-(A-B)*(A-B))$
3	$min(A,B)$	10	$10(A*B)$	17	$0.5*exp(-(A + B)*(A + B))$
4	$A*A$	11	A/B		
5	$B*B$	12	$sin(100\pi\ (A + B))$		
6	$A + B$	13	$cos(100\pi\ (A + B))$		

1) The *selection* of $K + 1$ (input) features (with feature indices, $\alpha1, \beta1, \ldots, \beta K \in [0, N-1]$),
2) Their weights ($0 \leq w_{\alpha1}, w_{\beta i} < 1$) and,
3) K operators (via indices, $\theta_{i\in[1,K]}$ enumerated in Table 10.1). As a result of this K-depth feature synthesis, the jth element of the d-dimensional feature vector can be generated as,

$$y_j = \Theta_K \left(w_{\beta K} x_{\beta K},\ \Theta_{K-1} \left(\ldots, \Theta_2 \left(w_{\beta2} x_{\beta2}, \Theta_1 \left(w_{\beta1} x_{\beta1},\ w_{\alpha1} x_{\alpha1} \right) \right) \ldots \right) \right) \qquad (10.1)$$

In other words, the inner most operator Θ_1, is first applied to the scaled features $x_{\beta1}$ and $x_{\alpha1}$ then operator Θ_2 is applied to the result of the first operation and the scaled feature $x_{\beta2}$, and so on until the last (outer most) operator Θ_K is applied to the result of the previous operations and the scaled feature $x_{\beta K}$.

Note that letting $K = N+1$ and fixing the operator Θ_K to a typical activation function such as *sigmoid* or *tangent hyperbolic*, and the rest, $\Theta_{i\in[1,K-1]}$, to "+" operator (*Operator*(6) in Table 10.1) makes the feature synthesis technique equivalent to a basic feed-forward ANN (or single-layer perceptron, SLP). Similarly, if more than one EFS runs are performed ($R > 1$), the overall scheme is equivalent to a typical MLP. In short, feed-forward ANNs are indeed a special case of the EFS technique, yet the most complex one due to the usage of *all* input features ($K = N+1$), which voids feature selection. Moreover, ANNs make it the most limited case, since it uses only two operators among many possibilities. Therefore, the focus is drawn to achieve a low complexity by selecting only a reasonable number of features (with a low K value) and performing as few MD PSO runs as necessary (with a low R value).

10.3.2.3 The Fitness Function

Since the main objective is to maximize the CBIR performance, a straightforward fitness function (to be *minimized*) is the inverse average precision (*-AP* or 1-*AP*) or alternatively, the average normalized modified retrieval rank (*ANMRR*), both of which can directly be computed by querying all images in the ground truth dataset and averaging individual precision or *NMRR* scores. This, however, may turn out to be a costly operation especially for large databases with many classes. An alternative fitness function that seeks to maximize discrimination among distinct classes can be a clustering validity index (CVI) where each cluster corresponds to a distinct class in the database. CVI can be formed with respect to two widely used criteria in clustering: intra-cluster compactness and inter-cluster separation.

For each potential EFS encoded in a MD PSO particle, the CVI computed over the d-dimensional synthesized features, $Z = \{z_p, \ z_p \in c_j\}$, for each class, $c_i, \ i \in [0, C - 1]$, can be computed as in Eq. (10.1).

$$f(\mu_{c,\Sigma}, Z) = FP\ (\mu_{c,\Sigma}, Z) + \left(Q_e(\mu_{c,\Sigma}, Z) / d_{\min}(\mu_{c,\Sigma})\right)$$

$$\text{where}\quad Q_e(\mu_{c,\Sigma}, Z) = \frac{1}{C} \sum_{i=1}^{C} \frac{\displaystyle\sum_{\forall z_p \in c_i} \|\mu_{c,i} - z_p\|}{\|c_i\|} \tag{10.2}$$

where $\mu_{c,\Sigma}$ is the centroid vector computed for all classes, and $Q_e(\mu_{c,\Sigma}, Z)$ is the quantization error (or the average intra-cluster distance). $d_{\min}$ is the minimum centroid (inter-cluster) distance among all classes and $FP\ (\mu_{c,\Sigma}, Z)$ is the number of *false positives* i.e., synthesized feature vectors which are closer to another class centroid than their own. So the minimization of the validity index $f(\mu_{c,\Sigma}, Z)$ will simultaneously try to minimize the intra-cluster distances (for better *compactness*) and maximize the inter-cluster distance (for better *separation*), both of which lead to a low $FP\ (\mu_{c,\Sigma}, Z)$ value or in the ideal case $FP\ (\mu_{c,\Sigma}, Z) = 0$, meaning that each synthesized feature is in the closest proximity of its own class centroid, thus leading to the highest discrimination.

In the second approach we adapt a similar methodology to the one used in ANNs, i.e., target vectors are assigned to synthesize the features from each class, the EFS system searches for a proper synthesis to get to this desired output, and the fitness is evaluated in terms of the mean square error (MSE) between the synthesized output vectors and the target output vectors. However, we do not want to fix the output dimensionality to C as in ANNs, but instead let the EFS system search for an optimal output dimensionality. Therefore, target output vectors are generated for all dimensionalities within the range of $\{D_{\min}, \ldots, D_{\max}\}$. While generating the target vector table, the two criteria are applied for a good error correcting output code (ECOC) suggested in [16], i.e.,

- *Row separation*: Each target vector should be well-separated in the sense of Hamming distance from each of the other target vectors.
- *Column separation*: Each column in the vector table should be well-separated in the sense of Hamming distance from each of the other columns.

Large row separation allows the final synthesized vectors to somewhat differ from the target output vectors without losing the discrimination between classes. Each column of the target vector table can be seen as a different binary classification i.e., those original classes with value 1 in the specific column form the first metaclass and those original classes with value -1 in that column form the second metaclass. Depending on the similarity of the original classes, some binary classification tasks are likely to be notably easier than others. Since the same target output vectors should be used with any given input classes, it is beneficial to keep the binary classification tasks as different as possible i.e., maximize the column separation.

There is no simple and fast method available to generate target vectors with maximal row and column separation, but the following simple approach is used to create target output vectors with row and column separations which are satisfactory for this purpose:

1. Assign *MinBits* as the minimum number of bits needed to represent C classes.
2. Form a bit table with *MinBits* rows where each row is the binary representation of the row number.
3. Assign the first *MinBits* target vector values for each class c_i equal to the ith row in the bit table.
4. Move the first row of the bit table to the end of the table and shift the other rows up by one row.
5. Assign the next *MinBits* target vector values for each class c_i equal to the ith row in the bit table.
6. Repeat the previous two steps until D_{max} target vector values have been assigned.
7. Replace the first C values in each target vector by a 1-of-C coded section.

This procedure will produce different binary classification tasks until the bit table is rotated back to its original state (large column separation) and simultaneously the row separation is notably increased compared to using only the 1-of-C coded section. While step 7 reduces the row separation, it has been observed that for distinct classes it is often easiest to find a synthesizer that discriminates a single class from the others and, therefore, conserving the 1-of-C coded section at the beginning generally improves the results. The target vectors for dimensionalities below D_{max} can then be obtained by simply discarding a sufficient number of elements from the end of the target vector for dimensionality D_{max}. Since the common elements in the target vectors of different lengths are thus identical, the FGBF methodology for EFS can still freely combine elements taken from particle positions having different dimensionalities.

The target output vector generation for a 4-class case is illustrated in Table 10.2. For clarity, the elements set to -1 are shown as empty boxes. D_{max} is set to 10 and for four classes *MinBits* is 2. Note that the 1-of-C coding is used for

Table 10.2 A sample target vector encoding for 4 classes, $c_1,..., c_4$

t_{c1}	1				1		1	1		
t_{c2}		1			1	1				1
t_{c3}			1					1	1	
t_{c4}				1		1	1		1	1

the first four elements, while the remaining elements are created from the (shifted) rows of a 2-bit representation table.

When the target outputs are created as described above, the fitness of the jth element of a synthesized vector (and thus the corresponding fractional fitness score, $f_j(\mathbf{Z_j})$) can simply be computed as the MSE between the synthesized output vectors and the target vectors belonging to their classes, i.e.,

$$f_j(\mathbf{Z_j}) = \sum_{k=1}^{C} \sum_{\forall z \in c_k} \left(t_{jc_k} - z_j\right)^2. \tag{10.3}$$

where t_{jc_k} denotes the jth element of the target output vector for class c_k and z_j is the jth element of a synthesized output vector. The most straightforward way to form the actual fitness score $f(\mathbf{Z})$ would be to sum up the fractional fitness scores and then normalize the sum with respect to the number of dimensions. However, we noticed that since the first C elements with 1-of-C coding are usually easiest to synthesize successfully, the MD PSO usually favors dimensionalities not higher than C, indicating a crucial local optimum for dimensionality $d = C$. In order to address this drawback efficiently, the elements from 1 to C in the target vector are handled separately and moreover, the normalizing divisor used for the rest of the elements is strengthened, i.e., with the power of $\alpha > 1$, to slightly increase the probability of finding better solutions in higher dimensionalities, $d > C$. As a result, the fitness function is formulated as follows,

$$f(\mathbf{Z}) = \frac{1}{C} \sum_{j=1}^{C} \sum_{k=1}^{C} \sum_{\forall z \in c_k} \left(t_{jc_k} - z_j\right)^2 + \frac{1}{(d-C)^\alpha} \sum_{j=C+1}^{d} \sum_{k=1}^{C} \sum_{\forall z \in c_k} \left(t_{jc_k} - z_k\right)^2. \tag{10.4}$$

with this fitness function, $D_{\min}$ will not be set below $C + 1$.

10.4 Simulation Results and Discussions

MUVIS framework [17] is used to create and index a Corel image database to be used in the experiments in this section. The database contains 1,000 medium resolution (384 × 256 pixels) images obtained from Corel image collection [18] covering 10 classes (*natives, beach, architecture, busses, dinosaurs, elephants, roses, horses, mountains*, and *food*). In order to demonstrate the efficacy of the EFS technique for CBIR, we used well-known low-level descriptors that have a limited discrimination power and a severe deficiency for a proper content description. The descriptors used are 64-bins unit normalized RGB and YUV color histograms and 57-D Local Binary Pattern (LBP) and 48-D Gabor texture descriptors. The synthesis depth, K, was set to 7 meaning that only 7 operators and 8 features were used to compose each element of the new feature vector. The parameter α in Eq. (10.4) was set to 1.1. Unless stated otherwise, the set of 18 empirically selected operators given in Table 10.1 was used within the EFS. All numerical results (discrimination measure, classification error, ANMRR, AP, output dimensionality) are presented in terms of basic statistics such as the mean and best value obtained during 10 separate EFS runs.

In all the experiments MD PSO parameters were set as follows: The swarm size, S, was set to 200 and the number of iterations, $iterNo$, to 2000. The positional search space range, $\{X_{min},...,X_{max}\}$, was set according to the number of operators in use, T, and the number of features in the input feature vectors, F. The positional velocity range, $\{V_{min},...,V_{max}\}$, was empirically set to $\{-(X_{max} - X_{min})/10,..., (X_{max} - X_{min})/10\}$. The dimensionality range, $\{D_{min},...,D_{max}\}$, was set to $\{11,...,40\}$ and the dimensional velocity range $\{VD_{min},...,VD_{max}\}$, to $\{-4,...,4\}$.

10.4.1 Performance Evaluations with Respect to Discrimination and Classification

Let us first concentrate on demonstrating the capability of EFS to improve the discrimination power of the original (low-level) features. The discrimination ability is evaluated in terms of the discrimination measure (DM) given in Eq. (10.2) (DM $= f(\mu_{c,\Sigma}, Z)$) and the same function is used as the fitness function in EFS. The DM along with the number of false positives (FP) for the original features is presented in Table 10.3. DM and FP statistics, when EFS is performed over the entire database, are presented in Table 10.4.

The results clearly indicate a significant improvement in the discrimination power achieved by EFS. Since the entire database was used as the EFS dataset, these results also represent the maximal capability of EFS to improve the discrimination power. However, in real life, the ground truth information of the entire database is rarely available; only a mere fraction of it is usually known. In this case, EFS is still expected to have a high level of generalization ability, i.e., its performance (over the entire database) should be as close as possible to the one obtained when the whole database ground truth data is used for evolving the EFS. In order to evaluate this EFS is now evolved by using the ground truth data over only a part of the database (45 %). The corresponding DM and FP statistics (computed over the whole database) are given in Table 10.5. The results indicate reasonably good generalization ability since the deterioration from DM and FP measures are limited within the range of 7–27 % and the synthesized features still achieve a superior discrimination compared to the original features. Henceforth in the rest of the experiments presented in this section, only 45 % of the ground truth data of the database shall be used.

Table 10.3 Discrimination measure (DM) and the number of false positives (FP) for the original features

Corel	RGB	YUV	LBP	Gabor
DM	431.2	384.6	984.2	462.7
FP	357	334	539	378

Table 10.4 Statistics of the discrimination measure (DM), the number of false positives (FP) and the corresponding output dimensionalities for features synthesized by the first run of EFS over the entire database

Corel	RGB	YUV	LBP	Gabor
Min/mean DM	179.9/203.9	187.2/201.3	306.8/334.7	299.2/309.5
Min/mean FP	161/181.7	167/179.9	281/307.5	272/283.4
Best/mean dim.	37/32.6	36/33.7	26/24.7	37/28.8

Table 10.5 Statistics of the DM, FP and the corresponding output dimensionalities for features synthesized by the first run of the EFS evolved with the ground truth data over 45 % of the database

Corel	RGB	YUV	LBP	Gabor
Min/Mean DM	245.8/259.1	239.2/261.7	384.5/408.4	336.6/363.9
Min/Mean FP	223/237.5	218/239.5	357/372.3	303/330.6
Best/Mean Dim.	33/33.4	37/31.4	30/24.0	31/29.4

Table 10.6 Test CEs for the original features

Corel	RGB	YUV	LBP	Gabor
Test CE	0.420	0.371	0.560	0.433

Table 10.7 Test CE statistics and the corresponding output dimensionalities for features synthesized by a single run of the EFS with 45 % EFS dataset

Corel	RGB	YUV	LBP	Gabor
Min/mean test CE	0.293/0.312	0.285/0.307	0.402/0.422	0.327/0.367
Best/mean dim.	33/34.4	37/31.4	30/24.0	29/29.4

While our main objective is to improve CBIR results, EFS can also be used for other purposes as long as a proper fitness function can be designed. A possible application may be the synthesis of such features that can be classified more efficiently. To demonstrate this potential of EFS, a K-means classifier is trained by computing the class centroids using 45 % of the database, and the test samples are then classified according to the closest class centroid. When it is applied over the original and synthesized features, the classification errors presented in Tables 10.6 and 10.7 are obtained. The results clearly indicate a clear improvement in the classification accuracy, leading to the conclusion that when EFS is evolved with a proper fitness function, it can synthesize such features that can be classified more efficiently and accurately.

10.4.2 *Comparative Performance Evaluations on Content-Based Image Retrieval*

In the following experiments, image features are synthesized using the ground truth data of only part of the image database (45 %). Then ANMRR and AP are computed by the batch query, i.e., querying all images in the database. In order to evaluate the baseline performance, ANMRR and AP performance measures obtained using the original low-level features are first computed, as given in Table 10.8.

Recall that a single run of EFS can be regarded as a generalization of a SLP. Therefore, to perform a comparative evaluation under equal terms, a SLP is first trained using PSO and used as a feature synthesizer in which the low-level image features are propagated to create the output (class) vectors that are used to compute (dis) similarity distances during a query process. Performing batch queries over class vectors, the final ANMRR and AP statistics are given in Table 10.9. It can be seen that, even though the SLP output dimensionality is notably lower than the vector dimensionalities of the original low-level features, ANMRR and AP statistics for the synthesized LBP and Gabor features exhibit a significant performance improvement. This is also true but somewhat limited for RGB and YUV histograms. It is, therefore, clear that feature synthesis performed by SLP improves the discrimination between classes, which in turn leads to a better CBIR performance.

To demonstrate the significance of each property of EFS, a series of experiments are conducted enabling the properties one by one and evaluating their individual impact. Let us start with a single run of the most restricted EFS version that has the highest resemblance to a SLP. In this version, fixing the output dimensionality to C, the same 1-of-C coding for target (class) vectors and the same MSE fitness function are used. As in SLP, PSO (with FGBF) process in this limited EFS searches for only the feature weights, selecting only between addition or subtraction operators (in SLP the weights are limited between $[-1, 1]$, i.e., we need also subtraction operation to compensate for the negative weights).

Table 10.8 ANMRR and AP measures using original low-level features. "*All*" refers to all features are considered

Corel	RGB	YUV	LBP	Gabor	All
ANMRR	0.589	0.577	0.635	0.561	0.504
AP	0.391	0.405	0.349	0.417	0.473

Table 10.9 ANMRR and AP statistics obtained by batch queries over the class vectors of the single SLP

Corel	RGB	YUV	LBP	Gabor	All
Min/mean ANMRR	0.542/ 0.573	0.526/ 0.554	0.544/ 0.555	0.482/ 0.495	0.367/ 0.390
Max/ mean AP	0.446/ 0.414	0.463/ 0.434	0.443/ 0.432	0.503/ 0.490	0.611/ 0.589

Table 10.10 ANMRR and AP statistics for features synthesized by a single run of the EFS when the output dimensionality is fixed to $C = 10$ and operator selection is limited to addition and subtraction

Corel	RGB	YUV	LBP	Gabor	All
Min/mean ANMRR	0.564/ 0.580	0.540/ 0.548	0.636/ 0.651	0.598/ 0.623	0.482/ 0.487
Max/ mean AP	0.419/ 0.404	0.447/ 0.439	0.350/ 0.334	0.385/ 0.360	0.498/ 0.492

Furthermore, the combined features are then bounded using *tanh* function. The bias used in SLP is mimicked by complementing each input feature vector with a constant '1' value. Thus note that, the only property different from the SLP is the feature selection of only 8 ($K = 7$) features for composing each element of the output vector. The retrieval result statistics obtained from the batch queries are presented in Table 10.10.

Even though the feature synthesis process is essentially similar to the one applied by the SLP with the only exception that the number of input features used to synthesize each output feature is notably reduced as a result of the feature selection, the retrieval performance statistics for synthesized RGB and YUV histograms are quite similar. This leads to the conclusion that, in these low-level features, the original feature vectors have many redundant or irrelevant elements for the discrimination of the classes, which is a well-known limitation of color histograms. The feature selection, therefore, removes this redundancy while operating only on the essential feature elements and as a result leads to a significant reduction in computational complexity. The statistics for the synthesized LBP features are close to the ones obtained from the original LBP features, while for the synthesized Gabor features they are somewhat inferior. This suggests that with these descriptors, and especially with the Gabor features, most of the original feature vector elements are indeed essential to achieve maximal discrimination between classes. However, feature selection is also essential to achieve the objective of reducing the overall optimization complexity and it may hence be a prerequisite for applying feature synthesis on larger databases. In the following experiments, it will be demonstrated that the feature selection no longer presents a disadvantage whenever used along with the other properties of EFS.

In the following experiment, the significance of having several operators and operator selection in EFS is examined. All operators are now used in the EFS process. The retrieval performance statistics are given in Table 10.11. Note that

Table 10.11 ANMRR and AP statistics for features synthesized by a single run of the EFS when the output dimensionality is fixed to $C = 10$ and all operators are used

Corel	RGB	YUV	LBP	Gabor	All
Min/mean ANMRR	0.550/ 0.587	0.517/ 0.562	0.499/ 0.509	0.486/ 0.516	0.380/ 0.398
Max/ mean AP	0.432/ 0.394	0.468/ 0.423	0.487/ 0.477	0.498/ 0.467	0.597/ 0.578

Table 10.12 ANMRR and AP statistics and the corresponding output dimensionalities for the synthesized features by a single EFS run using the fitness function in Eq. (10.4)

Corel	RGB	YUV	LBP	Gabor	All
Min/meanANMRR	0.485/0.500	0.475/0.505	0.507/0.519	0.506/0.520	0.346/0.357
Max/ mean AP	0.494/0.478	0.507/0.477	0.477/0.465	0.479/0.464	0.630/0.619
Best/mean dim.	34/35.7	25/34.7	37/35.3	14/27.0	

ANMRR and AP statistics for synthesized RGB and YUV histograms are quite similar with and without operator selection, while for LBP and Gabor features a significant improvement can be observed with operator selection.

In the next experiment, EFS is allowed to optimize the output dimensionality and the fitness function given in Eq. (10.4) is used; i.e., now all the properties of EFS are in use but only in a single EFS run. ANMRR and AP statistics along with the corresponding statistics of the output dimensionality (best/mean) of the synthesized feature vector (*dbest*) are presented in Table 10.12.

ANMRR scores obtained with the single run of the EFS are better than to the ones obtained with the original features and the features (class vectors) synthesized by the SLP. The only exception is the synthesized Gabor features for which the statistics are slightly worse than the ones synthesized with the SLP. As discussed earlier, this indicates the relevance of each element of the Gabor feature vector and selecting a limited subset may yield a slight disadvantage in this case.

Finally, several runs of the EFS (without exceeding 25) are performed until the performance improvement is no longer significant. The ANMRR and AP statistics and along with the corresponding statistics of the output dimensionality (best/ mean) of the final synthesized feature vector (*dbest*) are presented in Table 10.13. The average numbers of EFS runs for the synthesis of the RGB, YUV, LBP, and Gabor features were 19.6, 18.7, 22.8, and 24.5, respectively.

Note that when EFS is performed with all the properties enabled, the retrieval performance of the synthesized features has been significantly improved compared to EFS with a single run. It can be also observed that the average dimensionalities of the final synthesized feature vectors (*dbest*) become lower. This is not surprising since repeated applications of consecutive arithmetic operators can achieve a similar discrimination capability with fewer output feature elements. Dimensionality reduction in the synthesized features is also a desired property that makes the

Table 10.13 Retrieval performance statistics and the corresponding output dimensionalities for the final synthesized features by several EFS runs using the fitness function in Eq. (10.4)

Corel	RGB	YUV	LBP	Gabor	All
Min/mean ANMRR	0.365/ 0.385	0.369/ 0.397	0.372/ 0.396	0.408/ 0.428	**0.258/ 0.280**
Max/ mean AP	0.616/ 0.596	0.610/ 0.584	0.613/ 0.589	0.577/ 0.559	**0.716/ 0.694**
Best/mean dim.	12/12.4	18/17.8	11/11.2	11/11.0	

retrieval process faster. However, we noticed that the best retrieval results were usually obtained when a higher output dimensionality was maintained for several runs. This suggests that it could be beneficial to set the value of power α in Eq. (10.4) to a higher value than now used 1.1 in order to favor higher dimensionalities even more. However, this may not be a desired property especially for large-scale databases.

Figure 10.5 illustrates four sample queries each of which is performed using the original features, features synthesized by a single EFS run, and the features synthesized by four EFS runs. It is obvious that multiple EFS runs improve the discrimination power of the original features and, consequently, an improved retrieval performance is achieved.

In order to perform comparative evaluations against evolutionary ANNs, features are synthesized using feed-forward ANNs since several EFS runs are conceptually similar to MLPs or especially to multiple concatenated SLPs. For ANNs, MD PSO is used to evolve the optimal MLP architecture in an architecture space, $R_{\min} = \{N_i, 8, 4, N_o\}$, $R_{\max} = \{N_i, 16, 8, N_o\}$. Such 3-layer MLPs correspond to 3 EFS runs. The number of hidden neurons in the first/second hidden layer corresponds to the output dimensionality of the first/second EFS run (with MLPs the number of hidden neurons must be limited more to keep the training feasible). Naturally, the number of output neurons is fixed to $C = 10$, according to the 1-to-C encoding scheme. In order to provide a fair comparison, the number of MD PSO iterations is now set to $3 \times 2,000 = 6,000$ iterations. Table 10.14 presents the retrieval performance statistics for the features synthesized by the best MLP configuration evolved by MD PSO.

It is fairly clear that except for RGB histograms, the retrieval performance statistics for features synthesized by EFS even with a single run are better than the ones achieved by the best MLP. This basically demonstrates the significance of the feature and operator selection.

As each EFS run is conceptually similar to a SLP, the EFS with multiple runs in fact corresponds to the synthesis obtained by the concatenation of the multiple SLPs (i.e., the output of the previous SLP is fed as the input of the next one similar to the block diagram shown in Fig. 10.3). The retrieval performance statistics obtained for the features synthesized by the concatenated SLPs are given in Table 10.15.

Compared to the retrieval result statistics given in Table 10.9 for the batch queries over the class vectors of a single SLP, slightly better retrieval performances are obtained. However, similar to the results for evolutionary MLPs, they are also significantly inferior to the ones obtained by EFS, as given in Table 10.13. Therefore, similar conclusions can be drawn for the significance of the three major properties of EFS presented in this chapter, i.e., feature and operator selection and searching for the optimal feature dimensionality. Furthermore, EFS provides a higher flexibility and better feature (or data) adaptation than regular ANNs, since 1) it has the capability to select the most appropriate linear/nonlinear operators among a large set of candidates, 2) it has the advantage of selecting proper features

Fig. 10.5 Four sample queries using original (*left*) and synthesized features with single (*middle*) and four (*right*) runs. *Top-left* is the query image

Table 10.14 ANMRR and AP statistics for the features synthesized by the best MLP configuration evolved by MD PSO

Corel	RGB	YUV	LBP	Gabor	All
Min/mean ANMRR	0.392/ 0.442	0.527/ 0.558	0.513/ 0.545	0.490/ 0.547	0.307/ 0.348
Max/ mean AP	0.594/ 0.543	0.465/ 0.433	0.445/ 0.475	0.498/ 0.442	0.673/ 0.633
Best ANN	$\{N_i,9,5,N_o\}$	$\{N_i,9,5,N_o\}$	$\{N_i,8,6,N_o\}$	$\{N_i,14,5,N_o\}$	

Table 10.15 ANMRR and AP statistics for the features synthesized by the concatenated SLPs

Corel	RGB	YUV	LBP	Gabor	All
Min/mean ANMRR	0.498/ 0.535	0.495/ 0.525	0.494/ 0.503	0.459/ 0.474	0.346/ 0.365
Max/ mean AP	0.491/ 0.454	0.493/ 0.463	0.493/ 0.485	0.527/ 0.512	0.634/ 0.615

that in turn reduces the complexity of the solution space, and 3) it synthesizes features in an optimal dimensionality.

Note that the application of EFS is not limited to CBIR, but it can directly be utilized to synthesize proper features for other application areas such as data classification or object recognition, and perhaps even speech recognition. With suitable fitness functions, synthesized features can be optimized for the application at hand. To further improve performance, different feature synthesizers may be evolved for different classes, since the features essential for the discrimination of a certain class may vary.

10.5 Programming Remarks and Software Packages

Recall that the major MD PSO test-bed application is **PSOTestApp** where several MD PSO applications, mostly based on data (or feature) clustering, are implemented. The basics of MD PSO operation is described in Sect. 4.4.2 and FGBF in Sect. 6.4 whereas in Sect. 8.4, N-D feature clustering application for Holter register classification is explained. In this section, we shall describe the programming details for performing MD PSO-based EFS operation. For this, the option: "8. Feature Synthesis..." should be selected from the main GUI of the **PSOTestApp** application, and the main dialog object from **CPSOtestAppDlg** class will then use the object from **CPSOFeatureSynthesis** class in **void CPSOtestAppDlg:: OnDopso()** function to perform EFS operation over MUVIS low-level features stored in FeX files.

The API of this class is identical to the other major classes in the **PSOTestApp** application. The thread function, **PSOThread()**, is again called in a different thread where the EFS operations are performed. The class structure is also quite similar to **CPSOclusterND** class where there are several fitness functions are implemented each for different EFS objective. Table 10.16 presents the class **CPSOFeatureSynthesis**. The input file for this application is a MUVIS image database file in the form of "*.idbs." Using the database handling functions, a MUVIS database file from which its low-level features are retrieved, is loaded. For each low-level feature, an individual EFS operation (with multiple repetitions if required) is performed and the synthesized features are then saved into the original FeX file, while the FeX file with low-level features is simply renamed with the "FeX0" enumeration at the end. If there are more than one EFS blocks (runs), then the synthesized features from the latest run is saved into the original FeX file while the previous runs outcomes are saved into enumerated FeX files with "FeX1," "_FeX2," etc. For example, let *Corel_FeX.RGB* be the FeX file for the Corel database where RGB color histogram features are stored. Assume that 2 EFS runs

Table 10.16   The class **CPSOFeatureSynthesis**

```
class PSOTESTER_API CPSOFeatureSynthesis
{
public:
        ...
        // API Functions..
        void        Init (HWND hWnd) {m_hWnd = hWnd;}
        void        ApplyPSO(CQueueList<CFileName> *pFL, PSOparam psoParam, SPSAparam saParam);
        void        Stop();
        void        ShowResults();

protected:
        void        PSOThread();
        float       GetResults(FILE *fp);

        // Database handling functions..
        int         Load_idbs();
        int         LoadDbsFeatures(FILE *fp);
        void        Load_qbf();
        void        Load_FeX( int u, int& img_no, int& ftr_no, int& no_params, double*& params );
        void        SeparateTrainingFeatures();
        int         dt_to_FeX();

        // STATICS
        static double           FitnessFnAP(CSolSpace<CVectorNDopt> *pPos, int);
        static double           FitnessFnANN(CSolSpace<CVectorNDopt> *pPos, int);
        static double           FitnessFnCVI(CSolSpace<CVectorNDopt> *pPos, int);
        ...

protected:

        PSOparam                m_psoParam; // PSO parameters..
        SPSAparam               m_saParam; // SA parameters..
        CQueueList<CFileName>   *m_pFL; // The input filename queue.
        CPSO_MD<CSolSpace<CVectorNDopt>,CVectorNDopt>         *m_pPSO; // The current PSO obj..
        FITFN*                  m_fpGetScore; // Generic fn. pointer for Fitness Func..
        CFileName               *pFL; // the name of the current file to be EFS..
        ...
        int             m_noVisDbsFeat;
        CDbsFeature     *m_pVisDbsFeat; // Visual Feature parameter array for all features present in the database..
};
```

are performed where each run can be repeated more than one time and the EFS with the best fitness score is only kept while the others are simply discarded. At the end of 2 EFS runs, the following files are generated:

- Corel_FeX0.RGB (The original low-level features)
- Corel_FeX1.RGB (The synthesized features after the first EFS run)
- Corel_FeX.RGB (The synthesized features after the second –last– EFS run)

In this way, the synthesized features of the last EFS run can directly be used for a CBIR operation in MUVIS while the features from the previous runs as well as the original features are still kept for backup purpose.

Recall that EFS has certain similarities to a classification operation. For instance, the EFS is also evolved over the training dataset (the GTD) of a database and tested over the test dataset. Therefore the function, **Load_qbf()**, loads the GTD for this purpose and another function, **SeparateTrainingFeatures()**, separates the FVs of the training dataset.

Table 10.17 presents the first part of the function, **CPSOFeatureSynthesis::PSOThread()** with its basic blocks. Note that the first operation is to load the database ("*.idbs") file, and the "*.qbf" file where the entire GTD resides. Once they are loaded, the EFS operation is performed for each feature in the database (i.e., the first for-loop) while for each EFS operation, there can be one or more EFS runs concatenated (as shown in Fig. 10.3). For each EFS run, first the current features (either low-level features if this is the first EFS run, or the features synthesized by the previous EFS run) are loaded by the **Load_FeX()** function. Then the training dataset features that are used to evolve the synthesizer, are separated by the function, **SeparateTrainingFeatures()**. Then the MD PSO parameters, internal and static variables/pointers are all set, created and/or allocated. Finally, the EFS operation by the MD PSO process is performed and repeated by **m_psoParam._repNo** times. As mentioned earlier, the synthesizer, **best_synthesis**, obtained from the best MD PSO run in terms of the fitness score is kept in an encoded form by a **CVectorNDopt** array of dimension, **best_dim**, which is the *dbest* converged by the MD PSO swarm. Figure 10.4 illustrates the encoding of the **best_synthesis**, in an array format.

Table 10.18 presents the second part of the function, **CPSOFeatureSynthesis::PSOThread()** where the EFS, **best_synthesis**, is tested against the previous synthesizer or the original features in terms of the fitness score achieved, i.e., it will be kept only if there is an improvement on the fitness score. To accomplish the test, first a dummy synthesizer is created (**pOrigFeatures**), which creates an identical FV as the original one and then the fitness score is computed by simply calling the fitness function, **FitnessFnANN(pOrigFeatures, s_v_size)**. The following *if* statement compares both fitness scores and if there is an improvement, then the synthesizer, **best_synthesis**, generates new features in a pointer array, **SF[]**, and saves them into the *FeX* file while renaming the previous one with an enumerated number. Note that at this stage the synthesizer is applied over the entire database, not only over the training dataset. Therefore, it is still unknown

Table 10.17 The function: **CPSOFeatureSynthesis::PSOThread()** (part-1)

```
void CPSOFeatureSynthesis::PSOThread()
{
 Load_idbs();// Load feature info from the idbs
 Load_qbf();// Load class info from gbf file

 for( int u = 0; u < m_noVisDbsFeat; ++u )
 {
   for( int h = 0; h < EFS_NO; ++h )
   {
               // Load the input features
               int img_no, ftr_no, no_params = 0;
               double* params = NULL;
               Load_FeX( u, img_no, ftr_no, no_params, params );

               // Separate the training features from others
               SeparateTrainingFeatures();

               // Set MD PSO parameters and internal variables..
               ...
               // Create static parameters to be used in fitness func..
               SynthesizedFeatures = ( double**) calloc( s_no_train, sizeof( double* )); // array of synth. features..
               for( int i = 0; i < s_no_train; ++i ) SynthesizedFeatures[i] = ( double* ) calloc ( max_dim, sizeof( double ));
               ...
               for(int rep = 0; rep < m_psoParam._repNo; ++rep)
               { // Repeat MD PSO feature synthesis..
                   m_pPSO  =  new  CPSO_MD<CSolSpace<CVectorNDopt>,CVectorNDopt>  (m_psoParam._noAgent,
m_psoParam._maxNoIter, m_psoParam._mode);
                   m_pPSO->SetFitnessFn(m_fpGetScore); // Set Fitness Fn. here...
                   m_pPSO->SetFindGBDimFn(FindGBDim);
                   m_pPSO->SetSAparam(m_saParam); // Set SA params..

                   m_pPSO->Init(m_psoParam._dMin, m_psoParam._dMax, m_psoParam._vdMin, m_psoParam._vdMax,
psoPosMin, psoPosMax, psoVelMin, psoVelMax );

                   m_pPSO->Perform(); // Apply MD PSO for clustering..
                   ...
                   CSolSpace <CVectorNDopt>* best_result= m_pPSO -> GetBestPos( dbest, ind, score );
                   if( score < min_score )
                   {
                         min_score = score;
                         best_dim = dbest;
                         if( best_synthesis ) delete[] best_synthesis;
                         best_synthesis = new CVectorNDopt[ best_dim ];
                         CVectorNDopt *pCC = best_result -> GetPos();
                         for( int i = 0; i < best_dim; ++i )
                             best_synthesis[i] = pCC[i];
                   } // if..
                   delete m_pPSO; m_pPSO = NULL;
               } // for..
               // Compare with the original features
               ...
```

whether or not the CBIR performance is improved by the synthesized features, and this can then be verified by performing batch queries using this feature alone.

Table 10.19 presents the current fitness function used, **FitnessFnANN(pOrigFeatures, s_v_size)**. This corresponds to the second approach, which adapts a similar methodology to the one used in ANNs, i.e., using the ECOC scheme. The first step is to apply the synthesis with the potential synthesizer stored in the MD

Table 10.18 The function: **CPSOFeatureSynthesis::PSOThread()** (part-2)

```
...
// Compare with the original features.. First create a dummy (all pass) synthesizer..
CSolSpace <CVectorNDopt> *pOrigFeatures = new CSolSpace <CVectorNDopt> ( s_v_size, psoPosMin,
psoPosMax );
CVectorNDopt *pCC = pOrigFeatures->GetPos();
for(int d=0; d<s_v_size; d++)
{       pCC[d].m_pV[0] = min_feat_val +d;
        for(int op = 1; op < FS_DEPTH; ++++op)
        {
                pCC[d].m_pV[op] = min_feat_val + d + 1;
                pCC[d].m_pV[op+1] = min_op_val;
        } // for..
} // for..

float orig_score = FitnessFnANN( pOrigFeatures, s_v_size); // fitness score of the orig. features..
delete pOrigFeatures;

if( min_score < orig_score )
{
    // Synthesize the whole data using the best operations found...
    double** SF = ( double**) calloc( s_no_images, sizeof( double* )); // synth. Features..
    for( int i = 0; i < s_no_images; ++i ) SF[i] = ( double* ) calloc ( best_dim, sizeof( double ));
    ApplyFSadaptive( false, true, SF, best_synthesis, best_dim );

    // Rename old FeX file..
    ...
    // Write the synthesized features into a FeX file...
    ...
    for(int j=0;j<s_no_images;j++) free( SF[j] );
    free( SF );
} // if..
// Step 6: Clean-up locals and statics..
...
} // for h..
} // for s..
// Clean up Dbs Features..
...
```

PSO particle's position, **pPos** in dimension, **nDim**. The function call, **ApplyFS-adaptive(true, true, SynthesizedFeatures, pCC, nDim)**, will synthesize features stored in the static variable, **SynthesizedFeatures**. The fitness score is the normalized mean square error (MSE) between the synthesized features and the ECOC code in this dimension stored in the static variable, **mean_vec[c][d]**. While calculating the MSE in the last for-loop, note that the MSE per dimension is also stored as the individual dimensional fitness scores that are used for the FGBF operation.

Table 10.20 presents the feature synthesizer function, **CPSOFeatureSynthesis::ApplyFSadaptive()**, which synthesizes either the FVs of the entire database, **s_pInputFeatures** or the training dataset, **s_pTrainFeatures**. Recall that **nDim** is the dimension of the MD PSO particle, and also corresponds to the dimension of the synthesized feature (FV). Therefore, the particle's position vector is in dimension **nDim** and each dimensional component holds an individual synthesizer encoded in an array of fixed length, **FS_DEPTH**. As shown in Fig. 10.4, within

Table 10.19 The function: **CPSOFeatureSynthesis::FitnessFnANN()**

```
double CPSOFeatureSynthesis::FitnessFnANN(CSolSpace<CVectorNDopt> *pPos, int _nDim)
{
        CVectorNDopt *pCC = pPos->GetPos();
        int nDim = (pPos->GetDim() < _nDim) ? pPos->GetDim() : _nDim;

        if( !ApplyFSadaptive( true, true, SynthesizedFeatures, pCC, nDim ))
                return 1e9; // From original features Synth. nDim dimensional features..

        memset(class_intra_dist, 0, nDim*sizeof(float)); // init..
        for(int d = 0; d < nDim; ++d )
        {
                for(int c = 0; c < s_cp.no_train_class; ++c )
                {       // for an individual class..
                        for( int i = s_pTrainInfo[c]; i < s_pTrainInfo[c + 1]; ++i )
                                class_intra_dist[d] += (mean_vec[c][d] - SynthesizedFeatures[i][d] )*(mean_vec[c][d]
        - SynthesizedFeatures[i][d] );
                } // for..
        } // for..

        float VI =0; // Final Fitness value..
        for(d = 0; d < nDim; ++d )
        {
                pCC[d].m_bScore  = class_intra_dist[d];
                VI += class_intra_dist[d];
        } // for..

        return VI/nDim;
}
```

the *for-loop* of each dimension, the elements of each synthesizer array is decoded into FV component indices, operator indices, weights, and biases. For instance,

int feature1 = ((int) pCC[j].m_pV[0]) + feat_min;
float weight1 = ABS(pCC[j].m_pV[0] - ((int) pCC[j].m_pV[0]));

The variable, **feature1**, is the index of the first FV component (of the original feature) and the variable, **weight1**, is its weight. Note that within the following *for-loop*, the FV component index and the weight of the second FV are also decoded as well as the first operator index, **FSoperator**. Then for the entire dataset, the (*j*th) dimensional component of the synthesized features are computed using the associated operator, **oFn[FSoperator]**, as,

for(int i = 0; i < img_no; ++i)
    SynthesizedFeatures[i][j] =
        oFn[FSoperator](SynthesizedFeatures[i][j], weight2*pOriginalFeatures[i][feature2]) + threshold;

Note that the first term, **SynthesizedFeatures[i][j]**, was already assigned to the scaled (with **weight1**) version of the first feature, **feature1**. Therefore, the first term of the operator will always be the output of the previous operator, and the second term will be the new scaled (with **weight2**) component of the original FV. The final synthesized feature may then be passed through the activation function, *tanh*, to scale it within [-1, 1] range.

Table 10.20 The function: **CPSOFeatureSynthesis::ApplyFSadaptive()**

```
bool  CPSOFeatureSynthesis::ApplyFSadaptive(bool   bTrain,   bool   bAct,   double**   SynthesizedFeatures,
CVectorNDopt *pCC, int nDim )
{
        double** pOriginalFeatures = NULL;
        int img_no = 0;
        if (bTrain )
        {
                pOriginalFeatures = s_pTrainFeatures; // Synth. only train dataset FVs..
                img_no = s_no_train;
        }      else
        {
                pOriginalFeatures = s_pInputFeatures; // Synth. all dbs. FVs..
                img_no = s_no_images;
        } // else..

        int feat_min = s_v_size/2;
        int op_min = (1+no_oFn)/2;

        for( int j = 0; j < nDim; ++j )
        {
                int feature1 = ( (int) pCC[j].m_pV[0]) + feat_min;
                if( feature1 < 0 || feature1 >= s_v_size ) feature1 = s_v_size - 1;
                float weight1 = ABS(pCC[j].m_pV[0] - ( (int) pCC[j].m_pV[0]));

                for( int i = 0; i < img_no; ++i )
                        SynthesizedFeatures[i][j] = weight1*pOriginalFeatures[i][feature1] ;

                for(int op = 1; op < FS_DEPTH; ++++op)
                {
                        float weight2 = ABS(pCC[j].m_pV[op] - ( (int) pCC[j].m_pV[op]));
                        int feature2 = ( (int) pCC[j].m_pV[op]) + feat_min ;
                        if( feature2 <0 || feature2 >= s_v_size)
                                feature2 = s_v_size - 1;
                        int FSoperator = (( (int) pCC[j].m_pV[op+1]) + op_min) % no_oFn;
        //              float threshold = ABS(pCC[j].m_pV[op+1] - ( (int) pCC[j].m_pV[op+1]));
                        float threshold = 0;

                        for( int i = 0; i < img_no; ++i )
                                SynthesizedFeatures[i][j] =
oFn[ FSoperator]( SynthesizedFeatures[i][j], weight2*pOriginalFeatures[i][feature2] ) + threshold;
                } // for..

                if(bAct) // if activation function is used..
                        for( i = 0; i < img_no; ++i )
                                SynthesizedFeatures[i][j] = tanh(SynthesizedFeatures[i][j]); // apply Act. Fn..
        } // for j..

        return true;
}
```

The MD PSO operation for EFS is identical to the one for N-D feature clustering, as for both operation, the task is to find out certain number of (*dbest*) N-D arrays (for EFS carrying synthesizers in encoded form) or FVs (for N-D clustering, carrying the N-D cluster centroids). However, the FGBF operations differ entirely. As explained in Sect. 6.4.1, the FGBF operation for clustering operation is handled in the **FGBF_CLFn()** function whereas the other function, **FGBF_FSFn()**, is for

EFS. Note that this is the basic FGBF operation as explained in Sect. 5.5.1 and the implementation is identical to the one presented in Table 5.17.

References

1. A. Antoniou, W.-S. Lu, *Practical optimization, algorithms and engineering applications* (Springer, USA, 2007)
2. D. Goldberg, *Genetic Algorithms in Search, Optimization and Machine Learning* (Addison-Wesley, Reading, MA, 1989), pp. 1–25
3. S. Kirkpatrick, C.D. Gelatt, M.P. Vecchi, Optimization by simulated annealing. Science **220**, 671–680 (1983)
4. K. Price, R. M. Storn, J. A. Lampinen, *Differential Evolution: A Practical Approach to Global Optimization* (Springer, 2005) ISBN 978-3-540-20950-8
5. A. Unler, A. Murat, A discrete particle swarm optimization method for feature selection in binary classification problems. Eur. J. Oper. Res. **206**(3), 528–539 (2010)
6. L.-Y. Chuang, H.W. Chang, C.J. Tu, C.H. Yang, Improved binary PSO for feature selection using gene expression data. Comput. Biol. Chem. **32**(1), 29–38 (2008)
7. Y. Lin, B. Bhanu, Evolutionary Feature Synthesis for Object Recognition. IEEE Trans. Man Cybern: Part C **35**(2), 156–171 (2005)
8. Y. Liu, Z. Qin, Z. Xu, X. He, Feature selection with particle swarms computational and information science. Lect. Notes Comput. Sci. **3314**, 425–430 (2005)
9. K. Geetha, K. Thanushkodi, A. Kishore kumar, New Particle Swarm Optimization for Feature Selection and Classification of Microcalcifications in Mammograms, in *Proceedings of IEEE International Conference on Signal Processing, Communications and Networking*, pp. 458–463, Chennai, India, Jan. 2008
10. B. Bhanu, J. Yu, X. Tan, Y. Lin, "Feature Synthesis using Genetic Programming for Face Expression Recognition", *Genetic and Evolutionary Computation (GECCO 2004). Lect. Notes Comput. Sci.* **3103**, 896–907 (2004)
11. J. Yu, B. Bhanu, Evolutionary feature synthesis for facial expression recognition. Pattern Recogn. Lett. **27**, 1289–1298 (2006)
12. A. Dong, B. Bhanu, Y. Lin, Evolutionary Feature Synthesis for Image Databases, in *Proceedings of 7th IEEE Workshop on Applications of Computer Vision*, Breckenridge, CO, USA, Jan. 2005
13. R. Li, B. Bhanu, A. Dong, Coevolutionary Feature Synthesized EM Algorithm for Image Retrieval, in *Proceedings of 13th Ann. ACM Internation Conference on Multimedia (MM'05)*, pp. 696–705, Singapore, Nov. 2005
14. R. Li, B. Bhanu, A. Dong, Feature Synthesized EM Algorithm for Image Retrieval. ACM Trans. Multimedia, Commun. Appl. **4**(2), article 10, May 2008
15. R.S. Torres, A.X. Falcao, M.A. Goncalves, R.A.C. Lamparelli, A genetic programming framework for content-based image retrieval. Pattern Recogn. **42**(2), 283–292 (2009)
16. T.G. Dietterich, G. Bakiri, Solving Multiclass Learning Problems via Error-Correcting Output Codes. J Artif Intell Res **2**, 263–286 (1995)
17. MUVIS. http://muvis.cs.tut.fi/
18. Corel Collection/Photo CD Collection (www.corel.com)

Printed in the United States
By Bookmasters